The Real Cost of Cheap Food

The Real Cost of Cheap Food

Michael Carolan

from Routledge

First published by Earthscan in the UK and USA in 2011

For a full list of publications please contact:
Earthscan
2 Park Square, Milton Park, Abingdon, Oxon OX14 4RN
711 Third Avenue, New York, NY 10017

Earthscan is an imprint of the Taylor & Francis Group, an informa business

© 2011 Dr Michael S. Carolan. Published by Taylor & Francis.

Notices
Practitioners and researchers must always rely on their own experience and knowledge in evaluating and using any information, methods, compounds, or experiments described herein. In using such information or methods they should be mindful of their own safety and the safety of others, including parties for whom they have a professional responsibility.

Product or corporate names may be trademarks or registered trademarks, and are used only for identification and explanation without intent to infringe.

British Library Cataloguing in Publication Data
A catalogue record for this book is available from the British Library

Library of Congress Cataloging in Publication Data
Carolan, Michael.
The real cost of cheap food / Michael Carolan. — 1st ed.
  p. cm.
Includes bibliographical references and index.
1. Food supply—Social aspects. 2. Public health. 3. Agriculture—Costs. 4. Agriculture—Environmental aspects. I. Title.
HD9000.5.C258 2011
338.1'9—dc22
              2011003697

ISBN 978-1-84971-320-7 (hbk)
ISBN 978-1-84971-321-4 (pbk)

Typeset in Sabon by FiSH Books, Enfield, UK

Printed and bound in the United States of America by Edwards Brothers Malloy on sustainably sourced paper.

Contents

List of Figures, Tables and Boxes

Figures

Tables

Boxes

Acknowledgements

Writing this book has been a pleasure, in part because of those to whom I was introduced through the process. First: a tip of the hat (and big 'Thanks!') to commissioning editor extraordinaire Tim Hardwick. I've never worked with a more involved (in a good way) editor. From reviewer of draft manuscripts to part-time research assistant, Tim has been indescribably helpful. He was also a ringleader of sorts, bringing me into contact with a variety of food and agriculture specialists. They include:

- Geoffrey Lawrence (University of Queensland, Australia): your timely review was spot on, pointing me in many directions that improved the manuscript tremendously. Thank you, thank you!
- Gyorgy Scrinis (University of Melbourne, Australia): I hope I did justice to your concept of nutritionism. Thanks for your correspondence (and suggested readings); I look forward to many more.
- Adrian Williams (Cranfield University, UK): the suggestions you offered on the subject of life-cycle analysis (LCA) were a big help as I rewrote Chapter 6. Thanks also for promptly answering the occasional email looking for additional clarification on the particulars of a given study.
- Richard Bennett (University of Reading, UK): thanks for pointing me in the direction of seminal pieces within the economics of animal welfare literature.
- Anonymous reviewers of my book proposal: there were quite a few of you. I don't know who you are, but your initial input was invaluable. Thank you all.
- Tara Garnett (University of Surrey, UK) and the Food Climate Research Network (FCRN) mailing list: I have been diligently reading your postings with quiet excitement. They were an invaluable resource. To anyone interested in the ecological footprint of our food system, I recommend looking at their website (www.fcrn.org.uk).

I would also like to thank the Rural Economy Graduate Students' Association (REGSA) at the University of Alberta for inviting me to Edmonton to give their Keynote Address in January 2011. While the manuscript was already written, preparing for the lecture gave me the opportunity to tweak certain arguments, which undoubtedly improved the final product. Thanks also to my university

(Colorado State University), college (Liberal Arts) and department (Sociology) for finding money (when none is usually to be found) to grant me a sabbatical. That time away from my usual university duties allowed me to complete this manuscript in a timely fashion. A word of appreciation also to Cambridge University Press for allowing me to reproduce Table 10.2, found in Chapter 10. The table originally appeared in M. S. Carolan (2009) 'The costs and benefits of biofuels: A review of recent peer-reviewed research and a sociological look ahead', *Environmental Practice*, vol 11, pp17–24.

And, finally, to Nora: 'thanks' just doesn't seem to sufficiently convey my appreciation for all that you've done. You read previous drafts. You put up with my terribly long workdays. And just a few weeks ago, you gave birth to Joey. What a wild ride. Here's to many decades more.

June 2011

List of Acronyms
and Abbreviations

aka	also known as
AOC	*Appellation d'Origine Contrôlée* (controlled designation of origin)
BMP	best management practice
BOGO	buy one get one
BSE	bovine spongiform encephalopathy
CAFO	concentrated animal feeding operation
CEO	chief executive officer
CGIAR	Consultative Group on International Agricultural Research
CIA	US Central Intelligence Agency
CLA	conjugated lineolic acid
CO	carbon monoxide
CO_2	carbon dioxide
CR4	sum of the market shares of the top four firms
CRP	Conservation Reserve Program
CSA	community-supported agriculture
CSE	Centre for Science and Environment
CWRF	per capita water requirements for food
cwt	hundredweight (*or* centum weight)
Defra	UK Department for Environment, Food and Rural Affairs
DOJ	US Department of Justice
EEP	Export Enhancement Program
EU	European Union
FAO	Food and Agriculture Organization of the United Nations
FCRN	Food Climate Research Network
FDA	US Food and Drug Administration
FDI	foreign direct investment

FPEAK	Fresh Produce Exporters' Association of Kenya
FRC	US Federal Trade Commission
FSA	UK Food Standards Agency
GAO	Government Accountability Office
GAO	*Grupo de Agricultura Organica* (Cuban organic farming association)
GDP	gross domestic product
GE	genetic engineering/genetically engineered
GEM	Green Energy Madagascar
GHG	greenhouse gas
GM	genetically modified
GURT	genetic use restriction technology
GWP	global warming potential
HFCS	high fructose corn syrup
HLY	healthy life year
H_2O	water
IATP	Institute for Agriculture and Trade Policy
IMF	International Monetary Fund
IPM	integrated pest management
IRC	International Relations Center
kcal	kilocalorie
kg	kilogram
km	kilometre
kWh	kilowatt hours
LCA	life-cycle analysis
LDC	least developed countries
LED	light-emitting diode
LMC	large modern confinement (system)
m	metre
mg	milligram
MIRG	management-intensive rotational grazing
ml	millilitre
mmt	million metric tonnes
MST	*Movimento dos Trabalhadores Rurais Sem Terra* (Landless Workers' Movement)
NAFTA	North American Free Trade Agreement
NASA	National Aeronautics and Space Administration

NCGRP	National Center for Genetic Resources Preservation
NGO	non-governmental organization
NHANES	National Health and Nutrition Examination Survey
N_2O	nitrous oxide
NO_x	nitrogen oxide
NPK	nitrogen, phosphorus and potassium
OECD	Organisation for Economic Co-operation and Development
OLED	organo light-emitting diode
OSHA	Occupational Safety and Health Administration
oz	ounce
P	phosphorus
PCFFA	Pacific Coast Federation of Fishermen's Associations
PES	payments for environmental services
PL480	Public Law 480
PTSD	post-traumatic stress disorder
PVP	plant variety protection
R&D	research and development
REDD	reducing emissions from deforestation and degradation
ROPPA	Network of Farmers' and Producers' Organizations of West Africa
RWR	renewable water resources
SARS	Severe Acute Respiratory Syndrome
S-CHIP	State Children's Health Insurance Program
SO_x	sulphur oxide
SSE	Seed Savers Exchange
TC	traditional confinement (system)
UAE	United Arab Emirates
UFW	United Farm Workers of America
UK	United Kingdom
UN	United Nations
UNICEF	United Nations Children's Fund
UPOV	International Union for the Protection of New Varieties of Plants
US	United States
USAID	US Agency for International Development
USDA	US Department of Agriculture
VOC	volatile organic compound

W	watt
WFF	World Forum of Fish Harvesters and Fish Workers
WFFP	World Forum of Fisher Peoples
WFP	World Food Programme
WHO	World Health Organization
WTO	World Trade Organization

1

Introduction

The 'cheap food argument': I hear it weekly. There was the instance, at a recent global food security conference, where an audience member was lambasted by someone on an expert panel for questioning the logic of the current food system: 'I'm interested in feeding the world, in producing as much food as cheaply and as efficiently as possible. What do you have against cheap food?' The audience member responded with a look of indignation but no words. Agricultural professionals, consumers, politicians and producers make the point repeatedly: if there was a more efficient (aka *cheaper*) way to produce, process and distribute food, we would be doing it. As evidenced by that silent rebuttal, the cheap food argument is quite powerful. It's an ace in the hole for proponents of the status quo. Whenever pressed into a corner, it can be pulled out at a moment's notice. Like at that global food security conference: 'What do you have against cheap food?' Check and mate.

This book centres on something purportedly central to the cheap food argument: cost. Its conclusions will not please proponents of the current food system. The cheap food produced in today's food system, as I am about to detail, is actually quite expensive. Can we afford to stop producing cheap food? How can we afford not to?

I am employed by a Land Grant University (historically, a bastion of pro-industrial food production sentiments) (Beus and Dunlap, 1990): Colorado State University. Studying food and agricultural policy and its impacts routinely places me face-to-face with arguments as to why we cannot afford to shift the course of our food system juggernaut. Some of those faces are friends, such as colleagues in the College of Agriculture who tell me it is not the *spirit* of alternative food system visions that cause them concern, but rather their *impracticality*. They recognize the benefits of eating more whole foods and fewer of the highly processed artefacts that populate grocery stores, fast-food chain menus and kitchen cabinets. They question the long-term sustainability of a global food system heavily based upon corn (maize) and soybean production. And they admit that after decades of advancements in plant breeding, millions still starve.

But, alas, in a case of failing to see the forest for the trees, they ultimately refuse to talk about systemic change. While well intentioned, they tell me, alternative food system visions cannot compete with the dominant food system in one important respect. The dominant system is our best chance at global food security because the one thing that it does better than all the alternatives is supply us with the cheapest food possible.

We spend less of our annual incomes today on food than any previous generation. The percentage of disposable income spent on food within the US has steadily decreased since 1947 (Miller and Coble, 2007, p98). Between 1970 and 2005, the percentage of disposable income spent on all food in the US dropped from 13.9 to 9.8 per cent, a decrease all the more remarkable given that approximately half of what US consumers spend on food is spent eating away from home (up from 30 years ago when the figure was 34 per cent and 50 years ago when the figure was 25 per cent) (USDA, 2008). The same for Britons: in 1957 (three years after the end of rationing), the average household spent one third of its income on food; today the figure is 15 per cent (Wallop, 2008). And for many people, the most persuasive indicator is found no further than their local grocery store. Who hasn't noticed organic (insert your favourite fruit or vegetable here) commanding more per kilogram than the conventionally grown variety? And internationally: is not the growing global obesity epidemic (see, for example, James, 2008) proof that the food system is delivering on its promise of cheap, abundant food?

Yet, is *cheap* the same as *inexpensive*? Or to put it another way, can we afford this cheap food? I argue *no*, on both accounts. Using the concept of 'cost' against cheap food, this book attempts to change the parameters of the food debate. As long as 'cheapness' – a concept I define in a moment – remains something lauded and desired, the status quo is at an unfair advantage. This is because, for many critics of the status quo, cheap food is precisely what they *don't* want. Cheapness, rather, is part of the problem, not something to be praised and replicated around the world – an unsustainable example of market and social failures of the first order. We need inexpensive food, food we can afford, and food that affords us certain capabilities. Changing the debate over food from being about *cheapness* to about *affordability* allows us to seriously ask questions such as 'Can we afford cheap food?' and 'What exactly does cheap food afford us?'

So what do I mean by 'cheap' food? While standard uses of the term tend towards such meanings as 'inexpensive' (a favourite definition among cheap food proponents) and 'inferior' (a definition held by many cheap food critics), I have another designation in mind. Cheap food refers to the de-contextual-ization of food in its broadest sense (more about this in the next section). At one level, then, we can think about cheap food as a euphemism for myopic economic accounting practices, exemplified by the thinking that the price of a food item at the grocery store reflects its full cost. This is the most obvious angle from which to critique cheap food. Agro-food scholars, food and environmental activists, and ecological economists have been arguing for decades about how today's food system rests upon the market (and society) turning a blind eye to many of its costs (see, for example, Tegtmeier and Duffy, 2004; Carolan, 2006; Colman, 2008).

But the market does not a world make. To say that cheap food is only the effect of externalizing costs, minimizes – dare I say, cheapens – our understanding of the world. It assumes that there is a market solution to the problems that compel me to write this book. It assumes that if only we could assign the right value to those things currently externalized, we will have solved the problem of cheap food. If only it were that easy. The food system is more than markets. So when we talk about cheap food, it is important that we contextualize, *really* contextualize, the food system, not only in terms of its environmental additions (pollution) and withdrawals (resources extracted), but also in terms of what it affords physiologically, culturally, economically and politically.

We need a system that produces food and feeds people, especially those who are starving. And, yes, this food – since the price of food at the grocery store matters – needs to be priced so that people can buy it. But let us also remember that price is a context-dependent concept. Slightly higher prices will not hurt someone's pocketbook if their pocketbook grows faster than the price of food. This gets at another understanding of the *afford*-ability of food. I am thinking here of the term's original meaning, which originates from the Old English word *geforthian*: to carry out. Affordability, following this usage, speaks to an artefact's *enabling* ability. Just like the sun affords plants the energy to grow, I want a food system that affords people and nations the capabilities to develop and enhance their overall well-being. Thinking of the food system in this manner, as something that affords society and individuals certain in/capabilities, allows for a more honest discussion to take place about what we want from food and whether the current system can achieve those desired ends. As discussed in later chapters, rather than affording those in the developing world the ability to obtain greater food (and economic, political, cultural, etc.) security, cheap food has had just the opposite effect. It creates *dis*abilities.

This book works to shift the debate about what food ultimately is. Rather than viewing food as a thing, a noun, it shows food as a process, a verb – an artefact best understood as part of, rather than apart from, a broader system of interconnections. Food-as-a-thing: this is as good a place as any to start the discussion. Cheap food is shorthand for understandings that reduce food to dangerously unsustainable levels, giving us, for example, grocery store prices that don't reflect food's total costs or discussions about food security that centre exclusively on bolstering yields or a country's per capita calorie consumption, while neglecting broader social, political, nutritional, economic and ecological concerns.

Rethinking 'Food' Itself

Like so many of the things I write about, the idea for this book was born in the classroom. We were discussing food security. It was a spirited discussion. Some students were critical of the current food system for reasons related to its long-term human and environmental consequences. Expecting this to come up, since most have heard by now about how industrial food production externalizes certain costs, I had on hand calculations that attempt to give a value to these various impacts. One study, focusing on agricultural production in the US in the areas of natural resources, wildlife, biodiversity and human health, estimated these externalized costs to be between US$5.7 to $16.9 billion annually (Tegtmeier and Duffy, 2004, p1).

A remarkable figure. Yet, such analyses sit uneasily with me for a number of reasons. To start with: how can one place a value on something like biodiversity or wildlife? I believe it was Einstein who said: 'Everything that can be counted does not necessarily count; everything that counts cannot necessarily be counted' – a useful quote to keep in mind when wading through these attempts to quantitatively internalize the costs of our food. Also, *who* makes these evaluative determinations? Extending upon a question posed by Robert Starr Allyn (1934, p46), who asked: 'Pray tell me, what does an onion taste like?' to critique the idea that plant patents specify plants in their entirety, I might ask: 'What is the value of an onion's taste?' Focusing only on value that can be quantified is a slippery slope. As other less quantifiable aspects of food get pushed to the wayside through these internalization exercises, we risk sliding back to where we began: talking about cheap food (only now arrogantly thinking that *all* costs have been internalized).

I eventually asked my students the nebulous question: 'When is food?' As expected, I got a lot of confused looks (perhaps I am evoking such a look now). 'Don't you mean *what* is food?', someone asked. After explaining that I asked the question correctly, I waited, quietly, for about 30 seconds. Silence. I then told them the purpose of the question: of how it plays on something Yrjö Engeström (1990) asked over 20 years ago. In a paper entitled 'When is a tool?', Engeström illustrates how a tool is not a thing with fixed attributes, but an artefact that *becomes* a tool in practice. A tool, therefore, emerges *in situ*; it is an effect of practice, structures and cultural conventions (see also Star and Ruhleder, 1996). Asking '*what* is food?', I told my students, assumes too much. It presupposes a fixed essence of food; it views food as a thing, a noun. But, as Harris (1986, p13) reminds us, 'We can eat and digest everything from rancid mammary gland secretions to fungi to rocks (or cheese, mushrooms, and salt if you prefer euphemisms).' The question '*when* is food?' situates food as part of – rather than apart from – this broader context.

This little exercise seemed to help my class better see the *system* part of the food system. I then proceeded to explain the various ways in which cheap food is really quite expensive. And it is not just expensive for those in developing countries, but to everyone, including those living in the US, a country which prides itself on having the cheapest food in the world. They appeared to get it. Instead of focusing on un-contextual factors, like yield per hectare or percentage of disposable income spent annually on food, the remainder of the class period was used discussing global food security in a more interconnected, complex and sophisticated way. We talked about the costs of the current food system to, among other things, the world's smallholder farmers, global equality, human health and well-being, the environment, taxpayers in developed nations, animal welfare, and cultural systems. I hope this book elicits a similar understanding of cheap food among readers.

The Audience

When I write I like to imagine who I am writing for. When putting together this book, I had in mind an international audience. For this reason, I tried not to limit its focus to any one country in order to increase its appeal to, and impact upon, the global community. That said, certain chapters take a long hard look at the US, in terms of its policies and practices. This simply cannot be avoided given its pivotal role in the world system. Thus, when focus is on the US, the intent is not to make the discussion only about the food policies and practices of this country, but to shine light on *why* today's food system looks like it does in terms of creating global winners and losers. The reader can be assured that the trip ahead will take them around the world.

I also imagined my students. The writing style, the content covered, the examples given, and the use of tables and figures – all have been informed by my years in the classroom. Since we're all students in some way, I hope the end product resonates both inside and outside the university.

I also hope that my vocabulary, by framing the argument around the *costs* and *affordability* of our food system, is something that proponents of today's food regime can relate to. When writing this book, I have tried to imagine how someone sympathetic to the current food system might be persuaded to revisit their convictions. Those already critical of how food is produced, traded, processed, transported and consumed will find in this book plenty more reasons to feel as they do. Yet, I am more interested in those who think, such as those friends of mine referenced earlier, that today's food system, warts and all, remains our best option because of its ability to produce copious amounts of cheap food. This book, I hope, will cause them to re-evaluate their mental ledger that tells them that cheap equals inexpensive. It most certainly does not. The dominant food system socializes many of its costs, while simultaneously privatizing the majority of its benefits. This is not only egregiously unjust, but makes for bad policy when the goal is affordable food.

Chapter Overview

Chapter 2 focuses upon the idea that affordable food is food that *affords* people the capabilities to pull themselves out of poverty and develop along trajectories of their own choosing – affordances that also produce 'savings' for the developed world. Specifically, the chapter examines cheap food through the lens of international development. In most developing countries, poverty is concentrated in rural areas (Pinstrup-Andersen, 2002). A popular argument among development scholars is that rural poverty in poorer nations can be alleviated by increasing the productivity – whether by traditional breeding methods or genetic engineering (GE) techniques – of the world's small-scale farmers (see, for example, Mellor, 1966; Pinstrup-Andersen, 2002; Paarlberg, 2008; Rao and Dev, 2009). The so-called Green Revolution, in fact, has been justified heavily by this logic. I discuss whether or not the Green Revolution actually delivered upon these promises. At the same time, I highlight how this attention to productivity has also acted as a smokescreen, directing energies away from other equally important problems that have plagued the world's smallholder farmers. Without access to markets, for example, yield increases benefit no one. The lack of markets for the world's smallholder farmers is a major problem – a problem that has only been exacerbated by policies promising cheap food. The ideology of cheap food seduces affluent nations to erect programmes that have crippled millions of farmers from less developed parts of the world. The latter half of the chapter discusses how (and why) free trade is rarely fair, especially when it comes to food, an argument that is bolstered by discussing the subjects of government subsidies, tariffs and food aid.

Chapter 3 examines the links between cheap food and conflict. To quote Susan Rice (2006, p1), currently US Permanent Representative to the United Nations: 'global poverty is far more than solely a humanitarian concern. In real ways, over the long term, it can threaten US national security.' Among other things, poverty and food insecurity reduce the recruitment costs for extremists and 'non-friendly' militia groups. When facing a certain future of poverty for oneself and one's family, the promise of food and money, even a little, has been successfully used to enrol members, from foot soldiers to suicide bombers, into conflicts (UN Office for the Coordination of Humanitarian Affairs, 2008; al-Mukhtar, 2010). This gives new meaning to the slogan 'food not bombs'. The relationship between food security and national security is only beginning to be explored, though we've known of this link for quite some time – after all, the theme of the 1999 World Food Prize was 'Food, agriculture, and national security in a globalized world'. Another timely example of cheap food-related conflict is that stirred up over immigration. Elaborating upon this link by looking at the current situation of the US, the chapter explains why comprehensive immigration legislation, to be truly *comprehensive*, needs to have a well thought-out food policy component. The chapter also looks at the phenomena of food riots as well as the recent rise of what has come to be known as the global land grab.

Chapter 4 explores the links between cheap food, health and obesity. Some figures to contextualize this discussion:

- The Food and Agriculture Organization of the United Nations (FAO) has calculated that world agriculture produces enough to provide everyone in the world with at least 2720 kilocalories (kcal) per person per day (FAO, 2002, p9).
- A recent study estimates that over 23 per cent of the world's population is

overweight and an additional 10 per cent obese (it was further calculated that by 2030, 58 per cent of the world population will be obese) (Kelly et al, 2008).

- The number of 'undernourished' in the world rose to 1.02 billion people during 2009, even after food commodity prices declined from their earlier peak in 2008. This represents the highest level of chronically hungry people since 1970 (FAO, 2009, p11).
- We are witnessing the rise of the 'obesity-hunger paradox' in developed countries. In the words of a recent *New York Times* (Dolnick, 2010) article: 'the hungriest people in America today [those living with chronic food insecurity], statistically speaking, may well be not sickly skinny, but excessively fat'. That's right: we are now seeing obese individuals suffering from persistent malnutrition.

How could we let this happen, where one quarter of the world is at risk of dying from eating too much, another quarter at risk of dying from eating too little, and some at risk of dying from *both* obesity and malnourishment? The chapter attempts to provide some answers to this question. To provide the reader with a bit of a teaser (without spoiling anything): the chapter argues that cheap food policies rest upon a certain understanding of 'food' – namely, one that centres on elemental components such as calories, proteins, vitamins and the like: what has been called the ideology of nutritionism (Scrinis, 2008, p47). As long as nutritionism informs international food policy, we will never solve the problem of global malnourishment. The chapter concludes by listing some rather startling statistics about the costs to our health – being paid for by taxpayers – that come from eating all of this cheap food. Reading these statistics, one can't help but wonder if the 'cheapness' of cheap food is not partially a product of cost shifting from the food sector to the healthcare sector.

Chapter 5 examines cheap meat. According to a recent article in the journal *Science*, animals currently utilize, either directly or indirectly, up to 80 per cent of the world's agricultural land (although, as discussed in greater length in Chapter 10, some of this land is only suitable for grazing). Yet they supply just 15 per cent of all calories produced (Stokstad, 2010, p810). Then there is the fact that livestock (some worse than others) poorly convert grain (e.g. corn) and plant proteins (e.g. soybeans) into animal proteins. This led Francis Moore Lappe (1971, p62) to famously quip decades ago that a concentrated animal feeding operation (CAFO) cow is a 'protein factory in reverse'. Cattle tend to receive the brunt of criticism because, in a CAFO environment, they are the least efficient of the lot. As Jeffery Sachs, director of the Earth Institute at Colombia University, explained in a speech at the 1999 World Food Prize, in an attempt to summarize the various costs of cheap beef:

> ... a kilogram of final beef consumption requires up to 16 kilograms of grain input. The water use, the fertilizer use, the land use to produce that means that 40 percent of our grain production now is for animal feed.

> The animal feed is soaring because meat production is rising. The United States stands way off the charts in this, of course. And the nutritionists tell us, persuasively, that our beef consumption is so high that it is highly deleterious to our human health at the same time. (Sachs, 1999)

To be sure, animal agriculture has its place in an affordable food system. Yet, I do not see how much longer we can afford the expense of meat that's produced as cheaply as we produce it today (see, for example, D'Silva and Webster, 2010). From its health effects to its impacts upon international food security, its environmental footprint, and its costs to animal welfare, the affordability of cheap meat is placed in serious doubt in this chapter.

Chapter 6 focuses on some of the ecological costs associated with producing, processing, shipping, packaging, marketing and consuming cheap food. The concept of 'food miles' is discussed (and extensively critiqued). A fair amount of time is also spent reviewing life-cycle analyses (LCAs) as they relate to components of our food system. This is followed by a thorough treatment of water as it relates to food, where concepts such as 'virtual water' and 'water footprint' are introduced and discussed. The subject of food waste is then reviewed, followed by a rather harrowing account of cheap food's addiction to (cheap) phosphorus.

Chapter 7 is, in many respects, a continuation of the previous chapter. Much of its focus remains on the cost of cheap food to the environment. The big difference is that discussion centres on the actual *pricing* of those costs. This chapter, more than any other, deals with more hardnosed economic assessments of the cost of cheap food. It begins by offering some brief economic explanations for why 20th-century agriculture built itself up around the substitution of external inputs for internal ecological controls. Discussion then moves to reviewing how some economists are now working to value ecological processes, such as ecosystem services, as they attempt to paint a more accurate picture of the real costs of our food. Other topics discussed in this chapter include the true cost of pesticides and attempts by economists to calculate the real price of meat.

Chapter 8 examines the costs of cheap food policies to communities and culture. I examine the negative impact that industrial large-scale farmers have had upon rural communities. Also discussed are the links between biological monocultures and cultural monocultures, and how cheap food policies increase the likelihood of both. The chapter then focuses on what I call monocultures of tastes, which allows me to talk about how today's food system narrows our collective understandings of 'food' itself – a trend with clear negative implications to both cultural and biological diversity. This brings me to a fourth monoculture brought about by today's food system: a monoculture of geopolitical space. This space is hinted at earlier in the chapter when noting that as communities become surrounded by industrial farms, there is a decline in local control over public decisions. This phenomenon is explored in greater detail through the empirical entry point of 'anti-GE' (genetically engineered) laws.

So: who actually benefits from cheap food? Having detailed the various ways in which cheap food is predicated upon a socialization of costs, Chapter 9 highlights the winners of these policies and practices. The real winners of cheap food lie in the 'middle' of the food system hourglass – an exclusive, but very profitable and influential, club. This represents one of those chapters where the US case dominates the discussion. My rationale for this is simple: in order to fully investigate this 'hourglass' structure, I needed to pick one from a single country and stick with it throughout the chapter. I do, however, regularly reference examples from other countries to highlight that the US case is representative of structural changes occurring throughout much of the world.

The chapter begins by sorting out the winners from the losers among producers, although, as we'll see, the category of 'farmer' has become terribly blurred thanks to contract farming, specifically, and vertical integration, more generally. Next I discuss food processors and manufacturers. This part of the hourglass has seen remarkable market concentration during recent decades, giving remaining large firms remarkable buyer power. The monopsony conditions – where buyers rather than the market set prices – that arise from this concentration are then discussed. The remainder of the chapter focuses on the retail sector, where large-scale supermarket firms such as Walmart are located. Of all 'links' connecting farm to fork, this is arguably the one where the biggest winners reside.

Chapter 10 lays out suggestions on how to make food more affordable. The reader will find the usual fare discussed towards this end. Farmers' markets, community-supported agriculture and community gardens are all pieces to the puzzle of affordable food. Yet, if I am to follow my own advice and contextualize, really contextualize, food, I must also recognize the need to go beyond the usual suggestions if we hope to ever turn the juggernaut of cheap food around. I therefore also discuss, for example, subsidies for consumers (particularly for the most food insecure), vertical agriculture, the need to adjust producer subsidies to better incentivize other forms of agriculture, the adoption of conservation practices among farmers, as well as polycultures and organic agriculture. Any serious discussion of affordable food also requires that we revisit how corporations do business and our complicity in letting those unaffordable practices continue. I therefore call for a re-examination of the concepts of 'free trade' and 'efficiency'. The goal: to make these terms more amenable to food systems that afford food security, sustainability and fairness, and that are just to all parties (human and non-human) involved. In addition, the chapter highlights cases where food is being kept sociologically and culturally relevant in an attempt to stave off those monocultures discussed in Chapter 8. Finally, in light of Chapter 5, I would be remiss if suggestions were not offered on how to make meat more affordable.

References

Allyn, R. S. (1934) *The First Plant Patents: A Discussion of the New Law and Patent Office Practice*, Educational Foundations, New York, NY

al-Mukhtar, U. (2010) 'Poor women in Iraq easily recruited by insurgents', *Statesmen.com*, 22 February, www.statesman.com/opinion/al-mukhtar-poor-women-in-iraq-easily-recruited-268225.html, last accessed 2 April 2010

Beus, C. and R. Dunlap (1990) 'Conventional versus alternative agriculture: The paradigmatic roots of the debate', *Rural Sociology*, vol 55, no 4, pp590–616

Bezemer, D. and D. Headey (2008) 'Agriculture, development and the urban bias', *World Development*, vol 36, no 8, pp1342–1364

Carolan, M. S. (2006) 'Do you see what I see? Examining the epistemic barriers to sustainable agriculture', *Rural Sociology*, vol 71, pp232–260

Colman, D. (2008) 'Ethics and externalities: Agricultural stewardship and other behavior, Presidential Address', *Journal of Agricultural Economics*, vol 45, no 3, pp299–311

Dolnick, S. (2010) 'The obesity-hunger paradox', *The New York Times*, 12 March, www.nytimes.com/2010/03/14/nyregion/14hunger.html?src=me, last accessed 6 April 2010

D'Silva, J. and J. Webster (eds) (2010) *The Meat Crisis: Developing More Sustainable Production and Consumption*, Earthscan, London

Engeström, Y. (1990) 'When is a tool? Multiple meanings of artifacts in human activity', in Y. Engeström (ed) *Learning, Working and Imagining: Twelve Studies in Activity Theory*, Orienta-Konsultit, Helsinki, pp23–35

FAO (Food and Agriculture Organization) (2002) *Reducing Poverty and Hunger: The Critical Role of Financing for Food, Agriculture, and Rural Development*, FAO, International Fund for Agricultural Development, World Food Programme, www.fao.org/docrep/003/Y6265e/y6265e00.htm, last accessed 6 April 2010

FAO (2009) *The State of Food Insecurity in the World, 2009*, FAO, International Fund for Agricultural Development, World Food Programme, ftp://ftp.fao.org/docrep/fao/012/i0876e/i0876e.pdf, last accessed 30 September 2010

Harris, M. (1986) *Good To Eat: Riddles of Food and Culture*, Allen and Unwin, London

James, W. P. T. (2008) 'WHO recognition of the global obesity epidemic: WHO and the obesity epidemic', *International Journal of Obesity*, vol 32, ppS120–S126

Kelly, T., W. Wang, C. S. Chen, K. Reynolds and J. He (2008) 'Global burden of obesity in 2005 and projections to 2030', *International Journal of Obesity*, vol 32, pp1431–1437

Lappe, F. M. (1971) *Diet for a Small Planet*, Random House, New York, NY

Mellor, J. (1966) *The Economics of Agricultural Development*, Cornell University Press, Ithaca, NY

Miller, J. C. and K. Coble (2007) 'Cheap food policy: Fact or rhetoric', *Food Policy*, vol 32, pp98–111

Paarlberg, R. (2008) *Starved for Science: How Biotechnology Is Being Kept Out of Africa*, Harvard University Press, Cambridge, MA

Pinstrup-Andersen, P. (2002) 'Food and agricultural policy for a globalizing world: Preparing for the future', *American Journal of Agricultural Economics*, vol 84, pp1201–1214

Rao, N. C. and S. M. Dev (2009) 'Biotechnology and pro-poor agricultural development', *Economic and Political Weekly*, vol 44, no 52, pp56–64

Rice, S. (2006) 'National security implications of global poverty', Lecture at University of Michigan Law School, 30 January, www.brookings.edu/~/media/Files/rc/speeches/2006/0130globaleconomics_rice/20060130.pdf, last accessed 2 April 2010

Sachs, J. (1999) 'Food at the center of global crisis', The World Food Prize, 2009 Norman E. Borlaug International Symposium, Food, Agriculture, and National Security in a Globalized World, Des Moines, IA, 14–16 October, http://208.109.245.191/assets/Symposium/2009/transcripts/2009-Borlaug-Dialogue-Sachs.pdf, last accessed 8 April 2010

Scrinis, G. (2008) 'On the ideology of nutritionism', *Gastronomica*, vol 8, no 1, pp38–48

Star, S. and K. Ruhleder (1996) 'Steps to an ecology of infrastructure: Complex problems in design and access for large-scale collaborative systems', *Information Systems Research*, vol 7, no 1, pp111–134

Stokstad, E. (2010) 'Could less meat mean more food?', *Science*, vol 327, no 5967, pp810–811

Tegtmeier, E. M. and M. D. Duffy (2004) 'External costs of agricultural productivity in the United States', *International Journal of Agricultural Sustainability*, vol 2, pp1–20

UN Office for the Coordination of Humanitarian Affairs (2008) 'Afghanistan: Poverty pushing youth into arms of Taliban?', *IRINnews.com*, 27 February, www.irinnews.org/Report.aspx?ReportId=76986, last accessed 2 April 2010

USDA (US Department of Agriculture) (2008) *Total Expenditures*, Economic Research Service, www.ers.usda.gov/briefing/CPIFoodAndExpenditures/Data/, last accessed 27 March 2010

Wallop, H. (2008) 'Britons spend one-fifth of income on homes', *The Telegraph* 29 January, www.telegraph.co.uk/news/uknews/1576933/Britons-spend-one-fifth-of-income-on-homes.html, last accessed 28 December 2010

2

Cheap Food, Globalization and Development

Once I went to a house where a farmer took his life by drinking a toxic chemical because of his uncontrollable debts. I could do nothing but listen to the howling of his wife. *If you were me how would you feel?* ... I believe the situation of farmers in many other countries is similar. We have in common the problems of dumping, import surges, lack of government budgets ... I have been so worried watching TV and hearing the news that starvation is prevalent in many less developed countries, although the international price of grain is so cheap. (cited in Rosset, 2006, pxiii; emphasis in original)

These words come from a pamphlet, distributed on 10 September 2003 at the World Trade Organization (WTO) Ministerial Meeting in Cancun, Mexico. Its author, Lee Kyung Hae – a South Korean farmer, founder of a South Korean farmers' association, ardent WTO critic, and inspiration to individuals around the world – is now dead. He killed himself later that day. A sign bearing the slogan 'WTO Kills Farmers' in one hand, Lee thrust a red penknife into his chest while standing on top of a police barricade. Within a matter of days tens of thousands of smallholder farmers from all around the world – from Bangladesh to Chile, South Africa and Mexico – marched in memory of Lee and in protest to the current food system. Heard among their chants of solidarity was one poignant phrase: 'We are Lee' (Patel, 2009, p35).

Any system that compels a farmer to take a penknife to his heart – an action in turn symbolically amplified by thousands of others in their taking to the streets – ought to be questioned. As Lee indicated in his pamphlet, cheap food does not equal development. Policies aimed at making the international price of grains so cheap has hindered (and continues to hinder) progress for many nations and for most of the world's smallholder farmers, as I'll now explain.

The War on Small Farms

Classical development theorists never looked too kindly on peasant agriculture. It's too labour intensive. Those bodies could be more efficiently utilized working in factories. Farmers need to be freed from an unproductive farm sector and put to work in the factory, where the real wealth is said to be generated. At the same time, the farmers who remain need to produce food more efficiently with the help of technology, inputs and advancements in seed breeding. Cheap food would not only help to feed a growing non-farming population, but would prevent rising food prices, which would in turn help to offset the initial low wages offered in cities. This act will redistribute wealth from the countryside to urban areas. And as more of a country's population moves to the city – why wouldn't they if that's where the jobs are? – more of the population will experience an increase in living standards (Lewis, 1954; Johnston and Mellor, 1961). Presto: cheap food offers a path to development.

Unfortunately, we have a case here of something that sounds good in theory but which has failed to deliver the fruits promised when put into practice. Why has cheap food been unsuccessful at best – and, arguably, disastrous in some cases – as a development tool? Let's look into this question.

The world's smallholder farmers are still under attack. As one economist recently quipped: as people are 'freed from the shackles of unremitting toil on the land ... [t]owns and cities become teeming hives of small-scale activity' (Ellis, 2005, p144). Yet, the facts, for much of the world at least, say otherwise. The non-farming sectors of developing economies have rarely grown fast enough to absorb the surplus labour freed from the 'shackles' of farming. Building an industrial sector to absorb displaced farmers presupposes an already established infrastructure – among other things, roads, a steady and reliable supply of energy, and a water and waste disposal system. If an industry cannot supply its factories with, say, raw materials and electricity, or efficiently distribute products once manufactured, the depth of a country's surplus labour supply matters little. This strategy, of constantly squeezing the farming sector to build urban economic capacity (an act equivalent to robbing Peter to pay Paul), has made developing countries food insecure, economically fragile and heavily dependent upon major grain-exporting nations. Let's not forget: the small family farm in nations such as the US was not systematically wiped out in the name of 'development'. To be sure, their numbers have fallen dramatically over the last century. But this was a process decades in the making. In many developing nations, conversely, the structural adjustments that are said to represent the road to prosperity can (and have) put millions of smallholder farmers out of business within just a matter of years.

No longer a mere 'handmaiden of industrialization', greater attention needs to be placed on the role of 'agriculture *for* development rather than agriculture *in* development' (Byerlee et al, 2009, p17). With an estimated 2.5 billion people worldwide engaged in production agriculture – recognizing also, according to the World Bank (2007, p1), that 'three of every four poor people in developing countries live in rural areas' – it's hard to think of another developmental strategy that's more pro-poor and more interested in the needs of the developing world. Simply put: spending money on agriculture, even small-scale agriculture, is money well spent. Studies have consistently shown that public investments in agriculture in poor countries, in terms of research and extension, yield higher societal returns than expenditures in other productive sectors (Fan et al, 2000; Bezemer and Headey, 2008). It is therefore quite disturbing to find that investment in agricultural development in developing nations is actually on the *decline* – for example, in real 2008 dollars, US investment in this area dropped from US$400 million a year during the 1980s to US$60 million in 2006 (Bertini and Glickman, 2009, p97).

The neglect of the rural poor in developing countries has been pointed to as evidence of what is called the 'urban bias' in international policy. Popularized by Michael Lupton (1977), the term refers to a tendency in developmental and international agricultural policy circles, dating back to World War II, to under-allocate resources to, and extract surplus from, the rural class of poor countries. The urban bias, in other words, is just as it suggests: a developmental bias towards urban areas and the sectors of the economy that tend to be located there. So why have these policies been allowed to continue even though they undermine the livelihoods of billions? If cheap food policies are so bad for the farmers in (and the countries of) the developing world, why do they persist? Many of those countries are democratic. Do not democratic nations have mechanisms – namely, *democracy* – to keep policies that harm major segments of the voting population from enduring?

The problem is that poor rural populations are perhaps the world's most politically disenfranchised group (Dasgupta, 1998; Grindle, 2004). This helps to explain why, for example, in India – the world's largest democratic state – 80 per cent of the population live in rural areas but 80 per cent of government spending goes into urban areas (Patel, 2009, p241). Throughout much of Africa and Central and South America, rural populations are geographically isolated not only from centres of power, but also from each other, making collective mobilization difficult (Bezemer and Headey, 2008, p1348). The story in Asia is slightly different. Many Asian countries had, and still have, a high rural density, which greatly reduces the transaction costs of organizing rural pressure groups. The threat of a rural-based Communist insurgency – in countries such as South Korea, Taiwan, Malaysia and Indonesia – also made the political elite histor- ically more sensitive to the interests of their country's small farmers (Bezemer and Headey, 2008, p1348).

But this book is not about international development. It is about cheap food and its costs and consequences to the world. Having provided a brief overview of some of the logics driving international development policy since World War II, which have been of enormous consequence to the world's smallholder farmers, I now return to the topic of cheap food. I have discussed some of the rationale lying behind why cheap food policy is viewed as good for the developing world (for a summary of these major points, see Table 2.1) Now let's see whether cheap food lives up the hype.

Table 2.1 *Summary of some rationales for why cheap food benefits developing economies*

Labour efficiency
- A countryside populated by smallholders is an inefficient distribution of labour.

Production efficiency
- Small farmers can't product food as cheaply in part because of their heavy reliance on labour (their own), whereas large farms, substituting capital for labour, can afford to make less per unit because they are producing many more units.

Redistribution of wealth from rural areas (and agriculture) to urban areas (and industry)
- A surplus of cheap grain not only feeds the growing urban populations but also helps ensure that some of the wages earned there can be used to buy more than just food.

Free Trade Is Rarely Fair

Equal rules for unequal players are unequal rules. No one would think it fair if two people played a game of basketball but one was forced to have their hands tied behind their back while they played. In many respects, this is what we are asking of the developing world: to play in the global food economy even though they lack the same capabilities as affluent nations.

Before a developing country can gain access to the international market, they have to first agree to abandon any and all policies that discourage, distort or in any way distract from free trade. I would say that they have to be more like the developed West except that countries such as the US and regions such as the European Union (EU) have very little interest in real (or perhaps I should say 'fair') free trade. Most developed nations so aggressively protect their domestic agricultural sectors that I do not know how anyone in good faith can say that their policies match their free market rhetoric.

The urban bias against agriculture in developing countries is, in part, the product of a bias in favour of agriculture in the developed world (Bezemer and Headey, 2008, p1350; Kay, 2009, p114), which can spend between US$6000 and $10,000 per farm labourer per year (Bezemer and Headey, 2008, p1350). For comparison, the typical African government spends less than US$10 per farm worker per year on agriculture (Bezemer and Headey, 2008, p1350). As agricultural economist E. Wesley Peterson (2009, pxv) notes in his book *A Billion Dollars a Day*, approximately US$1 billion are spent each day globally supporting agriculture, the vast majority of which is due to the spending priorities of wealthy nations. Take tariffs, which continue to hamper farmers in developing countries. While tariffs on industrial goods coming from developing nations have slowly been on the decline since the 1970s, tariffs on many of their agricultural goods remain high. This not only perpetuates trade distortions in the short term, but erodes long-term investment in these countries as tariffs place them at a comparative disadvantage.

There's an interesting concept – comparative advantage. The idea assumes that countries are equal by being unequal; that all countries have something special about them – have some natural advantage – that allows them to do certain things better than others. Countries simply need to find out what that something is and do it. For centuries, comparative advantage referred to a 'natural' attribute, like annual rainfall patterns or soil type. More recently, comparative advantage refers to phenomena such as a country's cheap labour or lax environmental laws, making me wonder if comparative *disadvantage* might not be a more apt term. But with agricultural subsidies and tariffs, the levers that dictate who has a comparative advantage are being pulled by those who have already developed. In agriculture, it is becoming less important what advantageous internal attributes are held by a country (such as climate or cheap labour). More relevant is whether or not countries such as the US *choose* to give their farmers a comparative advantage over the rest of the world. Today, comparative advantages can be *produced* by using such things as direct payments, export subsidies and tariffs. In short, with affluence comes the ability as a nation to choose whether to have comparative advantage in the production of almost any agricultural commodity.

Why would a developing nation agree to open up its markets to cheap agricultural imports if doing so harms their farming sector? Before borders are opened to trade, agricultural exports typically account for approximately 10 per cent of the agricultural revenue generated in developing economies. The vast majority of this export revenue goes to a small handful of very large farmers. Conversely, most of the remaining 90 per cent of revenue finds its way into the hands of small-scale farmers (Rosset, 2006, p82). Becoming part of the global economy gives the country greater access to the affluent markets of wealthy nations. But this access really only benefits a small percentage of the country's farmers because only the largest grow for export.

What, then, are these nations giving up in exchange? They are giving up precisely what they are gaining from developed nations: market access. In exchange for giving their landed elite greater market access to wealthy nations, these governments are pitting the heavily subsidized, large-scale, technologically aided Western farmer in direct competition against the peasant. I don't need to tell you which of the two produces cheaper food. Many economists are okay with this, asking: 'Why should a farmer remain in agriculture if they cannot compete with other, more "efficient", producers?' Forcing peasants to compete in the open market will result in a redistribution of capital, resources and labour either within or across economic sectors, which, in the long run, will increase the overall efficiency and wealth of the economy – or, at least, so says the law of comparative advantage (Weis, 2007, p119).

There is also the belief that if the world's smallholder farmers were to become more market oriented, they could remain competitive – a point articulated recently, for example, by the World Bank (2007). The strategy is known as 'reconversion'. Smallholder farmers are told to shift from traditional forms of production to growing non-traditional crops aimed at the highly profitable export market (Kay, 2009, p126). Assuming that peasants are in a position to 'read' these market signals, which itself is a big assumption, resting the future of the world's smallholder farms on non-traditional export markets is a bad bet. It ignores the realities faced by most of the world's rural poor.

Take the experience of Mexican farmers following the signing of the North American Free Trade Agreement (NAFTA) and the structural adjustments that came with it. NAFTA is said to have displaced up to 15 million Mexican peasants (Bello, 2009, p49). As is too often the case, while neoliberal planners are happy to see the redundant labour force of peasant agriculture reduced, far less thought is given to what to do with this labour after it has been 'freed' from subsistence agriculture. Proponents of NAFTA recommended that small-scale Mexican farmers get into non-traditional agricultural exports – namely, fruits, vegetables and cut flowers (Bello, 2009, p49). These recommendations, however, gloss over basic realities. First, there is the problem of finance. While corn requires an investment of approximately US$210 per hectare, non-traditional commodities typically come with a considerably higher per hectare investment: melons, US$500 to $700; cauliflower, US$971; broccoli, US$1096; and snow peas (mangetout), US$3145 (Bello, 2009, p49). Yet, as has happened in all developing countries, following structural adjustments, government credit evaporates, making the financing of such ventures impossible for the rural poor.

Then there are the barriers associated with international technological and production standards. All food coming into the US from Mexico must pass through United States Department of Agriculture (USDA)-certified packing stations – an additional link in the food chain that increases the costs of doing business in the export market (Bello, 2009, pp49–50). Smallholder farmers around the world are also finding new private standards – set in place by major supermarket firms – an additional expense that many can ill afford (Clapp and Fuchs, 2009). These standards can dictate any number of things, from how something is produced (the specific methods of production) to aesthetic standards telling producers how their commodities must look.

Most certifying organizations, which oversee the following of these standards, are located in affluent nations. Primus Labs in the US, for instance, certifies for 68 per cent of the fruit and vegetables firms in northwest Mexico (Narrod et al, 2008, p361). This can add considerable expense for farmers in developing countries, who frequently need certain certifications to penetrate high-profit niche markets in countries and regions such as the US and the EU (Fuchs et al, 2009, p46). The costs to obtain certification can reach as high as US$850 per hour (Narrod et al, 2008, p361). Unable to afford necessary investments (e.g. equipment, buildings) that go along with satisfying production standards imposed upon them by organizations in other countries, hundreds of thousands of smallholders are expected to go out of business in Africa alone (save for, perhaps, Kenyan farmers; see Box 2.1) (Fuchs et al, 2009, p46).

Box 2.1 The curious case of Kenya

Kenya's exports of approximately 450,000 tonnes of vegetables, fruit and cut flowers to UK and European markets have become the East African country's fastest growing economic sector. Generating US$1.3 billion a year, Kenyan horticulture brings in more revenue than banking, telecommunications and tourism (Manson, 2009), a remarkable fact given that the industry hardly existed just a couple of decades ago. The report *Kenya's Flying Vegetables*, published by the Africa Research Institute, gives a first person account of this transformation. The report's author, James Gikunju Muuru (2009), tells of his experience as a smallholder farmer in central Kenya – one of 4.5 million in Kenya – who grows green beans, tomatoes, cabbage, sweet potato and baby corn on his 4 acres (1.6ha) and makes enough to support his wife and six children. Muuru claims to make seven times as much from growing green beans and other export crops as he would from growing a more traditional crop such as corn. It's a reasonable claim. A study from 2005 of smallholder farmers who produced for export found that net farm incomes were five times greater than smallholder farmers who did not grow horticultural products for export (Weinberger and Lumpkin, 2005, p10). Moreover, only 10 per cent of the total weight of food grown in Kenya is exported (yet this 10 per cent represents 50 per cent of the industry's total value). The remaining 90 per cent remains within the region or country (Manson, 2009).

Only time will tell if Kenya's farmers can sustain this success. While more than half the exports are produced by smallholders, the total number of smallholders producing for export is relatively small (Minot and Ngigi, 2004, pi). The benefits from export are therefore unevenly distributed. Moreover, recent research indicates that the fastest growing segment of Kenyan farmers growing for export is a new group of medium-sized (and ready to expand further) commercial operations managed by well-educated farmers (see, for example, Neven et al, 2009).[1] What the case of Kenya *does* show us is the folly of the urban bias. A country can make significant developmental strides through agriculture. Indeed, in the case of Kenya, this development has largely occurred through agriculture alone.

Free market ideology expects all rational farmers to read market signals and invest in crops in which they have a comparative advantage. The realities on the ground, however, betray this philosophy – and not because most farmers are irrational (I am not about to disparage this hard-working segment of the population by calling them 'irrational'). For those already well-off farmers, perhaps such actions are possible. Yet 'development' should not just be about improving the quality of life of those least in need. As researchers for the United Nations (UN) have pointed out: 'free market rules in a context of highly concentrated property and imperfect and missing markets leads to the marginalization of otherwise perfectly viable enterprises' (David et al, 2000, p1685).

As for government subsidies and tariffs: how can these blatant trade-distorting practices continue even as institutions like the WTO are charged with reining in their use? Domestic supports are assigned to three 'boxes' – amber, blue and green (see Table 2.2). The amber box is reserved for those measures deemed trade distorting – namely, those that send signals to increase production (like an input subsidy). These are to be reduced, though not eliminated entirely. The blue box contains measures that limit production. Blue box measures are not restricted. Finally, there is the green box. This box is for measures *said* to have no effect on production. These too are not restricted by the WTO.

Table 2.2 *The WTO's coloured boxes for domestic support in agriculture*

Amber box: Includes all domestic support measures believed to distort production and trade. These measures include policies to inflate commodity prices or subsidies that directly encourage production. Member states are required to reduce such support unless current levels fall below certain parameters. Developed countries can keep a total of amber box support that is equivalent to up to 5 per cent of total agricultural production, plus an additional 5 per cent on a per crop basis. In developing countries their amber box support can go up to an amount that is equivalent to 10 per cent of total agricultural production.

Blue box: Measures are directed at limiting production. There are currently no limits on spending on blue box subsidies.

Green box: Subsidies that do not distort trade or that at most cause a minimal distortion (like government monies for research and extension). Green box measures can provide support for things like environmentally sound farm management practices, policies defined as being for regional or rural development, pest and crop disease management, infrastructure, food storage, income insurance and even direct income support for farmers as long as that support is 'decoupled' from production. There are no limits on green box subsidies.

Compiled from World Bank (2005, p25)and Rosset (2006, pp84–86)

There is little to say about amber measures. Since they are frowned upon by the WTO, affluent nations – so they can proudly claim their policies are not trade distorting – have disguised any distorting policies in a rhetorical veil of green. Blue box measures are aimed at production reduction. The US abandoned production controls back in 1996, followed more recently by the EU. Other examples of blue box policies are conservation programmes, where farmers are paid to pull their land out of production for conservation purposes (Rosset, 2006, pp84–86). Green measures are the most interesting because this is the box where countries hide their trade-distorting subsidies.

The US, EU and Japan account for 87.5 per cent of the world's total green box expenditures (Maini and Lekhi, 2007, p176). These 'decoupled' payments

(payments said to be independent of production levels and therefore non-trade distorting) can exist at unlimited levels. Yet, let there be no doubt: these payments affect production levels. Unlike production controls (also known as coupled payments), which provided payments to farmers who agreed to limit their production, green box measures pay farmers under the guise of something like 'income insurance'. The payment is *decoupled* because producers are assured a pay cheque regardless of how much they produce. Yet, in practice these payments still shield producers from low prices – prices that would otherwise, in a less distorted market, send signals to farmers to produce less or something else. Green box policies therefore *do* distort markets. They make farmers deaf to market signals, allowing them to continue to (over)produce and profit even when the costs of production exceed what the market is willing to bear. We should not be surprised, then, when farmers in the developing world fail to produce food as cheaply as US or European farmers. This is because farmers in developed countries cannot produce food that cheaply either, not without help from the government. And because these trade-distorting measures are disguised as green box measures, they are allowed to continue.

The costs of tariffs in developed nations, directed at agricultural commodities coming from the developing world, have been estimated at approximately US$11 billion per year (in 1995 US dollars) (Anderson et al, 2006, pp168–169). One study estimates that a 50 per cent reduction in agricultural tariffs would lead to a US$40 billion increase in the collective gross domestic product (GDP) of developing nations (ABARE, 2001). As an article in the influential journal *World Development* argues, 'conventional trade biases within OECD [Organisation for Economic Co-operation and Development] countries are still a formidable source of underdevelopment in LDC [least developed countries] agriculture' (Bezemer and Headey, 2008, p1351).

Food Dependency Undermines Food Security

Cheap food has put tens, perhaps hundreds, of millions of small-scale farmers out of business. But we are assured that that's okay. Being food independent, you see, is unnecessary in a globalized world. Indeed, from the perspective of neo-liberalism, food independence is the result of gross inefficiencies in terms of labour and resource allocation. Better to allocate resources according to the law of comparative advantage.[2] Find out what you do best as a nation and do it. For many developing countries, however, given the subsidy-dependent comparative advantage that affluent nations have in this sector, food production is probably not in the cards. Former US Secretary of Agriculture John Block made just this point in 1986: 'The idea that developing countries should feed themselves is an anachronism from a bygone era. They could better ensure their food security by relying on US agricultural products, which are available in most cases at lower cost' (cited in Bello, 2008, p452).

It's not like developing nations have much of a choice. Trade liberalization inevitably leads to a flood of cheap food imports. Farmers in the developing world are being asked to compete against farms in countries that spend billions propping up their agricultural sector. They are also competing against farms that benefit from living in nations with an extensive infrastructure (which reduces, among other things, transportation costs), readily available credit (at least until recently, when global credit markets dried up), rich agricultural research traditions, and a strong history of agricultural extension. Cheap imports are therefore dumped throughout the developing world. Legally speaking, they are not 'dumped'. Dumping is illegal under well-established international rules (Annand, 2005). Nevertheless, in point of fact – if not in point of law – dumping is what is occurring.

Looking at US exports, the Institute for Agriculture and Trade Policy (IATP) made the following calculations as to the percentage of exports that US farmers sold at average prices *below* the cost of production between 1997 and 2003: 37 per cent of all wheat; 11.8 per cent of all soybeans; 19.2 per cent of all corn; 48.4 per cent of all cotton; and 19.2 per cent of all rice (IATP, 2005, p2). Unfortunately, there is not much that poor countries can do to combat this practice. From a legal standpoint, a country must prove that they are harmed by the action in question. This is not as easy as it sounds. Cheap grain imports certainly benefit some – such as processors – which complicates the calculation. Yet there is a more practical problem that presents sufficient disincentive to keep developing countries from making too much of a stink over having their market flooded with cheap food.

In addition to their growing food dependence upon affluent nations, developing countries are finding themselves increasingly reliant upon nations and economic regions such as the US and the EU for their markets. For example, according to statistics compiled by the UK government, about 75 per cent of exports for Bangladesh are concentrated in textiles and over 90 per cent of those exports are destined for US and European markets (due, in part, to Bangladesh's cheap labour) (British Council, 2009). This dependence, however, is not reciprocal. Bangladesh exports destined for the EU totalled 5.5 billion Euros in 2008, while EU exports destined for Bangladesh total a little over 1 billion Euros (European Union, 2009). Countries such as Bangladesh have more to lose from trade sanctions than the USs and EUs of the world, giving the latter tremendous leverage over the trade policies of the former (Rosset, 2006, p42).

The above discussion highlights an important principle of globalization. Trade liberalization globalizes not only free market principles but also market failures (see Box 2.2) (Perez et al, 2008, p6). As markets in different countries become increasingly interdependent, governments lose flexibility in establishing democratically informed domestic policy instruments, although, as just discussed, the degree of flexibility lost is far from evenly distributed. And, again, we have another example of just how unequal equal rules can be when players with different capacities are made to compete against each other.

Box 2.2 The globalization of Ugandan Nile perch

Before the late 1980s, Uganda's fishery sector served local and regional markets. This began to change during the late 1980s, thanks to structural adjustment programmes – trade liberalization requirements that countries must follow if they hope to obtain development loans from international agencies such as the International Monetary Fund (IMF). Once open for international business, investment in the country's fishery sector soared, earning US$90 million, $101 million, $142 million and $146 million in 2003, 2004, 2005 and 2006, respectively, while providing employment to some 500,000 people (Fulgencio, 2009, p433). The fish at the centre of it all: Nile perch. This massive fish (an adult can weigh in excess of 100kg) represents over 90 per cent of Uganda's total fish exports.[3] The fish stock, however, is becoming serious depleted. Acoustic surveys a decade ago already showed that Nile perch in Lake Victoria, the source of this fish bonanza, had dropped from 1.9 million tonnes in 1999 to 1.2 million tonnes in 2000. Two surveys in 2008 show that this figure has since plunged to 299,000 tonnes (Fulgencio, 2009, p434). In 2009, Uganda's export earnings dropped 35 per cent as a result of dwindling stocks (Biryabarema, 2009), the direct result of overexploitation for the international market (the Nile perch is considered a delicacy in Europe, fetching US$9 to $10 per kilogram). Declining fish stocks directly threaten the livelihoods of all those living around the lake as these fish represent not only their main source of income, but an important source of nutrition as well (Nunan, 2010).

Meanwhile, the small fishers who first profited from the fishery sector's growth are being pushed out. From an economic standpoint, full exploitation of the Nile perch, given its remarkable size, requires state-of-the-art fishing and processing equipment and methods. Processors therefore started to vertically integrate – that is, they began to purchase their own fishing equipment and crews to obtain a greater supply (Schuurhuizen et al, 2006). Thus, while the Nile perch industry makes up nearly all of the country's export earnings, little ends up actually trickling down to the people and communities surrounding the lake (van der Knaap and Ligtvoet, 2010).

The net result of all of this is food dependence. Between 1950 and 1970 the developing world went from taking in no grain imports to accounting for almost half of the world imports (Friedmann, 1990, p20). Harriet Friedmann (1992) discusses this growing dependency in the context of the global wheat trade. Before World War II none of the nations in Africa, Latin America or South Asia imported wheat. This changed drastically in a matter of decades. Nigeria, for instance, was entirely self-sufficient in food during the 1960s. In 1983, one quarter of Nigeria's total earnings was spent importing wheat (Jarosz, 2009). What makes this all the more remarkable is that between 1959 and 1961 – as countries started to become hooked on this commodity – wheat cost considerably more than either corn (25 per cent more) or rice (600 per cent more). Countries around the world, regardless of their traditional dietary profile, began to consume increasing quantities of wheat as the century progressed.

The addiction started with imports coming from the US. After World War II, the US had large stocks of surplus wheat due to New Deal price support programmes. The US government sought to unload this surplus without harming market prices (at the time, wheat was only second to petroleum in terms of volume traded in the international market). The US provided the developing world wheat through 'concessional sales', at a highly subsidized rate, under Public Law 480 (aka 'food aid'). During the 1950s, the US share of wheat exports grew from one third to over half of the market, with the majority destined for the developing world. Between 1950 and 1976, per capita consumption of wheat increased in the developing world over 60 per cent, while per capita consumption of cereals (minus wheat) increased 20 per cent and per capita consumption of root crops (a tradition staple in many developing nations) decreased by 20 per cent. Why did these nations choose not to invest to become food self-sufficient and decide instead to be food dependent upon countries such as the US? 'The answer', according to Harriet Friedmann (1992, p372), 'is both material and ideological'. She continues:

> First, the USA wanted to get rid of surplus stocks of wheat it accumulated through domestic farm programmes, and this conveniently coincided with a mix of foreign policy and humanitarian goals. The US government, through the Marshall Plan, had invented foreign aid as a mechanism to increase trade despite lack of dollars by prospective importers. In 1954 it adopted this mechanism to food aid through PL480. Second, Third World countries welcomed cheap food, that is, wheat imports subsidized by the US government, as an aid to creating an urban working class.

Directing monies away from sectors whose products can be readily substituted by cheap imports makes sense in the short term. But what happens when the river of cheap imports runs dry? Whether because of significant weather events, massive grain sales or the emergence of sectors that rely upon grain as a cheap input (such as the emerging biofuel industry), the market can never be trusted to deliver low-cost agricultural commodities year after year. Note what occurred when the Soviet Union purchased grain from the US in 1972 and 1973 on a scale never before witnessed. (Russia, coincidently, is poised to make a purchase of similar scale due to the hottest summer ever in the country's recorded history, 2010; *Moscow Times*, 2010.) This sale produced sudden global shortages as prices quadrupled and food aid contributions nearly disappeared. Countries dependent upon, up until then, cheap imports found themselves dependent upon food that was suddenly not so cheap. And because they no longer had the domestic capacity to feed themselves they had no choice but to pay the higher price.

The promises of reconversion have also been short lived. Just as developing countries learned to substitute cheap imports for domestically produced grains, affluent nations found ways to substitute their own products for tropical agricultural commodities. Thus, over the last few decades, export markets for commodities produced in the poorer nations have shrunk as a result of industrial substitution. Some of the tropical commodities most affected by this process include cane sugar, which has been replaced with high-fructose corn syrup; cotton, which has been replaced by synthetic fibres; and palm oil, which has been replaced by soy and canola oils (Weis, 2007, p102).

The dependency described, however, does not just occur on the import side. Trade liberalization is making developing nations dependent on more affluent nations for their markets, which they need if they hope to sell their non-traditional commodities. Take the case of the African country Burkina Faso, when its leaders took a page from the neoliberalism playbook and called upon growers to raise green beans for export rather than for domestic consumption (see Freidberg, 2003). Initially, the strategy proved profitable for the country and its farmers. Eventually, however, as farmers in other countries entered the market, conditions started to resemble those before the switch was made to non-traditional crops for exports. And the country and many of its residents returned to a precariously insecure position (see Box 2.3).

The problem – or at least one of them – with the non-traditional export market is that it cannot possibly remain profitable over the long term for the world's poor farmers. Eventually others, with more capital and better access to credit, who have witnessed this profitability from afar, will want to get in on the action. Peasant farmers thus often meet the same demise growing for the export market – the market they are asked to believe will save them – as they did growing traditional commodities for domestic consumption. Their socio-economic position virtually guarantees such a fate under the current system.

The free market does not care for small farmers. When a country's borders are opened and the free trade process is set into motion, the sudden drop in farm gate prices immediately begins putting indebted farmers out of business. Low prices increase the minimum area needed to support a family, driving out some farmers while driving up the average farm size. As the price per agricultural unit decreases, the farmer must make up the difference by producing more units (Cochrane, 1993, pp417–436). Farmers must either (or ideally both) buy (or lease) more land or adopt productivity-enhancing technologies (e.g. higher-yielding seed). Of course, not everyone is positioned to successfully pull off this strategy. Lacking credit and capital, poor smallholders have little option but to either remain on the land and become poorer or abandon that way of life for a job in the city that may not exist – remember, in most developing countries the industrial sector is not growing fast enough to absorb the millions displaced by trade liberalization. The already well-off farmers in developing nations therefore become even more well off, while small farmers face a bleak future.

We must also remember the pressure felt by farmers from the other 'side'. Unlike most other sectors of the economy, farmers sell their products at wholesale prices but pay retail for their inputs (Magdoff et al, 2000, p12). And over the decades, the price of those inputs – such as fuel, seed and fertilizer – has

Box 2.3 The *haricot vert* of Burkina Faso

Drought, crop failure and hunger plagued Western African countries during the 1960s. While the problems were nothing new to the region, the 'solution' pushed by agents both within and outside the country was. Aid monies began flowing into the region to fund the construction of reservoirs and irrigation canals. These projects sought to retain water from the rainy season and distribute it over time to maximize agricultural productivity. In addition to some rice, most of the aid was directed at producing green beans for export into the French market. The French have one of the highest rates of per capita green bean consumption in all of Europe. The *haricot vert* (aka 'green bean') is an ever-present side dish, especially during the holiday season. Located in the southern hemisphere, Burkina Faso was able to supply green beans to the French during the winter holidays, when it is too cold to grow them in France. The labour-intensive nature of picking green beans also served the country of Burkina Faso by creating more jobs than many other commercial crops that where highly mechanized.

Up until the 1980s this strategy served the country and the people of Burkina Faso well, giving peasants a reliable living wage and increasing the country's foreign exchange. By the mid 1980s, Burkina Faso was the second largest African green bean exporter, behind Kenya. Yet, as happens in agriculture, market prices have since tumbled as other countries have sought a piece of the high-value export pie. The increased competition has made conditions very difficult for Burkina Faso farmers. Prices have dropped. French importers no longer provide advanced financing. And endless stories are told of buyers abandoning contracted growers before harvest for someone willing to sell their beans for less, forcing farmers (or farmer co-operatives) to scramble to find buyers at the last minute before their beans spoil (Freidberg, 2003, pp454–455).

been on the rise. Farmers are therefore squeezed from both ends – by declining farm gate prices and increasing input costs.

Is food not sufficiently different from other commodities to justify policies that would exempt it from WTO requirements that everyone play under the same rules (Rosset, 2006; see also Hendrickson et al, 2001, p728)? Treating food differently, it seems to me, would make trade more fair. The hand of the free market already seems to be on holiday when it comes to production agriculture. When prices go down, a clear signal that society wants less of that commodity and more of something else (whose price at that moment is on the rise), farmers do not respond by producing something else. In fact, often they produce *more* of exactly what the market is telling them not to. The law of supply and demand has failed us when it comes to producing a self-regulating market in agriculture. In the meantime, millions have been thrown under the bus of trade liberalization for the benefit of a few.

This is not a matter of breaking a few eggs to make an omelette. The question of whether trade liberalization policies are actually correlated with growth and prosperity for developing nations is an empirical one. And the data are not encouraging. An article in the *Journal of Developmental Economics* spells things out quite clearly: 'contrary to the conventional view on the growth effects of trade barriers, our estimation results show that trade barriers are positively and, in most specifications, significantly associated with growth, especially for developing countries, and they are consistent with the findings of theoretical growth and development literature' (Yanikkaya, 2003, p57). Other research, looking at sub-Saharan African economies, found little evidence that trade liberalization and the resultant agricultural exports have had a positive effect on this part of the world (McKay et al, 1997, p129). Even former US Secretary of Agriculture Dan Glickman has admitted to the failings of free market capitalism when it comes to issues of food security and international development. Noting the hypocrisy of trade liberalization arguments coming from affluent nations, Bertini and Glickman (2009, p99) write:

> No one has found a way to make the United States' own small farms competitive in a free market without public subsidies of one kind or another. Nonetheless, for decades, the World Bank and many Western aid agencies preached a rather purist version of free-market capitalism, without subsidies, as the solution to the hunger problems of developing countries. To those countries' government officials, many educated in the United States and Europe, it must have sounded like a treatise on chastity penned by Casanova.

Food Aid

Famed agricultural economist Willard Cochrane (2000) made the following astute observation: whereas demand for food grows mainly with population growth, and is therefore inelastic because the stomach is inelastic, production grows much faster as new technologies increase yields and as additional land is brought into production (though, as discussed in Chapter 10, there are limits to this growth).[4] The net effect is that supply will continue to outstrip demand, resulting in declining prices in the long term. Setting the wisdom of this argument momentarily to the side, Cochrane overstates the slowness by which demand for agricultural commodities grows. The biofuel industry, for one, has greatly expanded demand for grains. In addition, while the human stomach may not be all that elastic, the stomach of, say, cattle are. With the help of hormones and large doses of antibiotics, livestock have been able to consume much more grain today than they did just a couple of decades ago and produce much more meat and milk. Finally, I am not entirely sold on Cochrane's premise that the stomach is inelastic. This position seems to ignore the empirical evidence that points to how our stomachs have, at least to a degree, grown over the last few decades.

According to the US government-funded National Health and Nutrition Examination Survey (NHANES), for example, US men increased their calorie consumption from 1971 to 2000 by an average of 150 calories per day, while women increased their consumption by over 350 calories (Taubes, 2007, p250).

Criticisms aside, Cochrane's underlying point about the ever-present threat of overproduction cannot be summarily dismissed. Surplus production has plagued farmers in developed nations for well over half a century. Threats to capital accumulation were accentuated as European countries began feeding themselves as they built production capacity following World War II. By the 1950s, the stresses of overproduction were particularly acute to the world's top exporter: the US. As Cochrane reminds us, there is only so much that can be done from a policy perspective to increase domestic consumption. Our stomachs are only so big. The US National School Lunch Program (created in 1946) is an example of a domestic policy aimed at eliminating some of this agricultural surplus by way of funnelling excess into the stomachs of US school children. What the US really needed, however, were alternative outlets for its surplus grain. Once the European agricultural sectors began to rebound and no longer needed (or wanted) cheap grain imports, US growers sought out new markets with the help of food aid.

In 1954, the US government created Public Law 480 (PL480) with the passing of the Agricultural Trade Development and Assistance Act. The Marshall Plan showed how foreign aid could be used to increase trade dependence despite a country's lack of money. PL480 provided a tool that could be used to flood the developing world with surplus grains without negatively affecting market prices.

I discussed earlier how a country gives up a significant degree of autonomy when it agrees to open its borders to the global market. In addition to reducing policies deemed trade distorting (e.g. tariffs), the country is often required to follow certain macroeconomic policy reforms, such as exchange rate devaluations (Peterson, 2009, pp217–218). An overvalued exchange rate makes imported goods cheaper, while increasing the price of the country's exports. Cheap food is politically popular, especially considering that peasant farmers – those with the most to gain from higher commodity prices – exist on the political margins in most countries. Then add food aid to the equation – food that is either donated or sold at prices below commercial rates. Overvalued exchange rates and food aid together exert tremendous downward pressure on domestic market prices, making it impossible for smallholder farms to make a profit. Like the import dependency mentioned earlier, this exceedingly cheap food sounds all well and good – until, that is, the food aid runs out.

For developed nations, food aid offers a means to dispose of price-suppressing surplus without causing market prices to tumble. Food aid can, however, affect the price within recipient countries. The real pay-off for donating countries, however, comes after the aid stops. Poor nations suddenly find themselves with a domestic grain deficit as fewer farms are producing what had previously been so cheap thanks to aid. Developing nations are often left with little choice but to continue their grain relationship with more affluent nations. Only now, this relationship is maintained through more formal market channels. Countries such as the US know that food aid can produce dependency. To quote Senator Hubert Humphrey in remarks to the Agriculture and Forest Committee of the US Senate in 1957: 'if you are looking for a way to get people to lean on you and to be dependent on you, in terms of their cooperation with you, it seems to me that food dependence would be terrific' (cited in Weis, 2007, p66).

There was an emerging consensus during the late 20th century among development practitioners that domestic food reserves, which could be drawn upon in the event of a food shortfall, were an inefficient distribution of resources. Better to have, it was believed, financial reserves, which could then be used to buy food, than food reserves, which require such additional capital outlays as storage and processing facilities (Jarosz, 2009, p2077). There are a number of practical problems with this strategy beyond the obvious one concerning why one would wish to make countries dependent upon others for something as fundamental as food. I could see, in the abstract, why such a policy might make sense. Financial reserves are, by definition, highly liquid – that is, they can be converted into many things very quickly. Moreover, unlike grain, money does not spoil. Unfortunately, seeing the world through abstract macroeconomic principles is like observing a country from an aeroplane at 30,000 feet: much ends up being missed that could add considerably to the situation one is trying to observe, understand and improve.

Many of the capital requirements for holding food on reserve are still needed for food aid. Countries that food aid helps the least are those lacking basic infrastructural and institutional capacities that allow for its storage and equitable distribution (Clay et al, 1999; Jayne et al, 2002). As I write these words, the one year anniversary for the catastrophic magnitude 7.0 Haiti earthquake is fast approaching.[5] The main topic of discussion on the situation in Haiti today, at least in the US, involves the snail's pace at which aid is trickling into the country (in addition to the disastrous cholera epidemic and conflict-ridden presidential elections). More than US$9 billion has been pledged from countries around the world, but only a fraction of that has been delivered and spent. The airwaves are abuzz with people trying to figure out why, and accusatory fingers are being pointed in all directions, from the Haitian government to non-governmental organizations (NGOs). What we need to realize, however, is that good will and money can only do so much when seeking to aid a country in distress. Without already existing capacity to process and distribute aid, how can anyone expect efficient and equitable distribution, especially in the case of Haiti where whatever capacity that might have existed was destroyed by the earthquake? The same applies to food aid. Giving food without also working to build infrastructural and institutional capacities is a bit like giving someone water but without a cup to drink it from or the means to apply it to their parched fields.

A study published in the journal *Food Policy* and authored by World Bank economists examines the degree to which food aid may create disincentives to not only domestic food production (by suppressing domestic grain prices), but also in public and private investments in food production. The authors wished to determine if the benefits of food aid in addressing severe short-term hunger might be offset by the cost of increased long-term food insecurity (Ninno et al, 2007, p414). They looked at India, Bangladesh, Ethiopia and Zambia. As it turns out, their differences in population, income levels and economic structure helped to explain their contrasting experiences with food aid and agricultural development. The large populations of India and Bangladesh (1.05 billion and 144 million, respectively) make their food consumption requirements substantially greater than either Ethiopia (with half the population of Bangladesh in an area twice as large) or Zambia (with a population density one fiftieth that of Bangladesh). The demographic characteristics of India and Bangladesh indicate a greater possibility for gains due to economies of scale in marketing and storage, and their high population densities point to lower food distribution costs. Ethiopia, in particular, suffers from lower per capita income (US$100 per person) compared to India (US$480 per person) and Bangladesh (US$360 per person), which places its residents in a precarious position during periods of heightened food insecurity.

The researchers concluded that Bangladesh has been only marginally negatively affected by food aid in the long term. While food aid flows of wheat had lasted over 25 years, the country still managed to double its rice and wheat output. India experienced gains as well, but not of the magnitude witnessed in Bangladesh. In contrast, Ethiopia – the leading recipient of food aid for the last 30 years – is no more food secure today than in 1980. Three key factors were highlighted that enabled India and Bangladesh to achieve greater food security and, thus, reduce their need for food aid:

> First, Bangladesh and India maintained a political will and had donor support for long-term investments in production, including agricultural research, extension, irrigation and rural roads ... Second, food aid flows were small relative to the size of total consumption ... In countries where the size of food aid shipments is large relative to the size of the markets, and especially where the food aid commodity is a close substitute for major domestically produced staples, the risk of adverse price effects on production incentives are especially great. Third, food aid inflows were channeled through a public distribution system, with adequate public storage and careful management of the timing of arrivals of food aid and the distribution of food. Food aid distributed mainly through emergency relief programs in Ethiopia has been less effectively managed. (Ninno et al, 2007, pp422–423)

In other words, India and Bangladesh attained some positive movement towards greater food security *in spite* of food aid, not because of it.

It is clearly an overstatement to proclaim that food aid, in all cases, creates food dependence. Some countries, as the aforementioned study suggests, have managed the perils of food aid better than others. I am also aware that the example of 'Shining India' is not without its critics. Citing work by economist Utsa Patnaik, Raj Patel (2009, p30) tells of how 'statistical sleights of hand have enabled India's poor to vanish since the 1970s' – namely, by lowering the official threshold of 'poverty' from 2400 calories during the 1970s to 1890 calories by 2000. As Patel (2009, p30) explains, 'around half a billion people have been written out of poverty, by the simple expedient of shifting the goalposts'.

Part of the problem with food aid rests in how it is distributed. When an emergency arises, surplus food from affluent nations is donated to relief organizations working closely with the United Nations World Food Programme. Another route is to donate food to the governments of the countries in need, who then sell the food at below market prices to raise money for general development programmes. In either case, farmers in countries receiving aid often get less for their products once the aid enters the country (Peterson, 2009, p115). Because of this, most countries, save for the US, give cash to the World Food Programme. This explains why the US leads all other nations combined in food aid shipments. From 1995 to 2005, the US accounted for approximately 60 per cent of all food aid. The remainder came from the EU (25 per cent), Japan (6 per cent), Canada (5 per cent), Australia (3 per cent), Norway (less than 1 per cent) and Switzerland (less than 1 per cent) (Hanrahan and Canda, 2005, p5). Countries other than the US are not necessarily any less giving, just giving differently – namely, cash rather than grain.

The practice of giving grain to later be sold to generate cash seems inherently inefficient on a number of levels. First, selling aid cheaply – one estimate (IATP, 2005) places the price of these sales at between 30 to 50 per cent below what the market would otherwise bear – to generate revenue for development projects minimizes taxpayers' bang for their buck. It makes little sense to sell something for between 50 and 70 cents on the dollar to generate aid revenue when the whole dollar (which previously was used to buy grain) could be given directly to those in need. Just *who* are we trying to help with food aid anyway? Are we looking to fill stomachs or pocketbooks?

Furthermore (and to respond to the two previous questions), the majority of money spent on food aid is wasted on transportation. The US Agency for International Development (USAID) food aid budget in 2005 was US$1.6 billion. Of that, only 40 per cent (US$654 million) went to paying for food. The rest was spent on overland transportation (US$141 million), ocean shipping (US$341 million), transportation and storage in destination countries (US$410 million) and administrative costs (US$81 million) (Dugger, 2005). The US government requires that 75 per cent of all food aid be transported by US flag carriers regardless of cost. Then there is the requirement that 25 per cent of the cargo must pass through Great Lakes ports. So, for example, wheat grown in Kansas might have to go first to Chicago, where it is put on a freight train, before being sent south to the Gulf of Mexico. According to the Government Accountability Office (GAO), these requirements cost US taxpayers an additional US$70 per tonne in 2007 (Martin, 2010). The rationale behind this policy is national defence. Having a significant presence of US flag carriers out on the open sea enhances US military readiness. That's the argument at least. In actuality, many of the US flag carriers are owned by foreign companies with little actual military value (Martin, 2010). Food aid can thus take as long as six months to get from a US storage facility to a foreign village (Martin, 2010). In 2007, the Bush administration proposed using a portion of PL480 funds to make direct cash transfers, which would allow food to be purchased either locally or in neighbouring countries. This, it was argued, would greatly reduce transportation costs – thus increasing taxpayers' bang for their buck – and expedite the allocation of aid. Congress rejected the proposal for a host of political reasons (Peterson, 2009, p115).

Former US Secretary of the Treasury Paul O'Neil expressed sentiments held by many when, referencing food aid to Africa, he exclaimed: 'We've spent trillions of dollars on these problems and we have damn near nothing to show for it' (cited in Ncayiyana, 2007). To this, Jeffery Sachs (2006, p310) responds: 'It is no surprise that there is so little to show for the aid to Africa, because there has in fact been so little aid to Africa!' Subtracting for fees for consultants from donor countries, money put towards servicing African nations' debt, and the costs of administrating and transporting food aid leaves a magnanimous sum of 6 *cents* for each person in the form of actual aid. It's hardly shocking we have gotten nothing to show for our actions when the actual aid, after all the deductions are made, turns out to be so miniscule.

Notes

1   Neven et al (2009) argue that Kenyan smallholders pushed out of farming still benefit from this export strategy through the labour market (i.e. they become farm labourers). Yet, what happens when those middle-sized farms begin investing in labour-saving devices? What happens when the price of those export commodities drop as farmers, in other countries enter into the market to take advantage of premium prices? It is a lot easier to get out of farming than to get back in.

2   Neoliberalism is a political economic worldview based on rebranded classical economics (hence, the moniker 'neo'), emphasizing the sanctity of free markets, free trade and free enterprise. Chapter 10 offers a closer look at this philosophy and the problematic understanding of 'freedom' upon which it is premised.

3   See www.ugpulse.com/articles/daily/news.asp?about=Research+says+Nile+perch+fish+reducing&ID=6955.

4   This point draws upon what is known in economics as Engel's Law, after the 19th-century statistician Ernst Engel. Simply put, Engel's Law states that given the natural limits of our stomachs, as household incomes rise the proportion of income spent on food falls, even though (as often occurs in developing economies) actual household expenditures on food rise.

5   The event occurred on 12 January 2010, killing hundreds of thousands and leaving over 1 million homeless.

References

ABARE (2001) *The Impact of Agricultural Trade Liberalization on Developing Countries*, ABARE Research Report No 01.6, Australian Bureau of Agricultural and Resource Economics, Canberra, Australia

Anderson, K., W. Martin and D. van der Mensbrugghe (2006) 'Distortions to world trade: Impacts on agricultural markets and farm incomes', *Review of Agricultural Economics*, vol 28, no 2, pp168–194

Annand, M. (2005) 'Why antidumping law is good for agriculture', in A. Schmitz (ed) *International Trade Disputes: Case Studies in North America*, University of Calgary Press, Calgary, Alberta, pp63–86

Bello, W. (2008) 'How to manufacture a food crisis', *Development*, vol 51, no 4, pp450–455

Bello, W. (2009) *The Food Wars*, Verso, New York, NY

Bertini, C. and D. Glickman (2009) 'Farm futures: Bringing agriculture back to US foreign policy', *Foreign Affairs*, vol 88, no 3, pp93–105

Bezemer, D. and D. Headey (2008) 'Agriculture, development and urban bias', *World Development*, vol 36, no 8, pp1342–1364

Biryabarema, E. (2009) 'Uganda's fish export earnings drop 35 pct in 2009', *Rutgers Africa*, http://af.reuters.com/article/investingNews/idAFJOE60S0FJ20100129, last accessed 31 December 2010

British Council (2009) *Bangladesh Market Introduction*, British Council, www.britishcouncil.org/eumd-information-background-bangladesh.htm, last accessed 30 June 2010

Byerlee, D., A. de Janvry and E. Sadoulet (2009) 'Agriculture for development: Toward a new paradigm', *Annual Review of Resource Economics*, vol 1, no 1, pp15–35

Clay, D., D. Molla and D. Habtewold (1999) 'Food aid targeting Ethiopia: A study of who needs it and who gets it', *Food Policy*, vol 24, no 4, pp391–409

Clapp, J. and D. Fuchs (2009) (eds) *Corporate Power in Global Agrifood Governance*, MIT Press, Cambridge, MA

Cochrane, W. (1993) *The Development of American Agriculture: A Historical Analysis*, University of Minnesota Press, Minneapolis, MN

Cochrane, W. (2000) *American Agriculture in an Uncertain Global Economy*, Minnesota Agricultural Economist, University of Minnesota Extension Service, No 700, http://ageconsearch.umn.edu/bitstream/13175/1/mae700.pdf, last accessed 8 July 2010

Dasgupta, P. (1998) 'The economics of poverty in poor countries', *The Scandinavian Journal of Economics*, vol 100, no 1, pp41–68

David, M., M. Dirven and F. Vogelgesand (2000) 'The impact of the new economic model on Latin America's agriculture', *World Development*, vol 28, no 9, pp1673–1688

Dugger, C. (2005) 'Africa food for Africa's starving is road blocked in Congress', *New York Times*, 12 October, www.nytimes.com/2005/10/12/international/africa/12memo.html?ex=1286769600&en=0de1afa6dd7990e7&ei=5090&partner=rssuserland&emc=rss, last accessed 19 February 2010

Ellis, F. (2005) 'Small farms, livelihood diversification, and rural–urban transitions: Strategic issues in sub-Sahara Africa', in *The Future of Small Farms: Proceedings of a Research Workshop*, International Food Policy Research Institute and Overseas Development Institute, Imperial College, London, pp135–149,

http://citeseerx.ist.psu.edu/viewdoc/download?doi=10.1.1.139.3719&rep=rep1&
type=pdf#page=142, last accessed 25 June 2010

European Union (2009) *Trade Has Always Been a Principal Driving Factor of
EC–Bangladesh Relations*, Delegation of the European Union,
www.delbgd.ec.europa.eu/en/trade/index.htm, last accessed 30 June 2010

Fan, S., P. Hazell and S. Thorat (2000) Government spending, growth and
poverty in rural India', *American Journal of Agricultural Economics*, vol 82,
no 4, pp1038–1051

Freidberg, S. (2003) 'French beans for the masses: A modern historical geography of
food in Burkina Faso', *Journal of Historical Geography*, vol 29, no 3, pp445–463

Friedmann, H. (1990) 'The origins of third world food dependence', in H. Bernstein,
B. Crow, M. Mackintosh and C. Martin (eds) *The Food Question: Profits versus
People*, Monthly Review Press, New York, NY, pp13–31

Friedmann, H. (1992) 'Distance and durability: Shaky foundations of the world food
economy', *Third World Quarterly*, vol 13, no 2, pp371–383

Fuchs, D., A. Kalfagianni and M. Arentsen (2009) 'Retail power, private standards,
and sustainability in the global food system', in J. Clapp and D. Fuchs (eds)
Corporate Power in Global Agrifood Governance, MIT Press, Cambridge, MA,
pp29–59

Fulgencio, K. (2009) 'Globalisation of the Nile perch: Assessing the socio-cultural
implications of the Lake Victoria fishery in Uganda', *African Journal of Political
Science and International Relations*, vol 3, no 10, pp433–442

Grindle, M. (2004) 'Good enough governance: Poverty reduction and reform in
developing countries', *Governance*, vol 17, no 4, pp525–548

Hanrahan, C. and C. Canda (2005) *International Food Aid: US and other Donor
Contributions*, CRS (Congressional Research Service) Report for Congress, US
Library of Congress, Washington, DC,
www.au.af.mil/au/awc/awcgate/crs/rs21279.pdf, last accessed 14 July 2010

Hendrickson, M., W. Heffernan, P. Howard and J. Heffernan (2001) 'Consolidation
in food retailing and dairy', *British Food Journal*, vol 103, no 10, pp715–728

IATP (Institute for Agriculture and Trade Policy) (2005) *Agricultural Export
Dumping Booms during WTO's First Decade*, Press Release, Minneapolis, MN, 9
February, www.iatp.org/iatp/press.cfm?refid=89731, last accessed 30 June 2010

Jarosz, L. (2009) 'Energy, climate change, meat, and markets: Mapping the
coordinates of the current food crisis', *Geography Compass*, vol 3, no 6, pp2065–
2083

Jayne, T., J. Strauss, T. Yamano and D. Molla (2002) 'Targeting of food aid in rural
Ethiopia: Chronic need or inertia', *Journal of Development Economics*, vol 69, no
2, pp247–288

Johnston, B. and J. Mellor (1961) 'The role of agriculture in economic development',
American Economic Review, vol 51, no 4, pp566–593

Kay, C. (2009) 'Development strategies and rural development: Exploring synergies,
eradicating poverty', *The Journal of Peasants Studies*, vol 36, no 1, pp103–137

Lewis, W. A. (1954) 'Economic development with unlimited supplies of labour', *The
Manchester School*, vol 28, no 2, pp139–191

Lupton, M. (1977) *Why Poor People Stay Poor: A Study of Urban Bias in World
Development*, Temple Smith, London

Magdoff, F., J. Foster and F. Buttel (2000) 'An overview', in F. Magdoff, J. Foster and
F. Buttel (eds) *Hungry for Profit: The Agribusiness Threat to Farmers, Food, and
the Environment*, Monthly Review Press, New York, NY, pp7–21

Maini, K. and R. Lekhi (2007) 'Implications of World Trade Organization on dairy sector of India', in R. S. Jalal and N. S. Bisht (eds) *Emerging Dimensions of Global Trade*, Sarup and Sons, New Delhi, pp174–180

Manson, K. (2009) 'Kenya's food miles', *Boise Weekly*, 9 September, www.boiseweekly.com/boise/kenyas-food-miles/Content?oid=1168196, last accessed 28 December 2010

Martin, S. (2010) 'Restrictions on US food aid waste time and money', *Tampa Bay Times*, 8 February, www.tampabay.com/news/world/restrictions-on-us-food-aid-waste-time-and-money/1070813, last accessed 13 July 2010

McKay, A., O. Morrissey and C. Vaillant (1997) 'Trade liberalization and agricultural supply response: Issues and some lessons', *The European Journal of Development Research*, vol 9, no 2, pp129–147

Minot, N. and M. Ngigi (2004) *Are Horticulture Exports a Replicable Success Story? Evidence from Kenya and Côte d'Ivoire*, International Food Policy Research Institute, Washington, DC, www.ifpri.org/sites/default/files/publications/eptdp120.pdf, last accessed 28 December 2010

Moscow Times (2010) 'Ministry says it will import feed grain', *Moscow Times*, 29 November, www.themoscowtimes.com/business/article/ministry-says-it-will-import-feed-grain/425153.html, last accessed 15 December 2010

Muuru, J. G. (2009) *Kenya's Flying Vegetables*, Africa Research Institute, London, www.africaresearchinstitute.org/files/policy-voices/docs/Kenyas-Flying-Vegetables-Small-farmers-and-the-food-miles-debate-0V6S400WZM.pdf, last accessed 28 December 2010

Narrod, C., D. Roy, B. Avendano and J. Okello (2008) 'Standards on smallholders: Evidence from three cases', in E. McCullough, P. Pingali and K. Stamoulis (eds) *The Transformation of Agri-Food Systems: Globalization, Supply Chains, and Small Farmers*, Earthscan, London, pp355–372

Ncayiyana, D. (2007) 'Combating poverty: The charade of development aid', *British Medical Journal*, vol 335, pp1272–1273, www.bmj.com/cgi/content/full/335/7633/1272, last accessed 3 August 2010

Neven, D., M. Odera, T. Reardon and H. Wang (2009) 'Kenyan supermarkets, emerging middle-class horticultural farmers, and employment impacts on the rural poor', *World Development*, vol 37, no 11, pp1802–1811

Ninno, C., P. Dorosh and K. Subbarao (2007) 'Food aid, domestic policy and food security: Contrasting experiences from South Asia and sub-Sahara Africa', *Food Policy*, vol 32, pp413–435

Nunan, F. (2010) 'Mobility and fisherfolk livelihoods on Lake Victoria: Implications for vulnerability and risk', *Geoforum*, vol 41, no 5, pp776–785

Patel, R. (2009) *Stuffed and Starved: The Hidden Battle for the World Food System*, Melville House Publishing, Brooklyn, NY

Perez, M., S. Schlesinger and T. Wise (2008) *The Promise and Perils of Agricultural Trade Liberalization: Lessons from Latin America*, White Paper, Washington Office on Latin America, Washington, DC

Peterson, E. W. (2009) *A Billion Dollars a Day: The Economics and Politics of Agricultural Subsidies*, Wiley-Blackwell, Malden, MA

Rosset, P. (2006) *Food Is Different: Why We Must Get the WTO Out of Agriculture*, Zed Books, New York, NY

Sachs, J. (2006) *The End of Poverty: Economic Possibilities of Our Time*, Penguin, New York, NY

Schuurhuizen, R., A. Van Tilburg and E. Kambewa (2006) 'Fish in Kenya: The Nile perch chain', in R. Ruben, M. Singerland and H. Nijhoff (eds) *Agro-Food Chains and Networks for Development*, Springer, The Netherlands, pp155–164

Taubes, G. (2007) *Good Calories, Bad Calories: Challenging the Conventional Wisdom on Diet, Weight Control, and Disease*, Knopf, New York, NY

van der Knaap, M. and W. Ligtvoet (2010) 'Is Western consumption of Nile perch from Lake Victoria sustainable?', *Aquatic Ecosystem Health and Management*, vol 13, no 4, pp429–436

Weinberger, K. and T. Lumpkin (2005) *Horticulture for Poverty Alleviation*, Working Paper No 5, World Vegetable Centre, Shanhua, Taiwan

Weis, T. (2007) *The Global Food Economy: The Battle for the Future of Farming*, Zed Books, New York, NY

World Bank (2005) *Agriculture Investment Sourcebook*, World Bank, Washington, DC

World Bank (2007) *World Development Report 2008: Agriculture for Development*, World Bank, Washington, DC

Yanikkaya, H. (2003) 'Trade openness and economic growth: A cross country empirical investigation', *Journal of Developmental Economics*, vol 72, no 1, pp57–89

3

Cheap Food and Conflict

What do the French Revolution, the Russian Revolution and World War II have in common, minus the obvious (such as the tremendous loss of life)? One common thread is food. A bread crisis further agitated the French masses by amplifying the inequalities between the Crown and Church and everyone else. Food protests in Russia had much the same effect. And Hitler's own need for conquest was rooted, in part, by a thirst for agricultural land – what Nazi propaganda continually referred to as *Lebensraum* (Cribb, 2010, p14). This chapter explores the links between cheap food policies and conflict. The conflict I am talking about, however, need not involve all-out civil war or an existential battle such as the one precipitated by Nazi Germany where the very existence of numerous nation states was on the line. The types of conflict discussed in this chapter may not be of the same magnitude as, say, either the French or Russian Revolutions (where 0.5 million and 9 million lives were lost, respectively). But they still involve the loss of life, the destruction of natural habitat, the pillaging of scarce resources and the weakening of (often already weak) states. Conflicts don't have to be of the scale of World War II to cast a long shadow over whoever is unfortunate enough to be living amongst the bloodshed.

This chapter corroborates the following hypothesis: cheap food policies increase the likelihood of inter- and intra-state conflict and thus increase the risk of global *dis*order. In today's integrated economy, instability in one part of the world can easily create worldwide ripple effects. Conflict is not cheap. Its costs are both numerable (we can count dead bodies and money spent to send UN peacekeepers into a region) and innumerable (like species extinction or the loss felt by a child after losing a parent). More than 3.6 million civilians died during internal conflicts in the 1990s, with children and teens accounting for over 50 per cent of all battlefield casualties (UNCTAD, 2004, p163). Estimates place 'typical' civil war costs at US$64.2 billion, a figure that includes the value of lost output, lost lives and healthcare (Addison, 2005, p2). Two civil wars, in other words, cost more than all the aid given annually to the developing world, which in 2009 totalled US$119.6 billion (despite the global recession this figure rose 0.7 per cent in real terms from the previous year) (OECD, 2010).

Moreover, we know conflict exacerbates food insecurity. Wars typically exact a high toll upon rural areas (Silberfein, 2008, pp85–86). Livestock and crops risk being plundered. Domestic agricultural markets can be disrupted, as can the infrastructure that transports commodities (Brückner and Ciccone, 2009, p1). And if the fighting is sufficiently widespread, the farmers who remain may be unable to raise anything as once cultivated fields become battlefields (Silberfein, 2008, pp85–86). A potentially very expensive spiral can thus ensue: cheap food policies, which reduce food security throughout much of the developing world, increase the likelihood of conflict, which further weakens food security, thereby prolonging the conflict (or increasing the chance of a flare-up the moment peace appears within grasp), and so on.

My point is not, however, to focus exclusively on war. War is only the most extreme form of conflict, and as the most extreme it is also the most expensive, which, after showing its links to cheap food policy, reveals some fundamental accounting errors in how we evaluate the costs of our food system. Other cheap-food-related conflicts discussed in this chapter include food riots and clashes brought on by the so-called global land grab. Even disputes tied to immigration – a hotly contentious issue in the US and throughout much of Europe – have roots deeply linked to cheap food policy.

The Scourge of Poverty

Here are some data to better contextualize the discussion:

- A country with US$250 gross domestic product (GDP) per capita has a 15 per cent risk of experiencing civil war within the next five years. In a country with US$5000 GDP per capita, that risk is 1 per cent (Rice, 2010, p7).
- The Central Intelligence Agency (CIA's) State Failure Task Force found that states with high levels of human suffering, as measured by high infant mortality, are 2.3 times more likely to fail and collapse into conflict than others (Rice, 2006).
- A study examining sub-Saharan Africa showed that countries produced, on average, 12.4 per cent less food per capita during war years than in peacetime. It was also estimated that peace would have added 2 to 5 per cent to sub-Saharan Africa's food production per capita per year (Messer et al, 1998).
- The Food and Agriculture Organization of the United Nations (FAO) calculated that from 1970 to 1997 the developing world experienced agricultural losses of US$121 billion due to conflict. In sub-Sahara Africa, the FAO estimated conflict-induced lost agricultural output during the 1980s and 1990s accounted for more than 50 per cent of all food aid received during that period and greatly surpassed foreign investment flows (FAO, 2000).

- Oxford University economist Paul Collier (2003) finds that a doubling of a country's per capita income halves the country's risk of internal war.

Norman Borlaug has been quoted as saying: 'we cannot build peace on empty stomachs' (cited in Pinstrup-Anderson and Shimokawa, 2008, p513). I have no reason to disagree with this statement. A country full of empty stomachs tells me that the country is poor and the state weak: a scenario ripe for conflict.

To quote a highly cited study looking into the links between food production, development and conflict: 'We conclude that the rehabilitation of agriculture [in the developing world] is a central condition for developing, reducing poverty, preventing environmental destruction – and for reducing violence' (de Soysa and Petter Gleditsch, 1999, p16). Poverty erodes state capacity and aids in recruitment for extremist groups. Take, as an example, Mali (a landlocked country in Western Africa). Like other poor countries, Mali's government lacks resources and the capacity to provide for its citizens. A significant portion of its population lacks access to clean water, basic healthcare, educational opportunities and jobs. Other organizations try to fill the many gaps in services left open by the state, from international aid groups to extremist Wahhabist charities. The latter offer not only food, clothing, medicine and education, but also the opportunity for young men to travel to Saudi Arabia for religious training (Rice, 2010, p9).[1] Food's fundamental nature makes it a particularly attractive recruiting tool. There is evidence, for example, that Islamic fundamentalists in Afghanistan and Pakistan have been using free food to attract hungry youths into radicalized *madrasahs* (Bertini and Glickman, 2009, p105).

It has long been believed that free trade is the ultimate peacekeeper (see, for example, Brace, 1879; Griswold, 2005). But a closer look at the data complicates this view. Trade reduces conflict incentives only if other political, economic and social stressors are absent and when trading partners are relatively equal in economic and military terms (Schneider et al, 2003). The latter point alludes to the argument, made in Chapter 2, that free trade cannot be fair when there is an asymmetrical power relationship between trading partners. And those asymmetries are only growing. An Organisation for Economic Co-operation and Development (OECD) study conducted during the 1990s shows a disturbing trend of *increasing* global inequalities. The income range between the most and least affluent countries has been growing at an alarming rate. In 1802 this range was just over 3:1; in 1870 the range increased to 7:1; by 1950 it hit 35:1; and in 1992 the ratio reached 72:1 (Maddison, 1995, p22).

To the point that trade promotes stability and peace: in many respects it has had just the opposite effect. Trade in agricultural commodities has been disruptive for newly 'opened' countries, contributing significantly to food insecurity and conflict. Declines in demand for sugar cane, for example, due to competition from cheap sweeteners (namely, high-fructose sugar) and US and European domestic market quotas have destabilized countries that depended heavily on revenue generated through its export (Messer and Cohen, 2006, p21). Overexpansion in the production of cotton and coffee, leading to rapid changes in global market prices, has also been conclusively linked to cases of intra-state conflict, especially in African countries where the problem of weak state capacity is particularly acute (Uvin, 1996; Daviron and Ponte, 2005).

We have to be careful, however, in how we respond to the links between food insecurity, development and conflict. For one thing, at least when it comes to terrorism, leaders are rarely poor. The 9/11 hijackers were predominately well-off educated men from Saudi Arabia. But poverty is unquestionably a conditioning factor when it comes to increases in conflict. Poverty reduces recruitment barriers, further erodes state capacity, and creates openings for scapegoating, which can be a useful recruiting tactic. We must not, however, make the poor *themselves* the problem, which opens the door for them being targeted militarily. For a number of reasons, governments find it easier to exercise their military might than to risk being perceived as intruding in the market by implementing redistributive measures. It is also politically easier to pass appropriation bills for the military than for foreign aid, even though the latter might avoid the expense of the former (Ferraro, undated, p6). Turning a war on poverty into a war on the poor justifies expanding the state's monopoly on violence and reduces international sympathy towards those caught in the middle who face poverty and hunger on a daily basis but who refuse to be recruited into the conflict.

Research looking at the political economy of war rarely examines the question of whether food insecurity contributes directly to conflict. This is understandable given the complexity of the variables involved, the difficulty in establishing causality, and the variability in how researchers conceptualize both 'conflict' and 'food security' (Messer and Cohen, 2006, p13). This is not to say that the literature is entirely silent on the subject. There is a well-documented relationship between conflict and proxy indicators of food insecurity, such as high infant mortality (Esty et al, 1998, p51), high childhood mortality (Pinstrup-Anderson and Shimokawa, 2008, pp513–514), high childhood malnutrition (Pinstrup-Anderson and Shimokawa, 2008, pp513–514) and high childhood under-nutrition (Pinstrup-Anderson and Shimokawa, 2008, pp513–514). There are also a number of studies that include food production as part of a collection of variables when modelling pathways to conflict. For example, Nafziger and Auvinen (2000) found that between 1980 and 1995, poverty, low agricultural output, high levels of inequality and large military budgets reduced a country's ability to respond to crises.

As mentioned earlier, free trade of agricultural commodities can enhance peace in a region, but only if other social, economic and political variables are present. High-value crops for export *can* contribute to poverty reduction and reduce the risk of conflict, but only *if* pro-poor policies, practices and institutions are present. In their absence, trade can have devastating effects on the poor. Pro-poor policies would include guaranteeing small farmers (especially women) access to land (women account for between 60 and 80 per cent of all peasant farmers), capital, credit, education, basic healthcare and information (Watkins and von Braun, 2003, p8; Messer and Cohen, 2006, p19). These programmes must be actively pursued to counteract cheap food's often harmful effects on small farms (see Box 3.1).

Box 3.1 Blood bananas

Chiquita Brands International (previously known as the United Fruit Company) has a history of allying itself with national elites and other powerful groups to control land, markets, wages, and domestic agricultural and trade policies within the countries where it owns land (Messer and Cohen, 2006, p20). Recent actions by the US Justice Department brought a rather startling example of this corporate influence to light (Associated Press, 2007). In 2007 the company agreed to a US$25 million fine after admitting its involvement with groups in Colombia that the US government designated in 2001 as terrorist organizations. One of the groups – the United Self-Defence Forces of Colombia – is responsible for a significant portion of the country's cocaine exports. Court documents indicate that Chiquita paid close to US$2 million between 1997 and 2004 to right-wing paramilitaries and leftist rebels in exchange for protection of its workers and land (Chiquita sold its Colombian banana operations in 2004). The US Justice Department found that the company had disguised the payments in company books and that no later than September 2000, senior executive know about its company's paramilitary links.

Take the case of Afghanistan. The US had hopes that a democratic Afghanistan would eventually make its agricultural sector rich by exporting such commodities as fruit, nuts and cotton (Messer and Cohen, 2006, p20). Unfortunately, given the depressed market prices of legal agricultural commodities, profits from opium have simply been too good to pass up for many Afghan farmers. The amount of revenue generated from 1 acre of wheat is one tenth of what that same acre would generate under opium cultivation (Addison, 2005, p11). Afghanistan's opium trade is said to be US$2.6 billion a year, which equals 60 per cent of the country's GDP (Addison, 2005, p11). This money is then used by the country's warlords for both peaceful and non-peaceful ends. A similar story can be told about Colombia. Cheap food (and coffee) places Colombian coffee growers in a precarious position at the best of times. So when the price of coffee in 2002/2003 hit an all-time low, many Colombian coffee growers turned to coca production. Few other profitable options were available as the state did little to help transition farmers to legal alternative markets (Keyzer and Van Wesenbeeck, 2004, p550; Messer and Cohen, 2006, p20). Colombia and Afghanistan are extreme cases. Yet they point to a fundamental truth: that export cropping in itself does nothing to promote peace; indeed, in some cases it might have just the opposite effect. 'Rather', as Messer and Cohen (2006, p31) observe, 'the structures of production and markets and the food and financial policy context [are what] ... determine local household incomes and peaceful or belligerent outcomes.'

Food Crises and Riots

Between 2006 and early 2008, global food prices rose by over 80 per cent (Loewenberg, 2008, p1209). By December 2008, almost 1 billion of the world's 6 billion inhabitants were chronically hungry (Lawrence et al, 2010, p1). In 2007, a one-day 'pasta strike' occurred in Italy in response to a 20 per cent increase in the price of this food staple (*The Guardian*, 2007). That same year, 75,000 people took to the streets in downtown Mexico City to protest about price increases for such food staples as tortillas and to demand changes in food policy (Grillo, 2007). In April 2008, Haitian protestors, demanding the resignation of President René Préval, attempted to break through the Presidential Palace gates in protest at rising food prices (they were ultimately driven back by Brazilian United Nations peacekeepers). The riots left four protesters dead and some 25 people injured (*Carib World News*, 2008). The price of basic foods such as rice, beans and condensed milk had increased by more than 50 per cent. In a country where 80 per cent survive on less than US$1 a day, and where the poor rely on biscuits made from mud to fill their stomachs (the mud is mixed with oil and sugar), such price increases pushed the vast majority of Haitians into a state of extreme food insecurity (Doyle, 2008). A mother working in the markets of Port au Prince was interviewed in 2008, explaining that before the food crisis, US$1.25 could buy vegetables, some rice, 10 cents of charcoal and a little cooking oil. At the height of the crisis, she was quoted as saying that with US$1.25 'You can't even make a plate of rice for one child' (Quigley, 2008).

A number of factors have been blamed for this drastic price spike: declines in productivity due to drought and land degradation; the utilization of agricultural commodities for the making of biofuels; increases in fuel costs (which increase both production and transportation costs); rising incomes in India and China, which lead to increases in meat consumption (meaning more grain is used as feed for livestock); and financial speculation in agricultural commodities. But what appear as causes are themselves symptoms of a deeper problem. Until we see food crises as products of a problematic system, and not as something for the system to solve, food crises and the riots, protests and civil disobedience that they evoke will most assuredly be normal occurrences in the future.

Timothy Wise (2010) recently wrote: 'Cheap food causes hunger.' Cheap food also causes food crises. How could an arrangement, which prides itself on supplying food that is both cheap and abundant, actually *cause* hunger to the point where, for example, mud biscuits become a food staple among the poor of a nation? Some of my colleagues view food riots as 'agential moments that can, in some cases, be understood as a movement toward an alternative best captured in the term "food sovereignty"' (Patel and McMichael, 2009, p11). Perhaps. It is encouraging to think that food riots, acts often promulgated by individuals whose pain and anger are matched only by the emptiness of their stomachs, 'provide us with opportunities to change and improve the way we do things' (Holt-Gimenez, 2010, pxxiii). Who wouldn't like to see food crises become a thing of past? If we ever hope to attain this goal, we have to first understand how cheap food is part of the problem, not part of the solution.

Take the case of the Philippines. Walden Bello observes that when it comes to food security, dictator Ferdinand Marcos had, sadly, a better track record than either the World Bank or the International Momentary Fund (IMF). As Bello explains (2008, p451):

> To head off peasant discontent, the regime provided farmers with subsidized fertilizer and seeds, launched credit schemes, and built rural infrastructure. During the 14 years of the dictatorship, it was only during one year, 1973, that rice had to be imported owing to widespread damage wrought by typhoons. When Marcos fled the country in 1986, there were reported to be 900,000 metric tons of rice in government warehouses. Paradoxically, the next few years under the new democratic dispensation saw the gutting of government investment capacity. As in Mexico, the World Bank and IMF, working on behalf of international creditors, pressured the Corazon Aquino administration to make repayment of the $26 billion foreign debt a priority.

Like other countries, the Philippine government was told by the IMF to abandon the practice of surplus storage. Nobel Prize-winning economist Paul Krugman (2008) explains that the shrinkage of these precautionary inventories of foodstuffs throughout the developing world occurred 'mainly because everyone came to believe that countries suffering crop failures could always import the food they needed'. The Washington consensus has been that a global food system would produce enough cheap food for everyone, which has lulled many into a false sense of (food) security. There is, however, a component missed by Krugman that further explains this shift away from surplus storage in the developing world. Bello (2008) alludes to it in his aforementioned quote. To be better positioned to repay their foreign debts, developing countries were pressured by such institutions as the World Bank to reduce expenditures. Yet doing this gutted the very domestic policies that previously helped to ensure food security for these nations, such as subsidies for small farms (for things like fertilizer and seed) and policies directed at creating government food reserves.

Back to Haiti. While only 966km (600 miles) from the US, Haiti is a world away from its neighbouring superpower to the north when it comes to food security. Having received immense sympathy from the international community after its capital city was decimated by an earthquake in January 2010, Haiti is still usually seen (at least when it comes to the issue of food security) as a victim of its own failed policies (see, for example, Dupuy, 2010; *Los Angeles Times*, 2010). It is true; no one forced Haitian officials to adopt the food and trade policies that were implemented over the last few decades. But cheap food doesn't run on force. An undying commitment to 'corporate liberalization' (Michael, 2009), masked under a thick layer of free trade rhetoric, has usually been sufficient to get countries to adopt policies that in the end proved not to be in their best interests.

After the dictator Jean Claude 'Baby Doc' Duvalier was overthrown in 1986, the IMF provided Haiti a US$24.6 million loan. Desperately in need of cash (Duvalier had emptied the treasury before fleeing the country), government officials accepted the loan as well as the conditions that it came with. Acceptance of the loan required Haiti to reduce tariff protections for their rice and other agricultural products and to open up their markets to imports. The US played a decisive role in dictating these requirements (Quigley, 2008), for reasons that will soon become clear.

Prior to 1986, Haiti imported very little rice. The agricultural sector was heavily protected through tariffs (of between 40 and 50 per cent) and import prohibitions (World Bank, 2002, p43). Trade liberalization brought down these protections – for example, tariffs on rice were reduced to 3 per cent. Not surprisingly, these neoliberal reforms flooded Haiti with cheap rice, largely from the US (in 2008, Haiti was the third largest importer of US rice). 'Within less than two years', according to a local Haitian doctor, 'it became impossible for Haitian farmers to compete with what they called "Miami rice". The whole local rice market in Haiti fell apart as cheap, US subsidized rice, some of it in the form of "food aid", flooded the market' (cited in Quigley, 2008). In 1985, Haiti imported only 7000 metric tonnes of rice; in 2004, that figure had increased to 225,000. Prior to the food riots of 2008, approximately one half of Haiti's hard currency was used to purchase food. It is therefore easy to see why, when food prices shot up in 2007, the Haitian government had a hard time feeding its people.

The flood of cheap imports did make food more affordable for the urban poor. Estimates by the World Bank (2002, p43) indicate that the real price of rice to consumers was reduced by about 50 per cent after trade liberalization in Haiti. So, yes, in the short run – and I stress *the short run* – cheap food made it easier for the Haitian urban poor to purchase (when available) food.

Yet cheap food policies tend to treat cheap food as an end in itself. Questions about how, for example, a population will earn money to buy cheap food are usually left out of the discussion (Patel and McMichael, 2009, p10). Indeed, the very existence of cheap food is often premised upon a population under-compensated for their labour. For them, cheap food is unjustly expensive. Oxfam Germany, for instance, documented in 2008 how German supermarkets were selling cheap fruit by contracting with farms in Costa Rica that were paying their workers US$0.75 an hour (Lang et al, 2009, p173).

The people of Haiti, like those in other poor countries, have paid dearly for policies transfixed on making cheap food, while assuming (incorrectly) that the food is affordable for those most in need. During the early 1960s, the agricultural sector employed 80 per cent of the labour force, which represented 50 per cent of the country's GDP, and made up 90 per cent of all exports. Unable to find jobs in other sectors, peasant farmers today have little choice but to continue to farm, which explains why approximately two-thirds of the country's population continue to depend on agriculture for a living (Arthur, 2009). Adopting policies that reduce wages for the majority of a country's population doesn't seem like a sound developmental strategy. Sure, such policies might reduce the price of certain foodstuffs. But what good is cheaper food when you have no money?

The bad news for Haiti doesn't end there. Cheap food policies have had a cascading effect on the macroeconomic stability of the country (this is why I stressed a couple paragraphs ago '*the short run*'). The state's capacity to govern eroded significantly in the aftermath of trade liberalization. Revenue was no longer generated through export. Income tax receipts fell as farmers lapsed further into poverty. And investments dropped as the country's infrastructure fell into disrepair, thus further undermining trade capacity and decreasing investment incentives.

While Haiti was forced to reduce its government support for agriculture – a sector that had long been the country's main economic driver – the US was doing just the opposite. According to Daniel Griswold (2006, p1), director of the CATO Institute's Center for Trade Policy Studies, direct taxpayer subsidies to the rice sector in the US have averaged US$1 billion a year since 1998. US consumers therefore pay for the rice programme twice, as taxpayers and again as consumers as tariffs on inbound rice artificially inflate its price at the grocery store. Thus, while the IMF was instructing Haiti to reduce its tariffs, the US was allowed to keep its rice-protecting tariff barriers (with rates as high as 24 per cent) (Griswold, 2006, p6). And so a hungry country became hungrier ... and violent.

According to Amartya Sen (1981), modern food crises have not been related so much to the absence of food as to the inability to buy it. Examining the 1943 Bengal famine, Sen found there to be plenty of food around. The problem, which a *Life* magazine reporter at the time estimated was taking the lives of more than 50,000 Bengalese a week (Fisher, 1943, p16), was not a lack of food but a lack of available food. Those who had food tended to hoard it, knowing that this action would only increase its market price and make them more money in the long run.[2] The Bengal famine revealed the limits of the market for delivering food during times of crisis. The British, because this was occurring during the British Raj, wrongly assumed that the market would justly and efficiently distribute food to the poor. Yet what the market delivered was not food but hunger.

Sen (1981) notes that many of the same variables were at work in the 1974 Bangladesh famine. Floods put farm labourers out of work and therefore reduced the incomes of landless labourers. The floods created expectations of impending rice shortages, leading to the hoarding of rice amongst those who could afford to hoard. As panic buying ensued, the price of rice was driven beyond the reach of the poor. Those with land and money easily survived the famine – indeed, for them there was no famine at all. As for the poor: the unregulated market showed them just how unfair the free market can be. Raj Patel (2009, p130) explains why the market sometimes fails to feed during times of perceived scarcity:

> Those who are in a position to control the distribution of grain will only
> do so if they're able to command a sufficiently high price. The only way
> that famine can be overcome is to guarantee rights to hungry people that
> trump those of grain-hoarders.

This, for Sen, is where democracy comes into play. Democracy, in theory, ought
to supply a massive counterweight – which, in the case of India, involves over
1 billion people – to counteract the interests of multinational corporations. For
Sen, it is not enough to have a regulated market. China experienced the Great
Leap famine (1958 to 1961), during which over 15 million lives were lost
(Xizhe, 1987, p639), with heavily regulated markets. Conversely, India managed
to avoid famine in 1965 and 1966 despite two consecutive years of failed crops
due to massive monsoon rains. 'The elected political leaders of India', Paarlberg
(2010, p50) explains, 'knew their own survival in office required a prompt
response to the crisis.' Indian officials were therefore quick to turn to outside
sources for aid – 10 million tonnes of grain from the US helped to fill the grain
deficit (Sud, 2009). The Indian government has also worked to expand public
food distribution systems to ensure that the food gets to those who needed it
during times of scarcity.

The Global Land Grab

The term 'global land grab' refers to transcontinental land deals whereby
corporations, investment firms or state-owned enterprises lease or purchase
large areas of land in other countries for agricultural purposes. Of the nearly 400
projects that have been inventoried (in 80 countries), 37 per cent are described
as for food production, while 35 per cent are for the production of biofuel crops
(GRAIN, 2010). Yet the true scope of this practice remains elusive. I say this
because the World Bank has tried and failed – yes, *the* World Bank – to conduct
a comprehensive study of 30 hotspot countries where land grabbing is rife due
to the unwillingness of the involved parties to talk (GRAIN, 2010). A recent
estimate by the International Food Policy Research Institute places the land
'grabbed' at 20 million hectares (49,421,076 acres) – an amount equivalent to
one fifth of all the agricultural land in the European Union (EU) with a value of
US$30 billion (Bladd, 2009).

Approximately 95 per cent of Asia's cropland has already been cultivated
and most of what can be farmed in the US and EU already is. Attention is
focusing in on Africa, where it is believed large quantities of 'idle' land are still
available to foreign investors (Rehman, 2010, p11). South Korea has purchased
1.3 million hectares in Madagascar and 690,000ha in Sudan. Saudi Arabia owns
equally massive tracks of land in Indonesia (1.6 million hectares) and Sudan
(10,117ha). The United Arab Emirates (UAE) has obtained 378,000ha in Sudan,
900,000ha in Pakistan and 3000ha in the Philippines. China, however, currently
leads all, having acquired 1.24 million hectares in the Philippines and 700,000ha
in Laos (Rehman, 2010, p12). China leases an additional 2,796,378ha in the

Democratic Republic of Congo, which will be used to build the world's largest palm oil plantation for biofuels (UPI, 2010). Other countries looking to acquire fertile farmland include Japan, Malaysia, India, Libya and Egypt. And they are looking in places beyond those already mentioned, such as Argentina, Brazil, Burma, Cambodia, Kazakhstan, Mali, Mozambique, Somalia, Thailand, Uganda, Ukraine and Vietnam.

Countries, however, are not the only ones grabbing land. Private investment firms are also looking at Africa and acquiring land at a brisk pace. For example, Philippe Heilberg, CEO of the New York-based investment fund Jarch Capital, has reportedly leased as many as 1 million hectares in southern Sudan from General Paulino Matip (Daniel and Mittal, 2009, p4; UPI, 2010). According to Heilberg, investing in war-torn areas makes financial sense, as long as you bet on the right horse: 'If you bet on the right shifting of sovereignty then you are on the ground floor [when the new regime gains power]' (cited in Daniel and Mittal, 2009, p5).

Where are the guardians of trade liberalization – why are they not denouncing these activities? The global land grab shows just how little faith free market proponents have in the free market when it comes to food security. For decades, developing countries have been told that the market will provide; the law of comparative advantage feeds all, regardless of whether or not you grow food. Some argue that the global land grab is an innovative way to create comparative advantage for food-dependent countries such as Japan, Saudi Arabia and South Korea (Rehman, 2010, p12). This logic, however, utterly guts the concept of any theoretical or empirical meaning and could ultimately be used to justify actions such as colonialism or the military occupancy of distant lands.

One of the things that makes Africa particularly attractive to foreign investors and states is its perceived abundance of land. The FAO-sponsored Global Agro-Ecological Assessment (Fischer et al, 2002) provides a compre- hensive survey of global agricultural potential. Based on satellite imagery from 1995 to 1996, it estimates that 80 per cent of the world's reserve agricultural land is concentrated in Africa. The report places total cultivable land in Africa at 807 million hectares, of which approximately 227 million hectares are under cultivation. It is not clear, however, how or if the study's authors adjusted for shifting cultivation patterns. Many small farms in Africa utilize a ratio of five plots under fallow to every plot under cultivation. Taking these practices into consideration, others have calculated that the range of total cultivated land in Africa could be anywhere between 227 million hectares to a maximum of 1182 million hectares, which greatly reduces the amount of land that's actually idle (Cotula et al, 2009, p60).

In many cases, land is practically or literally given away under these deals. Job creation and infrastructure development are viewed as the main economic benefits for developing countries. In Sudan, for example, a *feddan* (0.42ha) can be leased for as little as US$2 or $3 a year. In Ethiopia, rent was being paid in four out of the six projects examined in an FAO-sponsored study, with lease prices ranging from US$3 to $10 per hectare per year. A land deal in Madagascar between the state and the biofuel conglomerate Green Energy Madagascar (GEM) involves no rental fees for the leased 450,000ha. Instead, GEM promises to build up local infrastructures and hire around 4500 part-time workers (Cotula et al, 2009, pp79–80).

There is also little evidence that these deals increase public revenues through taxes due to the generous tax incentives provided by the government to attract foreign investment. International land deals granted by Ethiopia, for example, usually come with a profit tax holiday that lasts for a period of five years. It is estimated that this five-year tax exemption costs the Ethiopian government US$60,276,000 for *each* project (Cotula et al, 2009, p80).

Benefits are said to come in the form of other commitments. Qatar, for instance, is leasing 40,000ha of land in a fertile River Tana delta on the north coast of Kenya. In exchange for the lease, Qatar provided the Kenyan government a UK£2.4 billion loan to construct a port on the Indian Ocean island of Lamu (*The Telegraph*, 2008). But this raises the question: are these commitments enforceable? The enforceability of these promises depends upon the details of the contract and national legislation. Some promises, however, have proven empty. A good example of this is the announcement by GEM in Madagascar of their intent to increase mechanization, despite promises to pursue a labour-intensive business model and create jobs in exchange for 450,000ha. Other countries, such as Mali, have legislation allowing for the termination of leases if land grabbers fail to pay fees or maintain contractual obligations (Cotula et al, 2009, p82).

Yet, arguably the most problematic aspect of these deals involves their impact upon food security and, more specifically, food (and land) sovereignty. Many of the nations handing over land to foreign entities are incredibly food insecure. Reports have surfaced indicating that some investors are pushing for explicit provisions guaranteeing full repatriation of produce, including where this requires amending the national law of the host state. In other words, they are making the argument that 'If we grow it, we own it', even if it is being grown in a country suffering from starvation (Cotula et al, 2009, p87). Pakistan's Federal Investment Minister Waqar Ahmad Khan is quoted as saying that the government would ensure that the investors in corporate farming get the entire crop: 'We are negotiating with investors from Gulf states, particularly Saudi Arabia, for investment in corporate farming. Investors will be ensured repatriation of 100 per cent crop yield to their countries, even in the case of food deficit' (cited in *The News*, 2009). In the end, the Pakistani government had to revise its policy in response to public outrage and now call for 'reasonable percentages' of produce to be exported (Cotula et al, 2009, p87).

In light of the mounting evidence against current land-grabbing practices, the FAO has begun pushing investors to develop joint ventures with local farmers. Doing this would allow farmers to retain control of their land and be able to sell their produce at guaranteed prices. Some countries, such as Qatar, are at least publicly considering these alternative arrangements. Nasser Mohamed Al

Hajri, chairman of Hassad Food, an agricultural firm owned by the government of Qatar, is quoted as saying: 'We are not deleting the option of buying farmland but we don't feel like it is the right strategy. In many cases these deals are not win–win situations and we don't want to be in a situation where the rich are taking away food and land from the poor' (cited in Bladd, 2009).

In taking land, and lest we forget water (which is always included in these deals), away from local farms and pastoralists, these 'grabs' also risk exacerbating conflict within a region. The original and 1990s Mexican Zapatista rebellions involved, among other things, demands for land. Likewise, revolutions in El Salvador and Nicaragua were as much about struggles for land as they were about political philosophy (see, for example, Paige, 1996). And conflicts over water: an entire academic cottage industry has emerged during the last decade chronicling the rise of 'water wars' (see, for example, Shiva, 2002; Rand, 2003). One can be sure that such wars will follow opportunistic investors and states looking to grab land (and water) in states too poor (or weak) to say 'No!'

Food Policy as Immigration Policy

It is impossible for me, living in the US, to write this book and not carve out some space to talk about how cheap food relates to immigration. Food policy is invariably immigration policy in that it has shaped significantly the immigration debate that has gripped this country during recent years. In April 2010, the US state of Arizona passed controversial legislation (SB1070) giving its police the power to demand proof of citizenship among those detained who are suspected of being in the country illegally. More recent still, the US government (specifically the Department of Justice) has taken Arizona to court seeking to overturn this legislation. A massive debate is simultaneously under way about what exactly the federal government ought to do about immigration and with all the citizens living in the country illegally. Everyone, it seems, agrees that comprehensive immigration reform is needed. No one can agree, however, on just what this means.

The US government spends hundreds of millions of dollars annually 'securing the border', as it is called. It has also promised the Mexican government over US$1 billion to help fund its war on drugs and to better police its northern border. Then there is the Berlin wall of sorts – decked out with all the latest technology (night vision and motion detection cameras, ground sensors to detect tunnelling activity beneath the fence, etc.) – that is slowly being erected between the two countries.[3] The Great Wall of the US will one day stretch from the Gulf of Mexico to the Pacific Ocean. In addition to the many billions of dollars spent to erect the structure, the Government Accountability Office (GAO) estimates that it will cost taxpayers US$6.5 billion over the next 20 years to maintain (Wood, 2009). I am not suggesting that these costs are entirely attributable to food policy. But cheap food's fingerprints are among the more pronounced as we look for suspects to account for recent immigration trends, specifically with regards to those coming into the US from Mexico.

Taking effect in 1994, the North American Free Trade Agreement (NAFTA) further lowered trade barriers between Canada, Mexico and the US. NAFTA, it was promised, would give Mexico an early advantage relative to other less affluent nations when it came to trading with the world's largest consumer market: namely, the US (Perez et al, 2008, p7). With demand for fresh fruits and vegetables beginning to take off in the neighbouring country to the north, Mexico was poised to reap the benefits of trade. More than 16 years later, the results are in. Through NAFTA, Mexico saw more failures than successes. A series of studies assessing NAFTA's overall effect on Mexico was summarized at a recent event sponsored by the Carnegie Endowment for International Peace (a foreign-policy think-tank based in Washington, DC). According to the authors of these reports, NAFTA was not a total failure. Some positive economic trends were reported (Zepeda et al, 2009):

- Mexico's exports increased 311 per cent in real terms between 1993 and 2007, and non-oil exports increased 283 per cent.
- Foreign direct investment (FDI), mostly from the US, more than tripled between 1992 and 2006 (though this occurred at the expense of domestic investment).
- Inflation was reduced to below 5 per cent (from the 80 per cent levels recorded in the 1980s).
- Mexican firms saw an 80 per cent increase in productivity in the manufacturing sector.

These benefits, however, were insufficient to outweigh the 'profound shortcomings of the "NAFTA model"'. Some of the costs listed include the following (Perez et al, 2008, p8; Zepeda et al, 2009):

- The annual GDP per capita growth rate for Mexico was a mere 1.6 per cent between 1992 and 2007 (its average real per capita growth rate was 3.5 per cent between 1960 and 1979).
- Despite high FDI, domestic investment in Mexico has been greatly reduced, resulting in total investment levels (foreign plus domestic) of *below* 20 per cent (a net *loss* of investment).
- Macroeconomic vulnerability has increased as Mexico remains heavily dependent on oil exports for government revenues. Thanks, in part, to the tax-free zones (the '*maquila* zones'), located principally near the US–Mexico border, tax revenues today amount to less than 15 per cent of the country's GDP.
- Limited employment gains in manufacturing and service sectors of Mexico's economy have been offset by significant employment losses in the agriculture sector, which lost more than 2.3 million jobs between 1990 and 2008.
- Agricultural wages fell drastically in Mexico during the 1994 to 1995 peso crisis and have yet to reach pre-crisis levels in real terms.
- Mexico's agricultural imports from the US – most notably, corn – grew faster

than its exports, creating a negative trade balance for the sector.
- Mexican agricultural exports have become heavily dependent on large global corporate agri-food firms, leaving Mexican farms with expanded productivity but decreased power with production–distribution chains.
- NAFTA contributed to growing geographical inequality between Mexico's poorer southern and more affluent northern states.
- Environmental protection during the NAFTA era has been weak.

NAFTA has been particularly hard on Mexican farmers. The total value of agricultural exports from the US increased 280 per cent from 1992 to 2008 (Wise, 2009, p4). Ensuring the US's comparative advantage, agricultural subsidizes increased in the US just as the Mexican government drastically cut its support for agriculture. During the 1990s, the US also extended to Mexico US$3 billion in export credits to buy some of the corn that was spilling out of US grain elevators (Bello, 2008, p450). Between 1993 and 2004, pulled down by heavily subsidized products coming in from the US, the price of corn and soybeans in Mexico fell by approximately 50 per cent (the price of coffee, another key smallholder crop, also fell by 50 per cent during this period) (Perez et al, 2008, p8).

Beef producers in Mexico have also been hit hard by NAFTA. One study estimates that as many as 80 per cent of the cattle in Mexico are pasture (rather than grain) fed. Consequently, cheap corn and soybeans do not help Mexican beef producers to the same extent as beef producers in the US. Losses to the Mexican beef industry due to cheap US beef between 1997 and 2005 were estimated at US$1.6 billion (Wise, 2009, p26). Adding insult to injury, gains in the manufacturing sector have been outpaced by losses in agriculture, which has seen its total employment numbers drop from 8.1 million during the 1990s to 5.8 million in the second quarter of 2008 (Zepeda et al, 2008, p12). Not surprisingly, this surplus labour force is looking to the north – namely, the US – for employment.

Although migration to the US has deep historical roots, its pace has grown during the NAFTA years. As Wise (2009, p13) notes: 'In spite of the rising militarization of the US border, migration increased from about 350,000 per year before NAFTA to nearly 500,000 per year by the early 2000s.' Militarizing the border has, however, changed long-standing patterns set by seasonal demand in the US for temporary labour. Now, choosing not to risk annual return to Mexico, many immigrants stay in the US on a more permanent basis (Wise, 2009, p13).

Remittances – the money that migrant workers are sending home to families back in Mexico – are providing necessary capital to help small Mexican farms stay afloat. This explains why 3 million farms continued to grow corn in Mexico, in spite of their tremendous comparative disadvantage relative to US growers. As Laura Carlsen (2003), director of the Mexico City-based Americas Program of the International Relations Center (IRC), explains:

The remittances have a dual role. First, the money sustains agricultural activities that have been deemed nonviable by the international market but that serve multiple purposes: family consumption, cultural survival, ecological conservation, supplemental income, etc. Second, by sending money home, migrants in the US seek not only to assure [sic] a decent standard of living for their Mexican families but also to maintain the *campesino* identity and community belonging that continue to define them in economic exile.

In other words, remittances fill a hole left by the state by subsidizing smallholder agriculture while also conserving cultural identity. Again, Carlsen (2003):

The combination of these personal subsidies and subsistence tenacity account for the otherwise unaccountable growth in corn production in Mexico – despite the overwhelming 'comparative disadvantages' of a distorted international market. They reflect a deep cultural resistance to the dislocation and denial inherent in the free trade model.

It remains to be seen what effect a further militarization of the US–Mexico border will have on small farms in Mexico. If remittances from the US are, indeed, helping Mexican smallholders to make ends meet, then it is logical to assume that once those remittances disappear, so will some of those remaining 3 millions farms. Yet, there are not enough jobs to absorb Mexico's *current* surplus labour. Net losers from cheap food policies, Mexican farmers risk finding themselves with little choice but to go north for employment. Having already been breached well over 3000 times, requiring US$1300 for the average repair (Wood, 2009), it is clear that the wall is no substitute for well-fed stomachs and well-paying jobs. A focus on building food security through food sovereignty in Mexico ought therefore to be included in any comprehensive US immigration policy. Without addressing the drivers that send millions north, the so-called 'immigration problem' in the US will never be resolved.

I wish I could say that Mexico's experience over the past 20 years is an outlier case. I can't. Rather than the exception, what's happened in Mexico has been, unfortunately, the rule. I touched upon this in Chapter 2 in the limited context of rural out-migration, noting how neoliberal policies (and neoliberal global bodies, such as the World Bank) actively sought the removal of peasants from the countryside. In many cases, however, this ever-growing surplus labour force has few domestic (legal) employment opportunities, creating what Mike Davis (2006) has memorably called a 'planet of slums'. Many look to other countries, then, for employment; or, conversely, individuals may be recruited into less legitimate activities, involving anything from drugs to prostitution to terrorism. The consequence of all this inequitable social and economic upheaval is that, in the words of Geoffrey Lawrence and colleagues (2010, p4), 'people who once had access to food are no longer connected to the land and the food that it produced, a situation that is destroying food sovereignty ... and overall food security'.

Notes

1   At the height of the Iraq War it has been estimated that 25 per cent of the foreign terrorists recruited by Al Qaeda to fight the US-led coalition came from war-torn and poverty-ridden Africa countries (Schmitt, 2004).
2   I know some wheat farmers currently holding their wheat from the market because they expect its price to rise due to massive crop failures in Russia the previous year (2010). It will be interesting to see how this market-induced 'wheat shortage' will affect the global price of wheat.
3   According to a government report from 2009, the fence has already been breached over 3000 times (Wood, 2009).

References

Addison, T. (2005) *Agricultural Development for Peace*, Research Paper No 2005/07, United Nations University, January, www.wider.unu.edu/publications/working-papers/research-papers/2005/en_GB/rp2005-07/_files/78091751631815984/default/rp2005-07%20addison.pdf, last accessed 16 July 2010

Arthur, C. (2009) 'Haiti: New peasant alliance demands action on food crisis', *Inter Press Service*, 14 January, http://us.oneworld.net/article/359518-haiti-peasant-alliance-demands-action-food-crisis, last accessed 21 July 2010

Associated Press (2007) 'Chiquita admits to paying Colombia terrorists', *MSNBC.com*, 15 March, www.msnbc.msn.com/id/17615143/from/ET/#, last accessed 16 December 2010

Bello, W. (2008) 'How to manufacture a food crisis', *Development*, vol 51, no 4, pp450–455

Bertini, C. and D. Glickman (2009) 'Farm futures, bringing agriculture back to US farm policy', *Foreign Affairs*, vol 88, pp93–109

Bladd, J. (2009) 'Call for GCC "land grab" policy to stop', *Arabian Business.com*, 7 September, www.arabianbusiness.com/566961-call-for-gcc-land-grab-policy-to-stop—-experts, last accessed 22 July 2010

Brace, C. (1879) *Free Trade as Promoting Peace and Goodwill Among Men*, New York Free Trade Club, New York, NY

Brückner, M. and A. Ciccone (2009) *International Commodity Prices, Growth, and the Outbreak of Civil War in Sub-Sahara Africa*, FEDEA (Fundacion de Estudious de Economia Aplicada), Document no 2009-37, Madrid, Spain, www.antoniociccone.eu/wp-content/uploads/2009/08/civil-war-and-com-prices.pdf, last accessed 16 July 2010

Carib World News (2008) 'Haiti: Four riots leave four dead', *Carib World News*, 7 April, www.caribbeanworldnews.com/stories_one.php?sid=1062, last accessed 21 July 2010

Carlsen, L. (2003) 'The Mexican farmers' movement: Exposing the myths of free trade', International Forum on Globalization, Americas Policy Report, 25 February, www.ifg.org/analysis/wto/cancun/mythtrade.htm, last accessed 25 July 2010

Collier, P. (2003) 'The market for civil war', *Foreign Policy*, 1 May, www.foreignpolicy.com/articles/2003/05/01/the_market_for_civil_war, last accessed 16 July 2010

Cotula, L., S. Vermeulen, R. Leonard and J. Keeley (2009) *Land Grab of Development Opportunity: Agricultural Investment and International Land Deals in Africa*, Food and Agriculture Organization of the United Nations (FAO) and Laxenburg, International Institute for Applied Systems Analysis (IIASA), Rome, www.ifad.org/pub/land/land_grab.pdf, last accessed 23 July 2010

Cribb, J. (2010) *The Coming Famine: The Global Food Crisis and What We Can Do to Avoid It*, University of California Press, Berkeley, CA

Daniel, S. and A. Mittal (2009) *The Great Land Grab*, Final report, Oakland Institute, Oakland, CA, www.oaklandinstitute.org/pdfs/LandGrab_final_web.pdf, last accessed 22 July 2010

Daviron, B. and S. Ponte (2005) *The Coffee Paradox: Global Markets, Commodity Trade, and the Elusive Promise of Development*, Zed Books, New York, NY

Davis, M. (2006) *Planet of Slums*, Verso, New York, NY

de Soysa, I. and N. Petter Gleditsch (1999) 'To cultivate peace: Agriculture in a world of conflict', *Environmental Change and Security Project Report*, vol 5, summer, pp15–25

Doyle, L. (2008) 'Starving Haitians riot as food prices soar', *The Independent*, 10 April, www.independent.co.uk/news/world/americas/starving-haitians-riot-as-food-prices-soar-807016.html, last accessed 20 July 2010

Dupuy, A. (2010) 'Commentary beyond the earthquake: A wake-up call for Haiti', *Latin American Perspectives*, vol 37, pp195–204

Esty, D., J. Goldstone, T. Gurr, B. Harff, M. Levy, G. Dabelko, P. Surko and A. Unger (1998) *State Failure Task Force Report: Phase II Findings*, Science Applications International Corporation, McLean, VA, www.wilsoncenter.org/events/docs/Phase2.pdf, last accessed 17 July 2010

FAO (Food and Agriculture Organization of the United Nations) (2000) *The State of Food and Agriculture 2000*, Rome, Italy, www.fao.org/docrep/x4400e/x4400e00.htm, last accessed 17 July 2010

Ferraro, V. (undated) *Globalizing Weakness: Is Global Poverty a Threat to the Interests of States?*, Woodrow Wilson International Center for Scholars, Washington, DC, www.wilsoncenter.org/news/docs/ACF59B0.doc, last accessed 17 July 2010

Fischer, G., H. van Velthuizen, M. Shah and F. Nachtergaele (2002) *Global Agro-Ecological Assessment for Agriculture in the 21st Century*, Food and Agriculture Organization of the United Nations (FAO) and Laxenburg, International Institute for Applied Systems Analysis (IIASA), Rome

Fisher, W. (1943) 'The Bengal Famine: 50,000 Indians weekly succumb to disease and starvation in spreading catastrophe', *Life*, vol 15, no 21, pp16, 19–20

GRAIN (2010) *The World Bank in the Hot Seat over Land Grabbing*, GRAIN Policy Briefing, 7 May, www.commondreams.org/headline/2010/05/07-0, last accessed 22 July 2010

Grillo, I. (2007) '75,000 protest tortilla prices in Mexico', *The Washington Post*, 1 February, www.washingtonpost.com/wp-dyn/content/article/2007/01/31/AR2007013101343.html, last accessed 20 July 2010

Griswold, D. (2005) 'Peace on Earth? Try free trade among men', CATO Institute, 28 December, www.cato.org/pub_display.php?pub_id=5344, last accessed 17 July 2010

Griswold, D. (2006) *Grain Drain: The Hidden Cost of US Rice Subsidies*, CATO Institute, Trade Briefing Paper No 25, www.cato.org/pubs/tbp/tbp-025.pdf, last accessed 21 July 2010

The Guardian (2007) 'Pasta "strike" shocks Italy', *Guardian.co.uk*, 13 September, www.guardian.co.uk/world/2007/sep/13/italy, last accessed 20 July 2010

Holt-Gimenez, E. (2010) 'Forward', in G. Lawrence, K. Lyons and T. Wallington (eds) *Food Security, Nutrition and Sustainability*, Earthscan, London and Sterling, VA, ppxxiii–xxiv

Keyzer, M. and L. Van Wesenbeeck (2004) 'Changed market access in the North and the farm prices in the South: Some lessons from the war on drugs', *De Economist*, vol 152, no 4, pp543–560

Krugman, P. (2008) 'Grains gone wild', *New York Times*, 7 April, www.nytimes.com/2008/04/07/opinion/07krugman.html, last accessed 20 July 2010

Lang, T., D. Barling and M. Caraher (2009) *Food Policy: Integrating Health, Policy, and Society*, New York: Oxford University Press

Lawrence, G., K. Lyons and T. Wallington (2010) 'Introduction: Food security, nutrition and sustainability in a globalized world', in G. Lawrence, K. Lyons and T. Wallington (eds) *Food Security, Nutrition and Sustainability*, Earthscan, London and Sterling, VA, pp1–25

Loewenberg, S. (2008) 'Global food crisis looks set to continue', *The Lancet*, vol 324, no 9645, pp1209–1210, www.lancet.com/journals/lancet/article/PIIS0140-6736(08)61502-0/fulltext, last accessed 20 July 2010

Los Angeles Times (2010) 'Editorial: Haiti's plight', *Los Angeles Times*, 19 July, http://articles.latimes.com/2010/jul/19/opinion/la-ed-haiti-20100719, last accessed 20 July 2010

Maddison, A. (1995) *Monitoring the World Economy, 1820–1992*, OECD, Paris, France

Messer, E. and M. Cohen (2006) *Conflict, Food Insecurity and Globalization*, Discussion Paper 206, IFPRI (International Food Policy Research Institute), Washington, DC, www.ifpri.org/sites/default/files/publications/fcndp206.pdf, last accessed 17 July 2010

Messer, E., M. Cohen and J. D'Costa (1998) *Food from Peace: Breaking the Links between Conflict and Hunger*, IFPRI (International Food Policy Research Institute), 20/20 Vision Initiative, http://ageconsearch.umn.edu/bitstream/16205/1/br50.pdf, last accessed 16 July 2010

Michael, P. (2009) 'The world food crisis in historical perspective', *Monthly Review*, July/August, www.monthlyreview.org/090713mcmichael.php, last accessed March 14, 2011

Nafziger, E. W. and J. Auvinen (2000) 'The economic causes of humanitarian emergencies', in E. W. Nafziger, F. Stewart and R. Väyrynen (eds) *War, Hunger, and Displacement: The Origins of Humanitarian Emergencies*, vol 1, Oxford University Press for the World Institute of Development Economics, United Nations University, Oxford, UK

The News (2009) 'Corporate farming raises concerns among local growers', *The News*, 28 January, www.thenews.com.pk/print1.asp?id=159380, last accessed 23 July 2010

OECD (Organisation for Economic Co-operation and Development) (2010) *Development Aid Rose in 2009 and Most Donors Will Meet 2010 Aid Targets*, OECD Press Release, Paris, France, www.oecd.org/document/11/0,3343,en_21571361_44315115_44981579_1_1_1_1,00.html, last accessed 16 July 2010

Paarlberg, R. (2010) *Food Politics: What Everyone Needs to Know*, Oxford University Press, New York, NY

Paige, J. (1996) 'Land reform and agrarian revolution in El Salvador', *Latin American Research Review*, vol 31, no 2, pp127–139

Patel, R. (2009) *Stuffed and Starved: The Hidden Battle for the World Food System*,

Melville House Publishing, Brooklyn, NY

Patel, R. and P. McMichael (2009) 'A political economy of the food riot', *Review: A Journal of the Fernand Braudel Center*, vol 12, no 1, pp9–35

Perez, M., S. Schlesinger and T. Wise (2008) *The Promise and Perils of Agricultural Trade Liberalization: Lessons from Latin America*, White Paper, Washington Office on Latin America, Washington, DC

Pinstrup-Anderson, P. and S. Shimokawa (2008) 'Do poverty and poor health and nutrition increase the risk of armed conflict onset?', *Food Policy*, vol 33, pp513–520

Quigley, B. (2008) 'The US role in Haiti's food riots', *Counterpunch*, 21 April, www.counterpunch.org/quigley04212008.html, last accessed 20 July 2010

Rand, H. (2003) *Water Wars: A Story of People, Politics and Power*, Xlibris, Bloomington, IN

Rehman, S. (2010) 'The Obama administration and the US financial crisis', *Global Economy Journal*, vol 10, no 1, pp1–22

Rice, S. (2006) 'National security implications of global poverty', Lecture at University of Michigan Law School, 30 January, www.brookings.edu/~/media/Files/rc/speeches/2006/0130globaleconomics_rice/20060130.pdf, last accessed 2 April 2010

Rice, S. (2010) 'The national security implications of global poverty', in S. Rice, C. Graff and C. Pascual (eds) *Confronting Poverty: Weak States and US National Security*, Brookings Institute Press, Washington, DC, pp1–22

Schmitt, E. (2004) 'As Africans join Iraqi insurgency, US counters with military training in their lands', *New York Times*, 10 June, www.nytimes.com/2005/06/10/politics/10military.html?_r=1, last accessed on 19 July 2010

Schneider, G., K. Barbieri and N. Petter Gleditsch (2003) 'Does globalization contribute to peace? A critical survey of the literature', in G. Schneider, K. Barbieri and N. Petter Gleditsch (eds) *Globalization and Armed Conflict*, Rowan and Littlefield, Lanham, MD, pp3–30

Sen, A. (1981) *Poverty and Famines: An Essay on Entitlement and Deprivation*, Oxford University Press, New York, NY

Shiva, V. (2002) *Water Wars: Privatization, Pollution, and Profit*, South End Press, Cambridge, MA

Silberfein, M. (2008) 'The impact of conflict and resources on agriculture', in M. Taeb and A. H. Zakari (eds) *Agriculture, Human Security, and Peace: A Crossroad in African Development*, Purdue University Press, West Lafayette, IN, pp77–99

Sud, H. (2009) 'Poor rainfall not a threat to India's food supply', *UPI Asia.com*, 28 August, www.upiasia.com/Economics/2009/08/ 28/poor_rainfall_not_a_threat_to_indias_food_supply/9794, last accessed 21 July 2010

The Telegraph (2008) 'Qatar to lease 100,000 acres in Kenya in return for port loan', *The Telegraph*, 3 December, www.telegraph.co.uk/news/worldnews/middleeast/qatar/3543887/Qatar-to-lease-100000-acres-in-Kenya-in-return-for-port-loan.html, last accessed 23 July 2010

UNCTAD (United Nations Conference on Trade and Development) (2004) *The Least Developed Countries Report 2004: Linking International Trade with Poverty Reduction*, UNCTAD, Geneva

UPI (2010) 'Food and water drive African land grab', *UPI.com*, 29 April, www.upi.com/Science_News/Resource-Wars/2010/04/29/Food-and-water-drive-Africa-land-grab/UPI-11891272566080, last accessed 22 July 2010

Uvin, P. (1996) 'Tragedy in Rwanda: The political ecology of conflict', *Environment*, vol 33, no 3, pp6–15, 29

Watkins, K. and J. von Braun (2003) *Time to Stop Dumping on the World's Poor*, IFPRI (International Food Policy Research Institute) Annual Report 2002–2003, International Food Policy Research Institute, Washington, DC, www.ifpri.org/sites/default/files/publications/ar02e1.pdf, last accessed 17 July 2010

Wise, T. (2009) *Agricultural Dumping Under NAFTA: Estimating the Costs of US Agricultural Policies to Mexican Producers*, GDAE (Global Development and Environment Institute at Tufts University) Working Paper No 09-08, Medford, MA, www.ase.tufts.edu/gdae/Pubs/wp/09-08AgricDumping.pdf, last accessed 25 July 2010

Wise, T. (2010) 'The true cost of cheap food', *Resurgence*, vol 259, March/April, www.resurgence.org/magazine/article3035-the-true-cost-of-cheap-food.html, last accessed 20 July 2010

Wood, D. (2009) 'Billions for a US–Mexico border fence, but is it doing any good?', *Christian Science Monitor*, 19 September, www.csmonitor.com/USA/2009/0919/p02s09-usgn.html, last accessed 24 July 2010

World Bank (2002) *Global Economic Prospects and the Developing Countries: Making Trade Work for the World's Poor, 2002*, World Bank, Washington, DC

Xizhe, P. (1987) 'Demographic consequences of the great leap forward in China's provinces', *Population and Development Review*, vol 13, no 4, pp639–670

Zepeda, E., T. Wise and K. Gallagher (2008) *Rethinking Trade Policy for Development: Lessons from Mexico Under NAFTA*, Policy Outlook, Carnegie Endowment for International Peace, Washington, DC, www.carnegieendowment.org/files/nafta_trade_development.pdf, last accessed 25 July 2010

Zepeda, E., K. Gallagher, R. Stumberg and T. Wise (2009) *The Future of North American Trade Policy: Lessons from NAFTA*, Carnegie Endowment for International Peace, Washington, DC, 9 December, www.carnegieendowment.org/events/?fa=eventDetail&id=1495, last accessed 25 July 2010

4

Cheap Food, Hunger and Obesity

We have all heard the stat: 'The American farmer feeds X people.' Based upon what I have recently read and heard, the number today seems to be between 125 and 150 people. A popular figure used to tout American exceptionalism in the area of cheap food, the statistic was employed heavily during the Cold War. I recently stumbled upon an advertisement for the American farmer in a 1963 issue of *Life* magazine.[1] Paid for by the Chicago Board of Trade and entitled 'Red China's Greatest Weakness', the advertisement compares the American farmer, who at the time 'feeds himself and 29 other persons', with farmers from China and Russia, who are reported as feeding three and seven people, respectively. The advertisement then proclaims that the American farmer 'puts on our tables the best and cheapest food in the world' (*Life Magazine*, 1963, pR2).

I do not disagree with that statement. As earlier chapters make clear, it is difficult to compete against US farmers, who clearly produce some of the cheapest food in the world. I fail to find those farmer/eater ratios very informative, however, recognizing that the farms I grew up among in the Midwest do not produce food at all but feed (for livestock), feedstock (raw material for industrial processes), and foodstock (inputs for processed foods). It is a bit disingenuous so say that the average American farmer is feeding anyone at all, directly at least. Most of the farmers I know can't even feed themselves, which is just to say that they are as dependent upon the nearest grocery store as those of us who live in town. But cheap food is not concerned with things like edibility. Not always, at least. Cheap food, like field corn, need not be *food* at all – at least that's the position of some food writers and activists, such as Michel Pollan and Alice Waters. Others think it insane to deny 'food' status to these commodities. I remember how a family friend (and former officer for the Iowa Corn Growers Association) responded to the question of whether field corn was actually food: 'That's stupid; of course corn is food!' I too have grappled with this question: cheap food – to be food or not to be food? The conclusion I have arrived at, which both sides will probably find unsatisfactory, is that it depends on what we mean by 'food'.

Cheap food harkens to a specific understanding of food. It does not just speak of food that is, well, cheap. It also speaks of a *cheapening* of food – an understanding of food that is reductionist, de-contextualized and highly impoverished. I will talk much more about this in due course. For now, just remember that cheap food deals in a currency that is discrete, quantifiable and exchangeable. Cheap food is about maximizing things such as calorie production and carbohydrate and protein availability. When understood in these terms, we find that farms in countries such as the US make, as the *Life* advertisement proclaims, some of the best cheap food in the world. This has come, however, at great cost to our health.

What is wrong with wanting to maximize world calorie production? It's a fair question. The logic, I admit, is seductive. Reducing food to calories, at face value, makes sense when millions still go to sleep hungry every night. But sufficient calories alone do not a healthy body make. This point is glaringly evident today in the face of the hunger–obesity paradox. We are witnessing an increasing number of people who are both food insecure *and* obese.

The following section discusses how cheap food rests upon a cheapened understanding of food. After briefly discussing its roots, I highlight how this understanding of 'food' informed 20th-century international food policies and programmes (such as the Green Revolution), policies that continue to this day. In terms of maximizing calorie production, today's food system has been an unqualified success. The global obesity epidemic is testament to this fact. And there's the rub: cheapened food might be cheap – relative to, say, whole foods; but is it nourishing? We are seeing globally the substitution of *mis*-nourishment for *mal*nourishment. And this misallocation of calories comes with a pretty big price tag.

The Calorie-ization of 'Food'

If you do not know how the farmer/eater ratio is calculated, you need not worry. I have had a hard time finding anyone who does. Although I consider myself a rather savvy Google user, I have repeatedly been unable to locate a coherent description of the calculation online. I had to turn to friends and colleagues – folks such as agriculture economists and plant and soil scientists – for this information, though even most of them had no idea how the ratio was calculated. Those who could answer the question provided multiple formulae. There was, however, a common denominator: the calorie. One of the more common methods involved calculating how many calories the average US farmer produces each year (looking at things such as total bushels of grains and pounds of meat produced) and dividing that figure by the total number of producers. This gave me the number of calories the average farmer produces. Finally, take this calorie per producer figure and divide it either by the annual caloric consumption of the average US citizen or (because this figure is much higher than what's necessary for optimal health) a figure reflecting a more modest level of food consumption. Voilà: 125 to 150 people are fed annually by the average US farmer.

'At its surface the ratio is a measure of productivity – of yields and the like – but at its heart it's about calories', a colleague once told me. That's what I meant when I said edibility is not necessarily a concern of cheap food. It also explains why, by some standards, cheap food may not be food at all, while by others, such as those that conflate food with calories, it is.

When did we start thinking about 'food' in this way? Its roots extend back at least a century, born out of a desire to order food and to further expose understandings of nutrition and health to scientific treatment. Accounts from the late 1800s speak of a sense of confusion about how to order food without a standardized measuring stick. For example, one visitor commented after visiting the Agricultural Building (a 7.6ha glass-domed building holding foods from around the world) at the 1893 World's Fair in Chicago that 'the terms of comparison were unclear. Although care was taken to impose a taxonomic order, the effect on the viewer was of a culinary Babel, as the animal, vegetable and gastronomic novelties of five continents jostled in mystifying profusion' (cited in Cullather, 2007a, p2).

At the turn of the last century, nutrition was dominated by the likes of John Harvey Kellogg and Horace Fletcher, who let moral and aesthetic criteria guide their prescriptions on heath and diet. Quantifiable criteria took a back seat during this period to other less objective measures. A standardized figure, which would make all food comparable and substitutable, revealed itself in the waning years of the 19th century. I am talking about the calorie.

'The work of rendering food into hard figures', Nick Cullather (2007b, p340) explains, 'began just after breakfast on Monday, March 23, 1896, when Wilbur O. Atwater sealed a graduate student into an airtight chamber in the basement of Judd Hall on the Wesleyan University campus.' The student was placed in a small room reminiscent of a meat locker. Through an airlock the subject was fed precise quantities of bread, baked beans, Hamburg steak, milk and mashed potatoes. When not eating, A. W. Smith (the occupant), performed physical (weightlifting) and mental (such as reading German treatises on physics) tasks. As this was happening, Atwater and his staff measured the movement of heat, air and matter into and out of the chamber. The measurements were so precise that the heat generated from winding a watch from within this copper-lined room was said to have been detected (Mudry, 2009, p37). Smith was in a calorimeter.

Atwater did not invent or discover the calorie. Europeans first measured it in 1883 (Nichols, 1994, p1724S). Yet, whereas in Europe the calorie attracted little attention, in the US it quickly established itself as a major guiding policy principle. Those in agriculture especially liked the concept. Like the atom in physics, the calorie came to represent the fundamental irreducible base of the agricultural sciences. For example, the esteemed early 20th-century animal scientist from Pennsylvania State College (now Pennsylvania State University) Dr Henry Armsby frequently equated food with energy. The following is taken from a 1911 article written by Dr Armsby for *Popular Science*:

> Now the problem of food supply is in essence a problem of energy supply.
> Food yields the energy which operates the bodily mechanism and upon the
> regularity and sufficiency of this energy supply depends absolutely all
> human endeavor. To produce those carriers of energy which we call foods
> is the chief function of the farmer. (Armsby, 1911, p469)

A 1913 issue of *Popular Mechanics* offers a similar argument. Written by a
professor of physiology from Cornell University Medical College, the article
conceptually reduced food to nothing more than fuel. Just like no one wishes to
be cheated when purchasing fuel to heat their machines or homes, the author
argues, so we should be equally diligent when assessing the caloric value of
different foods:

> An engineer who wishes to supply a certain amount of power must know
> the heat value of certain kinds of fuel and the waste from each. From these
> he reckons the net cost of his power. Any one who cares to do so can make
> the same sort of a computation for his body. If the engineer pays $7 for a
> ton of coal, he sees to it that he gets $7 worth of heat. Why is it not just
> as reasonable when a person pays a certain price for food to expect a
> certain amount of food value? To demand the worth of one's money in
> heat units when the fuel in the house is under considerable is a plain
> proposition, and when housewives generally understand food values it
> will be a plain proposition in respect to fuel for the body. A properly
> educated public opinion will demand from manufacturers such
> information in regard to the food on which its energies depend. (Murlin,
> 1913, p345)

With the calorie, food can be talked about without reference, context or form.
Taste, culture, culinary tradition and variety: all secondary (and arguably
superfluous) qualities when compared to their shared status as carriers of energy.
With the calorie, Atwater and others began ranking foods. Grains, meat and
dairy goods ranked well compared to fruits, leafy vegetables and fish. Some
items in the latter group possessed so few calories 'that they could scarcely be
classified as food' (Cullather, 2007b, p345). The calorie also washed away
cultural differences in diet. Scientists argued that, for example, the potatoes and
cheese that made up so much of the Irish diet were identical, except perhaps in
quantity consumed, to the rice and *ghee* eaten by those from India and Southeast
Asia (Cullather, 2007b, p345). By the time of his death in 1907, Atwater used
the calorimeter to analyse approximately 8000 food items, publishing his
findings in various US Department of Agriculture (USDA) yearbooks and
bulletins (Mudry, 2009, p39).

With this understanding of 'food' in tow, officials began rethinking the
provisioning of army, prisons and schools. The intent of these reforms was to
maximize money spent on foods high in calories, while de-emphasizing diets
scoring poorly in Atwater's tests. In terms of improving national diets, it was

believed the people of Asia had the most to gain from this new knowledge as their diet lacked the type of energy-dense foods – most noticeably, red meat – common to North Americans (Cullather, 2007b, p341). By the 1920s, some politicians went as far as to argue for a system of agricultural trade based on the calorie (Cullather, 2007b, p352). The imperial army and navy of Japan began reforming rations during the 1920s, adding 'Western' recipes to boost the caloric intake of its soldiers while cutting costs. Added rations of beef, pork, wheat and fried batter (*tempura*) reflected an early arms race of sorts to place Japanese troops on an equal footing with their American and British rivals (Cullather, 2007b, p357).

The US government used the calorie as a tool of soft power and repeatedly employed it throughout the Cold War. General Lucius Clay, while military governor of Germany immediately after World War II, said in 1946 that food shortages meant 'no choice between becoming a Communist on 1500 calories and a believer in democracy on 1000 calories' (cited in Major, 1997, p236). Anxiety that West Germans would side with the Soviets simply to obtain more calories precipitated the Berlin airlifts, where food and other goods were flown into Berlin from June 1948 to May 1949. Later, Public Law 480 (passed in 1954), which gave birth to US food aid policy, was used early on to similar effect as the Berlin airlifts – namely, to stop communism from spreading in poor countries on account of calories. It also lent truth to the Eisenhower administration's claim that 'free men eat better' and the Kennedy administration's argument that 'wherever Communism goes, hunger follows' (cited in Cullather, 2007b, p363).

While the discovery of phenomena such as vitamins and minerals eventually undermined the calorie's identity as the single variable upon which all food is to be measured (and greatly helped to rehabilitate the value of fruits and vegetables), the reductionist ideology that the calorie helped to spur remains. While still central to how we think about food and agriculture – remember, the farmer/eater ratio is fundamentally a caloric ratio – we now reduce food in many ways, not just one. This ideology has been referred to as 'nutritionism'. Nutritionism refers to a 'quantitative logic' that 'obscures the broader cultural, geographical, and ecological contexts in which foods, diets, and bodily health are situated' (Scrinis, 2008, p44). Rather than seeing food as a part of a larger socio-material whole, nutritionism sees in food a bunch of smaller unconnected material components. To show the extent to which this highly reductionist, de-contextual view of food pervades agricultural policy, let's turn our attention to the Green Revolution.

Hidden Hunger and the Green Revolution

The Green Revolution: a series of strategies developed during the mid to late 20th century to combat starvation by expanding the global production of staple food crops through crop breeding. Was it a success? Yes. I know that answer will raise the ire of some. But what was the *primary* goal of the Green Revolution?

Reducing poverty and hunger? Both were secondary objectives, though it was assumed that the problems of poverty and starvation would naturally work themselves out if the first goal was satisfactorily achieved. The Green Revolution was fundamentally a revolution in calories. As a food calorie revolution, the Green Revolution was an unqualified success.

Global hunger is not the result of a lack of calories in the world. The global obesity epidemic is testament to that fact. Global hunger, rather, is a *distributional* problem. It is not something that can be solved by soil, crop or animal scientists. It is a socio-economic problem that desperately requires socio-economic remedies – for example, while the percentage of the world's population facing continual food insecurity has decreased, absolute numbers of those who are food insecure have increased (Weis, 2007, p11).

Then there is the Green Revolution's link to micronutrient malnutrition: also known as hidden hunger. Hidden hunger is a particularly pernicious form of malnutrition. Who hasn't seen the heart-wrenching infomercials depicting starving (near skeleton-like) children from some far away land. Watching from our comfortable couches, the images seem almost other worldly. *That's* hunger, as those of us in affluent countries have come to know the term. Hidden hunger doesn't look like that. In fact, it is often much harder to see, though no less life threatening and spirit crushing. So it's easier to miss ... or ignore.

In a peer-reviewed paper co-authored by a USDA plant scientist, the Green Revolution was described rather candidly as making a 'push toward food (i.e. calorie or energy) security' (Welsh and Graham, 1999, p3). Food security *equals* calorie security. When I ask students whether these two types of security – food and calorie – ought to be conflated, their initial response is overwhelmingly 'yes', until the practical implications of this conflation start sinking in. Someone will ask, for example, if a community would be considered food secure if they had all the corn (and, thus, calories) they wanted but nothing else to eat; or, perhaps reminded of Morgan Spurlock's 2004 documentary *Supersize Me* (on the consequences of relying entirely on McDonald's for food for thirty days), I occasionally hear the question of whether it's possible to live on Big Macs alone.

What are some of the reasons that help to explain this policy, which has made the Green Revolution a revolution in global caloric output? The market bears a lot of the responsibility – another case where pocketbook democracy has failed the least food secure. Improving the nutritional value of staple food crops is of secondary concern for higher income consumers who have, at least relative to the poor, a diverse diet (Unnevehr et al, 2008, p1). Moreover, the fortification of industrial foods in the developed world has largely eradicated any interest in the nutritional effects of plant breeding. If the food can be fortified during processing, I once had a crop scientist tell me, why go through all the expense of breeding for higher nutritional content and risk reducing the plant's yield? The sad truth is that the market historically rewards higher yield rather than nutritional content (Morris and Sands, 2006, p1078).

Even with all these available calories, micronutrient malnutrition plagues billions around the world. As stated in a report prepared by the United Nations

Children's Fund (UNICEF), the World Health Organization (WHO) and the United Nations World Food Programme (WFP):

> Deficiencies of micronutrients are a major global health problem. More than 2 billion people in the world today are estimated to be deficient in key vitamins and minerals, particularly vitamin A, iodine, iron and zinc. Most of these people live in low income countries and are typically deficient in more than one micronutrient. (UNICEF et al, 2007, p1)

Children are a particularly sensitive population to micronutrient malnutrition due to having higher nutritional requirements per kilogram of body weight. Almost two-thirds of all deaths of children globally are attributable to nutritional deficiencies (Caballero, 2002, pp3–4). A consensus is emerging that the growth lost in early years due to malnutrition is, at best, only partially regained during childhood and adolescence with dietary improvements (Martorell et al, 1994, pS45; Alderman et al, 2003, p7). One study, examining children who were between 12 and 24 months in age at the peak of the drought in rural Zimbabwe during 1994 to 1995 had average heights well below that of comparable children not affected by this event when measured at ages 60 to 72 months (Hoddinott and Kinsey, 2001). Another study notes that among children in rural Zimbabwe exposed to the 1982 to 1984 droughts, average heights at late adolescence were reduced by 2.3cm compared to children not exposed to this nutritional shock (Alderman et al, 2006). Noting the importance of iron intake for brain development and that more than 40 per cent of children aged zero to four in developing countries suffer from anaemia, Grantham-McGregor and colleagues (1999) argue that hidden hunger shares some of the blame for poor schooling outcomes in the developing world.

There is no disputing that the Green Revolution has occurred at the expense of dietary diversity (Blatt, 2008, pp81–85). Crops bred for traits associated with improved yields or to withstand mechanization have displaced traditional crops that are high in iron and other micronutrients. In South Asia, for example, while cereal production has increased more than fourfold since 1970, production of pulses (a high protein legume) has dropped 20 per cent (Welsh and Graham, 1999, p2; Gupta and Seth, 2007, p436). There is a well-documented relationship between Green Revolution cropping systems and a decline in the density of iron in the diets of people in South Asia (Seshadri, 2001; Welsh and Graham, 2002). Nor is it merely a matter of *what's* being raised. *How* the cereals of the Green Revolution are processed and consumed has also affected the dietary health of millions. Cereals and rice, in particular, are consumed primarily after milling – a process that removes micronutrients. Pulses, conversely, are traditionally consumed whole after cooking. Moreover, whole cereal grains contain relatively high levels of anti-nutrients, which are known to lower the absorption of micronutrients, and lower levels of substances that promote the bioavailability of micronutrients (Cordain, 1999, p28; Welsh and Graham, 1999, p6).

The costs associated with the calorie revolution (and the resultant rise in micronutrient malnutrition) of the 20th century are rather startling. The US Agency for International Development (USAID) estimates that iron deficiency costs India and Bangladesh about 5 and 11 per cent, respectively, of their annual gross domestic product (GDP) (Sanghvi, 1996). Another analysis, also looking at India, estimates that each year, 4 million healthy life years (or HLYs) are lost due to iron deficiency, an additional 2.8 million due to zinc deficiency, and still 2.3 million more due to vitamin A deficiency. When taken together and assigned a monetary value, these deficiencies are said to reduce the country's gross national income by between 0.8 to 2.5 per cent (Stein, 2006, pvi). Looking at the whole of South Asia, another analysis places the costs of iron deficiency at US$4.2 billion annually for losses in economic productivity (Horton and Ross, 2003, p71). The literature is clear: hidden hunger costs a lot. It impairs national development efforts, reduces economic output, lowers educational attainments, negatively affects school enrolments, increases mortality and morbidity rates, and raises healthcare costs (Sanghvi, 1996; Popkin, 1998; Horton and Ross, 2003).

To be fair, more attention has been paid to micronutrient malnutrition during recent years. Take the rise of bio-fortification, which involves breeding (or, more recently, genetically engineering) plants to have higher nutritional content. Yet, I am suspicious of any 'solution' based upon the same thinking that created the problem in the first place. The calorie-ization of food was problematic because it de-contextualized food. Bio-fortification similarly abstracts 'food'. Whereas calorie-zation reduces food to energy, bio-fortification views agricultural commodities as carriers of nutrients. Calorie-ization, bio-fortification: both are examples of nutritionism.

Golden Rice – rice genetically engineered to contain high levels of vitamin A – is an excellent example of bio-fortification. Yet, as a 'solution' to a diet deficient in beta-carotene, Golden Rice is terribly short sighted. Vitamin A is fat soluble, which means that its uptake within the body is dependent upon a level of fat in the diet. In less developed parts of the world, however, levels of dietary fat are often insufficient. And simply increasing a population's daily intake of rice, by itself, adds none of this life-sustaining fat to their diet. Moreover, Golden Rice entirely sidesteps the deeper 'why' question: namely, why is vitamin A deficiency increasing throughout the developing world, thus creating the need for such rice in the first place? Golden Rice is directed at symptoms rather than the root cause of vitamin A deficiency.

Bio-fortification strategies also do nothing to increase dietary diversity, recognizing that such diversity is not only widely recommended by nutritionists (Ruel, 2003; Johns and Sthapit, 2004), but something those in poor countries expressly desire (Bouis, 2003). Bio-fortification may, in fact, have the opposite effect. Concentrating more nutrients in a few staple foods risks further shrinking the global dietary diversity of those in the developing world. There is also the concern that fortification programmes will draw attention away from the more important issue of *food access*. As Tripp (2001, p258) notes, bio-fortification schemes embrace, whether implicitly or explicitly, a 'technical fix' attitude that 'can tempt governments to believe they don't have to worry about nutrition because the plant breeders are handling this'.

Bio-fortification schemes, by nature of their de-contextualized understanding, also do little to address the culturally rooted aspects of food. Bio-fortification can alter the cooking and storage properties of a plant in addition to its taste, odour, colour and texture – all of which are properties valued by small-scale farmers in developing countries (John and Eyzaguirre, 2007, p15). Food is much more than just a means to physical satiation. As John and Eyzaguirre (2007, p16) explain: 'farmers may prefer particular species or varieties for use in religion- or social-defined roles, recognize subtle benefits in reduced labour requirements, or attribute to them particular health properties that might outweigh the new values attributed to biofortified varieties'.

I am not suggesting that bio-fortification should not be done. If it is, however, it must be done with great care. We must be vigilant, for example, in making sure not to harm the very population whom we are trying to help, such as children. Safety issues ought to be a top concern when modifying a traditional food staple that accounts for a large share of a child's daily caloric intake (Unnevehr et al, 2008, p8). Compared to industrial fortification, variations in the nutrient content of a bio-fortified crop may possibly be more difficult to control since nutritional levels could vary depending on the agricultural practices of the farmer and the environmental conditions of the farm (Unnevehr et al, 2008, p9).

Nor am I suggesting that nutritionism only negatively affects those in developing countries. The ideology of nutritionism also runs rampant in affluent countries. On the subject of how far we in the West have managed to reduce food, humorist Dave Barry (2002, p128) makes the following observation:

> When I purchase a food item at the supermarket, I can be confident that the label will state how much riboflavin is in it. The United States government requires this, and for a good reason, which is: I have no idea. I don't even know what riboflavin is. I do know I eat a lot of it. For example, I often start the day with a hearty Kellogg's strawberry Pop-Tart, which has, according to the label, a riboflavin rating of 10 percent. I assume this means that 10 percent of the Pop-Tart is riboflavin. Maybe it's the red stuff in the middle. Anyway, I'm hoping riboflavin is a good thing; if it turns out that it's a *bad* thing, like 'riboflavin' is the Latin word for 'cockroach pus', then I am definitely in trouble.

The marketing of food today often focuses on the absolute or relative quantities of one or two nutrients in food (Scrinis, 2008, p45). This allows, say, the Kellogg Corporation to focus on how the Pop-Tart contains 10 per cent of the recommended daily intake of not only riboflavin but also vitamins A and B6, niacin and folate. Doing this draws attention away from not only the overall nutrient profile of a food and its ingredients, this obsession with the nutritional bits and pieces of a food also distracts from deeper (and, I would argue, more important) questions such as 'What is the real cost of a Pop-Tart?' and 'Is having nutritional value the same as being nutritious?'

'Nutritionism', as argued by Gyorgy Scrinis (2008, p47), 'has adapted and aligned our minds and bodies to the nutritional marketing strategies and nutritionally engineered products of the food industry'. Indeed, far from threatening the dominant (cheap) food system, nutritionism further secures its place in history. No longer can industrial food be as easily criticized as representing 'just empty calories'. It was fair criticism when food was equated merely with calories. Food companies can now point to specific nutritional values and thus deflect questions about whether or not their products promote health. Indeed, spending just a couple of extra cents per unit (box, can, bag, etc.) on fortification not only allows food companies to make health claims. A fortified food, even if it contains just a few pennies worth of vitamins or minerals, can be sold at a premium. Food historian Marion Nestle tells of when General Mills started fortifying its boxed cereal during the late 1970s. The addition of 7 cents worth of nutrients allowed the company to charge 36 cents more (a sizeable mark-up when a box sold for US$1.29) for a product that was otherwise identical to its non-fortified version (Nestle, 2007, p305).

The Political Economy of Mis-nourishment

'Hungry': a category that over half of the world's population falls into (Dybas, 2009, p646). Unlike in the past, however, the term no longer exclusively applies to those with distended stomachs (a product of the body turning upon itself in search of life-giving proteins). Hungry stomachs still stick out, but increasingly for other reasons – namely, *too many* calories. The obesity–hunger paradox is a phenomena where the hungriest people in some parts of the world 'may well be not sickly skinny, but excessively fat' (Dolnick, 2010).[2]

I will begin by addressing a question that is as topical as it is controversial: do agricultural subsidies lead to obesity? Among those answering this question in the affirmative include best-selling food writer Michael Pollan (e.g. Pollan, 2008, p186) and a number of nutrition and food policy scholars, such as Barry Popkin (2007, p60) (professor of nutrition at the Carolina Population Center of the University of North Carolina at Chapel Hill) and James Tillotson (2004, p629) (professor of food policy and international business at Tufts University). In the words of Pollan (2007), talking about the US farm bill:

> This resolutely unglamorous and head-hurtingly complicated piece of legislation ... sets the rules for the American food system – indeed, to a considerable extent, for the world's food system. Among other things, it determines which crops will be subsidized and which will not, and in the case of the carrot and the Twinkie, the farm bill as currently written offers a lot more support to the cake than to the root. Like most processed foods, the Twinkie is basically a clever arrangement of carbohydrates and fats

teased out of corn, soybeans and wheat – three of the five commodity crops that the farm bill supports, to the tune of some $25 billion a year. (Rice and cotton are the others.) For the last several decades – indeed, for about as long as the American waistline has been ballooning – U.S. agricultural policy has been designed in such a way as to promote the overproduction of these five commodities, especially corn and soy.

The logic underlying the above quote is fairly easy to follow. Agricultural subsidies have made certain food staples (most famously, corn) cheap and abundant. Making these food staples cheap and abundant provides an incentive for their use (substituting them for less cheap commodities), leading to artefacts such as high fructose corn syrup (HFCS). And so we find ourselves consuming an overwhelming amount of cheap and abundant industrial products derived primarily from a handful of subsidized food staples.

The argument seems almost self-evident. When first confronted by it I admit to having also been seduced by its sound, linear simplicity. It is not that Pollan and others are wrong when linking government agricultural subsidies to obesity. They admittedly smooth over the complexity of the relationship. But they are correct to label current agriculture subsidies as being largely antithetical to public health.

Many, however, are sceptical (and some, downright hostile) to this argument. The more thoughtful of these sceptics (see, for example, Alston et al, 2008; Beghin and Jensen, 2008, p480) acknowledge the undeniable growth of sweetener use in the food system and recognize that subsidies, especially early on (the 1970s), contributed to HFCS's meteoric rise in the US. Where they differ from those mentioned earlier is on the question of how much blame to place upon agricultural subsidies when looking to explain phenomena such as obesity and the obesity–hunger paradox. Pollan, for instance, is comfortable drawing a fairly thick causal arrow pointing from subsidies to the obesity pandemic. Others, however, argue that Pollan overstates the relationship between things such as cheap corn and bad diets.

Note already the 'slippage' in the debate. The earlier mentioned camp, which includes the likes of Michael Pollan, draws a thick causal arrow from agricultural and economic policies *as a whole* to obesity. The latter camp, conversely, takes a far more narrow interpretation of the debate, looking at the relationship between *specific* agricultural and economic policies (such as corn subsidies) and obesity. Let me address the more narrow interpretation first.

Critics of the subsidies equal obesity argument make a number of important points. They note, for example, that the cost of farm commodities as ingredients represents only a small share of the cost of retail food products – on average, approximately 20 per cent. And for foods such as soda (soft drinks), fully prepared meals and meals away from home, this percentage is often considerably less (Alston et al, 2008, p473). They therefore contend that subsidies do not affect the retail price of food all that much. In one recent analysis, it was estimated that a 20 per cent subsidy for corn processors in 1975 would have lowered food prices by 1.8 per cent. During more recent years, as food becomes increasingly processed, that same 20 per cent subsidy for corn users is estimated to decrease retail food prices a mere 0.3 per cent (Berghin and Jenson, 2008, p485). This recently led a trio of agriculture economists to assert the following: 'Hence, a very large percentage change in commodity prices would be required to have an appreciable percentage effect on food prices' (Alston et al, 2008, p473).

Today, HFCS represents over 40 per cent of the caloric sweeteners added to food. Per capita average use of all sweeteners in the US increased from 123 pounds in 1966 to 151 pounds in 1999 (Berghin and Jenson, 2008, p481). Per capita consumption of HRCS increased during that same period by 235 per cent (Harvie and Wise, 2009, p3). For a number of years the US has been converting 0.5 billion bushels of corn a year into HFCS. Corn subsidies are an undeniable factor driving this move away from sugar. Between 1963 and 2005, corn prices fell much faster than sugar prices, meaning that HFCS could historically be purchased for 20 to 70 per cent *less* than sugar (Harvie and Wise, 2009, p3). Coca-Cola reportedly saved US$70 million in its first year after switching to HFCS in 1983. This cost saving strategy was adopted by Pepsi one year later (Warner, 2006).

Yet, even if the price of corn marginally affects the unit cost of, say, a can of soda (recognizing that corn accounts for approximately 1.6 per cent of the total cost of soft drink manufacturing), given the total amount of soda sold and consumed annually, does this not still translate into tremendous taxpayer-funded savings for those manufacturing HFCS and soda?[3] According to a pair of researchers at Tufts University, Alicia Harvie and Timothy Wise (2009), the answer to this question is an unqualified 'yes'.

Thanks to subsidies, wet millers who refine HFCS purchased corn for 27 per cent less than its production cost from 1997 to 2005. If corn's value represents 44 per cent of HFCS production costs, then the wet milling industry spent US$18.4 billion producing HFCS over the period analysed, with corn accounting for US$8.1 billion of the total cost. Harvie and Wise (2009) calculate that had corn been priced to reflect its full production cost, HFCS would have cost almost 12 per cent more. Between 1997 and 2005, corn subsidies therefore saved HFCS producers US$2.19 billion. Harvie and Wise then looked to see what these savings meant to soda manufacturers. Their findings indicate that had corn been priced at its full production cost, soda manufacturers would have spent an additional US$873 million during that nine-year period. Using data from an earlier study that examines the savings to corn buyers between 1986 and 1996 (Starmer et al, 2006), Harvie and Wise (2009) estimate that subsidies saved HFCS manufacturers and soda companies roughly US$2 billion and $800 million, respectively. Combining these two estimates, 1986 to 1996 and 1997 to 2005, they calculate taxpayer-funded savings to HFCS producers of more than US$4 billion and to soda companies of US$1.7 billion.

To put this in perspective, the soda industry generates approximately US$50 billion a year in revenue in the US alone (Abramovitz, 2002, p140). So, yes, as a fraction of the costs and revenue generated, US$1.7 billion over a 20-year period is not much. But it is still almost US$2 *billion*. When I ask students what they think about giving the Cokes and Pepsis of the world this level of taxpayer-funded support, even when spread over 20 years, many mouths drop. Perhaps one has right now.

This brings us to another important point: *who* do subsidies really help? Although this question is discussed in greater detail in Chapter 8 (so I won't labour the point), it deserves brief attention here. When Coke switched over to HFCS, saving US$70 million that first year, who benefited? Consumers? Did the retail cost of a bottle of Coke suddenly drop? I have been unable to find any evidence of a substantial retail price change. What changed were the firm's margins. Farm subsidies translated into increased profits for soda companies. These profits include the approximately US$2 billion from 1986 to 2005 in addition to the tremendous amount of money saved by not having to use the more expensive sugar. Trying to dispel the link between subsidies and obesity, Larry Mitchell, former CEO of the US Corn Growers Association, is quoted as saying: 'When you examine the data, it doesn't support the theory. The fact is, farmers are capturing less and less of the total food dollar' (cited in Fields, 2004, p821). Mr Mitchell's point about farmers is correct; but it does nothing to dispel the link between subsidies and current US dietary practices. If anything, the fact that farmers are capturing less of every dollar spent on food shows that the real benefactors of agricultural subsidies are agro-processors and food and beverage manufacturers.

I have a theory about *how* agricultural subsidies link up with obesity. Subsidies make processed foods extremely profitable – at least significantly more profitable than whole (unprocessed) food. This profitability creates deep pockets to fund expensive advertising campaigns. In 2003, the USDA allocated US$333 million for nutrition education (Schoonover and Muller, 2006, p11). Compare this to the US$10 billion to $15 billion spent annually on food and beverage advertising aimed at children (Eggerton, 2007). PepsiCo's annual profits for 2010 were slightly less than US$6 billion (and $43.2 billion in full-year revenues) (Fredrix, 2010). Some of its product-specific advertising expenditures include US$162 million for Gatorade, US$145 million for Pepsi Cola, US$27 million for Tostitos, US$14 million for Doritos and US$11 million for Fritos (Nestle, 2010). Not included in these figures is money that Pepsi spent on things such as lobbying, supporting the American Beverage Association's efforts to fight soda taxes, a Pepsi-funded obesity study at Yale, or marketing to children and adults throughout the developing world (Nestle, 2010).

In truth, it is next to impossible to know the precise amount that these large companies spend on advertising to get us to buy their products. PepsiCo, for example, made the decision years back to hide some of this money in other budgets. The budget of the marketing department only constitutes a fraction of what the company actually spends in their quest to mould our minds and taste buds. Tens of millions of dollars spent on promoting their brand are hidden as event sponsorships (stadiums, buildings, etc.), for instance, allowing those dollars to be categorized as long-term capital expenditures. A former employee of PepsiCo estimates that the company spends as much as US$1 billion a year on direct and indirect consumer marketing (Nestle, 2010).

I do not mean to suggest that we are all mindless dolts. But neither are these corporate giants. I am well aware that there are many reasons why we buy and consume the foods we do (see Carolan, 2011). Clearly, PepsiCo would not be spending as much as US$1 billion a year on ads if that money had no effect. Rather than talk about the individual influence each company has on our diets, James Tillotson (2004, p634) suggests that we look at the 'composite influence of the total food industry'. Food companies spend many billions of dollars a year to get us to want their products, with the help of an army of social, behavioural and marketing scientists. I therefore refuse to brush this composite influence away as inconsequential.

And the data seem to indicate that this influence is having, well, an influence. US caloric consumption increased from 2234 calories per person per day in 1970 to 2757 calories in 2003. The distribution of these added calories are as follows: total per capita consumption of added fats and oils rose by 63 per cent, grain by 43 per cent, vegetables by 24 per cent, and sugar and sweetener by 19 per cent (though, as mentioned earlier, HFCS per capita consumption increased more than 200 per cent during this period) (Farah and Buzby, 2005). Studies have consistently found a positive correlation between number of television hours watched by children and frequency of requests, purchases and servings eaten of advertised food (see, for example, Claney-Hepburn et al, 1974; Taras et al, 1989; Frances et al, 2003). According to the *Guinness Book of Records*, the McDonald's arches are more widely recognized in the world than the Christian Cross. Children consume approximately 167 additional calories for each hour of TV watched (Wiecha et al, 2006). A child's risk of obesity increases by 6 per cent for every hour of TV watched on average each day (Robinson et al, 2001). One 30-second commercial can influence the brand preferences of a child (Borzekowski and Robinson, 2001). In one study, preschool children report that food in McDonald's wrappers taste better than food in plain wrappers even though *the same* food item was consumed (Robinson et al, 2007). And what are governments doing about all of this? In the US, at least, thanks to the Internal Tax Revenue Code of 1986, they are giving *tax breaks* to companies for advertising and marketing unhealthy food to kids.

In the US, advertising is currently considered a business expense as it can be used by corporations to reduce their federal tax liability. With the current corporate income tax rate at 35 per cent, elimination of the tax deductibility of food advertising costs would be the same as increasing the price of advertising by 54 per cent. A study by the Institute of Medicine (2004) estimates that eliminating the tax break would result in the reduction of fast food advertising messages by 40 per cent for children and 33 per cent for adolescents. Until that time, US taxpayers are helping to subsidize food advertisements, even for those foods that their government acknowledges must be eaten 'sparingly'.

If You Are What You Eat, Avoid Poverty

Accessibility to supermarkets is negatively correlated with income, which means that if you live in a low-income neighbourhood, your choice in supermarkets – and, thus, in healthy foods – is likely very limited (MacDonald and Nelson, 1991; Chung and Myers, 1999; Morland et al, 2002b). A widely cited article in the *American Journal of Preventative Medicine* finds the increased presence of supermarkets to be associated with a lower prevalence of obesity in the surrounding community, while the increased presence of convenience stores is associated with a higher prevalence of obesity (Morland et al, 2006). Deficiencies in food access seem to be further exacerbated in communities composed primarily of people of colour (Morland et al, 2002b; Zenk et al, 2005). Grocery stores in African American neighbourhoods, for example, have been reported to stock fewer fresh fruits and vegetables and healthier versions of standard foods, even though residents of the community report wanting greater access to these goods (Sloane et al, 2003). Proximity to a supermarket has also been shown to influence diet. The closer residents in low-income neighbourhoods live to a supermarket, the higher their fruit and vegetable consumption (Rose and Richards, 2004). A similar relationship has been reported among pregnant women: the closer they live to a supermarket, the larger the role that fruits and vegetables play in their diet (Laraia et al, 2004). In another study, fruit and vegetable consumption increased 32 per cent for each additional supermarket located in predominately African American neighbourhoods (Morland et al, 2002b).

On the subject of rural food security: a report by the Center for Rural Pennsylvania (2005) found that rural school districts in Pennsylvania had a greater percentage of overweight children than urban school districts and that the proportion of overweight children in rural schools was growing faster than those in urban districts. Similar findings have been reported in North Carolina (McMurray et al, 1999) and South Carolina (Felton et al, 1998). As rural communities erode (a process, as discussed further in Chapter 8, that is itself partially a product of cheap food policies), they risk reaching an economic and population threshold where it becomes difficult for local businesses, such as grocery stores, to remain open. Then there is the problem of rural public transportation – namely, there isn't any. Transportation limitations further complicate the issue of food accessibility. Having to travel greater distances than their urban counterparts for food, many remote rural areas are becoming food deserts (Morton et al, 2005; Schafft et al, 2009).

One of the largest risk factors for obesity is being poor (save for those, obviously, in abject poverty). National health surveys out of Australia, for example, consistently show highest obesity rates among populations with the least education and lowest income (Friel and Broom, 2007, pp152–161). There are a number of factors for explaining the unequal distribution of social conditions that enhance and weaken the health of individuals and populations (see, for example, Dixon and Broom, 2007; Lake et al, 2010). I can't go into them all here. What I will talk about, however, is how food policy has given us an abundance of cheap (and empty) calories.

A US study from 2004 examined how many calories US$1 could buy. Rather than use the dollar to buy items located along the periphery of the store (where milk products, meat and fish, fruits and vegetables, and fresh baked breads tend to be located), the biggest bang for one's buck was in the middle of the store, among the walls of soda, chips, candy and other non-expirable goods (Drewnowski and Specter, 2004). Similar results were obtained in a French study. Controlling for total caloric intake, the researchers found that a 100 gram increase in fats and sweets is associated with a 6 to 40 cent *decrease* in daily diet costs, whereas a 100 gram increase in fruits and vegetables is associated with a 22 to 35 cent *increase* (Maillot et al, 2007).

What makes this all the more remarkable is the innumerable value-added steps that these 'cheap' products have gone through. I counted more than 50 different ingredients in the Chocolate Fudge Pop-Tart. Looking online, I learned that I could buy a six-pack of this iconic American food (recently Pop-Tart World opened in Times Square, New York, joining the Hershey Store and M&M World at this landmark location) for US$3.99. Compare this to the red bell peppers I recently purchased at the grocery store for US$2 apiece. A remarkable feat: producing a box of frosting-covered pastries with a seemingly eternal shelf life, over 50 ingredients and a hefty annual advertising budget (after all, it has its own World) for the same price as two peppers.

We have a century of agricultural policies to thank for this. At most, one tenth of 1 per cent of all domestic agriculture subsidies in the US goes to supporting fruit and vegetable crops (Fields, 2004, p822). Why can't we do more to subsidize healthy foods? As Richard Atkinson, professor of medicine and nutritional sciences at the University of Wisconsin-Madison (and president and co-founder of the non-profit American Obesity Association), explains: 'There are a lot of subsidies for the two things we should be limiting in our diet, which are sugar and fat, and there are not a lot of subsidies for broccoli and Brussels sprouts' (cited in Fields, 2004, p823). Out of one side of its mouth the USDA encourages people to eat vegetables and fruits, many of which are not subsidized and are therefore unnecessarily expensive, while making it easier for consumers, thanks to subsidies, to buy the very foods it says they should eat sparingly (as illustrated in Figure 4.1).

The market is not going to help us. It will only make us fatter. Cheap food proponents say that the market is only responding to consumers. When consumers want to eat more fruits and vegetables, according to this logic, they will signal the market and the market will faithfully respond. But there is a big problem with this argument: consumers already want more food access and greater diet diversity (see, for example, Morland et al, 2002a; Laraia et al, 2004; Morton et al, 2005). The market is not responding to consumers' needs. That is because decades of cheap food policies have so greatly incentivized our consumption of things such as Pop-Tarts, Cokes and Big Macs that the transaction costs for many consumers to eat differently – to eat better – is simply

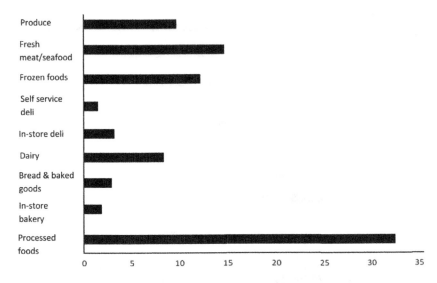

Figure 4.1 *Share of supermarket sales (percentage)*

Source: based on data from USDA ERS (2002)

too great. Some have the financial resources to buy, for example, more fresh fruits and vegetables (and possess sufficient time to prepare and cook these foods). Many, however, do not. Moreover, while fats and sweeteners have been getting cheaper, fruits and vegetables are getting more expensive (as illustrated in Figure 4.2). Looking at food items sold at major retail supermarket chains in Seattle, Washington, a recently published article finds nutrient density negatively associated with energy density and positively associated with cost (Monivais et al, 2010). The researchers also note that, at least from 2004 to 2008 when the study was conducted, the price disparity between healthy and unhealthy foods *increased* in favour of the latter. In other words: the more 'junky' the food, the cheaper it is and the cheaper it is becoming relative to more nutrient-dense foods.

No wonder so many eat as they do, given the economic incentives (and disincentives) that shape consumer 'choice'. Like the cheap food policies that it relies upon, the food industry operates according to logics that have little interest in public health. Food companies are motivated to continuously increase consumption of their food products, which arguably places their interests in opposition to those of our own (Tillotson, 2004, p634). 'Did consumers', Richard Denniss (2007, p23) rhetorically asks, 'demand larger portion sizes at restaurants, or did restaurants offer larger portions as an effective way to market themselves as providing "good value for money"' (while, I might add, widening their profit margins)?

Rather than waiting for consumers to signal the market by buying more whole foods, we might think about starting at the other 'end', with producers. We can nudge producers along by incentivizing the production of things such as

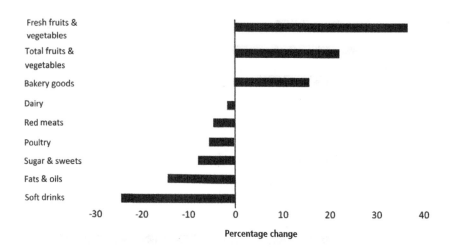

Figure 4.2 *Change in food prices, 1985–2000 (converted to real 2002 US$)*

Source: based on data from Putman et al (2002)

fruits and vegetables. These incentives must include not only subsidies for these crops, but also programmes that provide credit so farmers can purchase necessary equipment and that improve markets access so producers will have someone to sell their crops to.

Consumers will eat healthier foods if they are made more affordable (or if unhealthy foods are more correctly priced). In one study the prices of low-fat snack foods were reduced in 12 workplace and 12 secondary school vending machines. Relative price reductions of 10, 25 and 50 per cent produced sales increases of 9, 39 and 93 per cent, respectively (French, 2003). A similar study looked at a restaurant. The price of three low-fat restaurant foods – a grilled chicken sandwich, a salad with grilled chicken and a vegetable soup – were lowered between 20 and 30 per cent. The sales of the three low-fat target foods increased more than 30 per cent over the test period (Horgen and Brownell, 2002). Other researchers found that reducing the price of low-fat items in vending machines by 50 per cent increased the proportion of low-fat snacks purchased during the four-week test period from 25.7 to 45.8 per cent. When prices returned to normal after the test, the sale of low-fat items dropped 22.8 per cent, indicating that price was influencing consumers' decision in what they bought (French et al, 1997). Price increases of junk food through additional taxes, on the other hand, may be a suitable policy tool since the average consumer is responsive to price changes. Nevertheless, caution should be exercised to minimize any penalty on low-income and minority groups. As we've seen with cigarettes, such 'sin' taxes do not change everyone's behaviour and can have a disproportional negative impact upon certain populations, especially those living in a food desert where fast food and convenience store foods might be their only option.

You Call This Cheap?

Obesity is not just an individual problem. It's a societal problem brought upon us by policies of our own making – policies that now strip many of our so-called rights as sovereign consumers to 'choose' what we eat. There is no choice when one lives in a food desert. Nor is it fair to say that people choose Pop-Tarts, Cokes and Big Macs over bell peppers, lettuce and carrots when the consumption of the former has been so greatly incentivized over that of the latter. Overwhelming evidence links cheap food policy to mis-nourishment (see, for example, Hayne et al, 2004; Tillotson, 2004; Popkin, 2007; Wallinga et al, 2009). In light of this link, and all the health ailments associated with this risk factor, it makes sense to include many of the healthcare expenditures associated with the obesity pandemic in calculations of the 'cost' of the dominant food system. I conclude the chapter by listing some figures. While reading these figures, keep the following question in mind: is the 'cheapness' of cheap food, at least in part, an artefact of cost shifting, from one industry (food) to another (healthcare)?

I'll begin with the US, the world capital of both cheap food and obesity:

- Annual US obesity-attributable medical expenditures are estimated at US$75 billion in 2003 dollars, with approximately half of these expenditures paid for by Medicare and Medicaid (both taxpayer-funded programmes) (Finkelstein et al, 2004).
- Health expenditures due to obesity costs every US citizen US$150 annually (Bhattacharya and Sood, 2005).
- Direct and indirect costs of obesity in the US have been placed at US$117 billion (Frazao, 1999).
- Poor diet is a risk factor for four of the six leading causes of deaths in the US: heart disease, cancer, stroke and diabetes. When combined with obesity, these diseases have been estimated to cost US$556 billion per year (Wallinga et al, 2009).

Moving to Europe and the UK: though Europeans have long poked fun at their overweight neighbours to the West, they are starting to eat (and look) more like Americans with each passing year:

- It is estimated that total costs linked to obesity in England could increase by as much as 240 per cent between 2007 and 2025. Healthcare alone costs taxpayers UK£10 billion annually, while the wider costs to society and business are estimated to be close to UK£49.9 billion per year (Foresight, 2007, p5).
- It is estimated that treatment for obesity-related illnesses costs Germany

about UK£16 billion a year (which explains the occasional chatter among German politicians about implementing a so-called 'fat tax') (Hall, 2010).

Australia:

- In 2005, overweight and obese Australian adults cost the nation's economy AU$21 billion in direct healthcare (e.g. ambulatory services, hospitalization and prescription medication) and direct non-healthcare costs (e.g. special food and dietary support), plus an additional AU$35.6 billion in government subsidies (e.g. payments for disability, sickness allowances and unemployment benefits) (Colagiuri et al, 2010).

And, finally, Canada:

- A January 2011 issue of *Bloomberg Business Weekly* (2011) reports that the total economic cost of overweight and obese individuals in Canada is approximately Cdn$300 billion a year: Cdn$127 billion in healthcare; Cdn$72 billion in loss of productivity due to total disability; Cdn$49 billion in loss of worker productivity due to higher rates of death; and Cdn$43 billion in loss of worker productivity due to disability of active workers.

What about developing economies? Although many developing countries do not (yet) have obesity rates comparable to those in the US and England, neither do they (or their citizens) have the resources to pay for the exorbitant healthcare costs that go along with eating cheaply. According to a recent Organisation for Economic Co-operation and Development (OECD) report published in *The Lancet*, seven in ten Mexican adults are overweight or obese, while approximately half of all adult Brazilians, Russians and South Africans are also in this category (Cecchini et al, 2010, p1). Mexico is just behind the US when it comes to these statistics, where 74 per cent of adults are overweight or obese. In the case of Mexico especially, the fingerprints of cheap food policy are all over the country's obesity epidemic. Since it enacted the North American Free Trade Agreement in 1994, imports of processed food and drinks have increased rapidly. Mexico now consumes more Coca-Cola products per capita than any other country on the planet: a total of 635 eight-ounce bottles per person annually (three times the amount consumed in 1988) (Grillo, 2009).

The aforementioned OECD report (Cecchini et al, 2010) recommends that less affluent countries act now to combat obesity. Suggestions include media campaigns promoting healthier lifestyles, taxes on nutritionally empty foods, subsidies to increase consumption of nutritionally dense foods, and restrictions on food advertising for junk food. An earlier report in *The Lancet* revealed that if nothing is done to reduce the rates of chronic diseases associated with obesity, heavy losses of both human life and economic production can be expected. For example, for 23 low-income and middle-income countries, doing nothing will come at a cost of an estimated 250 million deaths and US$84 billion of lost national output between 2006 and 2015 (Abegunde et al, 2007, p1929).

Fifty years ago, Americans spent over 17 per cent of their income on food, while roughly 5 per cent of national income was spent on healthcare. Today, those numbers are just the reverse. The average US citizen spends less than 10 per cent of their income on food, while the cost of healthcare now tops 16 per cent of national income (Pollan, 2008, p110). In 2008, European Union (EU) countries spent, on average, 8.3 per cent of their GDP on healthcare; in 1998, that figure was 7.3 per cent (OECD, 2010, p12). Healthcare costs in Europe are rising in lockstep with obesity rates. The rate of childhood obesity is rising throughout EU member countries, where currently one child in seven, on average, is either overweight or obese. In southern European countries (namely, Malta, Greece, Portugal, Italy and Spain) the figure is much higher: one in five (OECD, 2010, p64). Over half of all the adults in the EU are overweight or obese, a statement that holds true in 15 of the 27 EU countries. In The Netherlands and the UK, obesity more than doubled between 1988 and 2008 (OECD, 2010, p11).

I am not denying that there are many reasons for the increasing burden that healthcare places on national and household economies. Cheap food isn't the entire cause ... but I'm convinced it's one.

Notes

1   Specifically, the message seeks to improve 'understanding of the American farmer and his dynamic role in helping provide an ever increasing standard of living and in contributing importantly to our national security'.

2   Recognizing how world 'food consumption patterns are showing signs of convergence towards a Western diet' (Pingali, 2006, p281), the developing world could learn from the experiences of affluent nations on this point before the obesity–hunger paradox replaces old-fashioned hunger in these countries.

3   HFCS represents only 3.5 per cent of the total cost of soft drink manufacturing, with corn accounting for 44 per cent of the total cost of HFCS production.

References

Abegunde D., C. Mathers, T. Adam, M. Ortegon and K. Strong (2007) 'The burden and costs of chronic diseases in low-income and middle-income countries', *The Lancet*, vol 370, pp1929–1938

Abramovitz, J. (2002) *Vital Signs 2002: The Trends that Are Shaping Our Future*, Norton, New York, NY

Alderman, H., J. Behrman and J. Hobbinott (2003) *Nutrition, Malnutrition, and Economic Growth*, Pan American Health Organization Report, www.idbgroup.org/res/publications/pubfiles/pubS-867.pdf, last accessed 9 August 2010

Alderman, H., J. Hoddinott and B. Kinsey (2006) 'Long term consequences of early childhood malnutrition', *Oxford Economic Papers*, vol 58, no 3, pp450–474

Alston, J., D. Sumner and S. Vosti (2008) 'Farm subsidies and obesity in the United States: National evidence and international comparisons', *Food Policy*, vol 33, pp470–479

Armsby, H. (1911) 'The conservation of the food supply', *Popular Science*, vol 79, pp469–501

Barry, D. (2002) *Dave Barry Hits below the Beltway*, Random House, New York, NY

Beghin, J. and H. Jensen (2008) 'Farm policies and added sugars in US diets', *Food Policy*, vol 33, pp480–488

Bhattacharya, J. and N. Sood (2005) *Health Insurance and the Obesity Externality*, National Bureau of Economic Research, Working Paper 11529, July, www.nber.org/papers/w11529, last accessed 18 August 2010

Blatt, H. (2008) *America's Food: What You Didn't Know about What You Eat*, MIT Press, Cambridge, MA

Bloomberg Business Weekly (2011) 'Cost of obesity approaching $300 billion per year', *Bloomberg Business Weekly*, 11 January

Borzekowski, D. and T. Robinson (2001) 'The 30-second effect: An experiment revealing the impact of television commercials on food preferences of preschoolers', *Journal of the American Dietetic Association*, vol 101, no 1, pp42–46

Bouis, H. E. (2003) 'Micronutrient fortification of plants through plant breeding: Can it improve nutrition in man at low cost?', *Proceedings of the Nutrition Society*, vol 42, pp403–411

Caballero, B. (2002) 'Global patterns of child health: The role of nutrition', *Annuals of Nutrition and Metabolism*, vol 46, pp3–7

Carolan, M. (2011) *Embodied Food Politics*, Ashgate, Burlington, VT

Cecchini, M., F. Sassi, J. A. Lauer, Y. Lee, V. Guajardo-Barron and D. Chisholm (2010) 'Tackling of unhealthy diets, physical inactivity, and obesity: Health effects and cost-effectiveness', *The Lancet*, vol 6736, no 10, pp1–10, www.oecd.org/dataoecd/31/23/46407986.pdf, last accessed 29 December 2010

Center for Rural Pennsylvania (2005) *Examining Demographic, Economic, and Educational Factors: Overweight Children in Pennsylvania*, Center for Rural Pennsylvania Research Brief, Center for Rural Pennsylvania, Harrisburg, PA

Chung, C. and S. Myers (1999) 'Do the poor pay more for food? An analysis of grocery store availability and food price disparities', *Journal of Consumer Affairs*, vol 33, pp276–296

Claney-Hepburn, K., A. Hickey and G. Nevill (1974) 'Children's behavior responses to TV food advertisements', *Journal of Nutrition Education*, vol 6, pp93–96

Colagiuri, S., C. Lee, R. Colagiuri, D. Magliano, J. Shaw, P. Zimmet and I. Caterson (2010) 'The cost of overweight and obesity in Australia', *Medical Journal of Australia*, vol 192, no 5, pp260–264

Cordain, L. (1999) 'Cereal grains: Humanity's double-edged sword', *World Review of Nutrition and Dietetics*, vol 84, pp19–73

Cullather, N. (2007a) 'American pie: The imperialism of the calorie', *History Today*, vol 57, no 2, pp1–7

Cullather, N. (2007b) 'The foreign policy of the calorie', *American Historical Review*, vol 112, no 2, pp337–364

Denniss, R. (2007) 'The commodified environment: How the economy feeds obesity', in J. Dixon and D. Broom (eds) *The Seven Deadly Sins of Obesity: How the Modern World Makes Us Fat*, University of New South Wales Press, Sydney, Australia, pp20–34

Dixon, J. and D. Broom (eds) (2007) *The Seven Deadly Sins of Obesity: How the Modern World Makes Us Fat*, University of New South Wales Press, Sydney, Australia

Dolnick, S. (2010) The Obesity-Hunger Paradox, *The New York Times*, March 12, www.nytimes.com/2010/03/14/nyregion/14hunger.html?src=me, last accessed 6 March 2011

Drewnowski, A. and S. E. Specter (2004) 'Poverty and obesity: The role of energy density and energy costs', *American Journal of Clinical Nutrition*, vol 79, no 1–6, pp6–16

Dybas, C. (2009) 'Report from the 2009 AIBS Annual Meeting: Ensuring a food supply in a world that's hot, packed, and starving', *BioScience*, vol 59, no 8, pp640–646

Eggerton, J. (2007) 'Food-marketing debate heats up: Congress to join FCC and FTC in pressing for action', *Broadcasting & Cable*, May 21, www.broadcastingcable.com/article/CA6444875.html, last accessed 14 March 2011

Farah, H. and J. Buzby (2005) 'US food consumption up 16 percent since 1970', *Amber Waves*, November, www.ers.usda.gov/AmberWaves/November05/Findings/usfoodconsumption.htm, last accessed 15 August 2010

Felton, G., R. Pate, M. Parsons, D. Ward, R. Saunders, S. Trost and M. Dowds (1998) 'Health risk behaviors of rural sixth graders', *Research in Nursing and Health*, vol 21, pp475–485

Fields, S. (2004) 'The fat of the land: Do agricultural subsidies foster poor health?', *Environmental Health Perspectives*, vol 112, no 14, pp820–823

Finkelstein, E., I. Fiebelkorn and G. Wang (2004) 'State level estimates of annual medical expenditures attributable to obesity', *Obesity Research*, vol 12, no 1, pp18–24

Foresight (2007) *Tackling Obesity: Future Choices*, UK Government's Foresight Programme, Government Office for Science, London, www.bis.gov.uk/assets/bispartners/foresight/docs/obesity/obesity_final_part1.pdf, last accessed 29 December 2010

Frances, L., Y. Lee and L. Birch (2003) 'Parental weight status and girl's television viewing, snacking, and body mass indexes', *Obesity*, vol 11, pp143–151

Frazao, E. (1999) 'High costs of poor eating patterns in the United States', in E. Frazao (ed) *America's Eating Habits: Changes and Consequences*, USDA Economic Research Service, Washington, DC, Agriculture Information Bulletin no 750, pp5–32, www.ers.usda.gov/publications/aib750/aib750a.pdf, last accessed 18 August 2010

Fredrix, E. (2010) 'PepsiCo 4th Quarter Profit Rises on Snacks Unit', *USA Today*, www.usatoday.com/money/companies/earnings/2010-02-11-pepsico_N.htm, last accessed 17 December 2010

French, S. (2003) 'Price effect on food choices', *American Society for Nutritional Sciences*, vol 133, no 3, ppS841–S843

French, S., R. Jeffery, M. Story, P. Hannan and M. Snyder (1997) 'A pricing strategy to promote low-fat snack choices through vending machines', *American Journal of Public Health*, vol 87, no 5, pp849–851

Friel, S. and D. Broom (2007) 'The social distribution of obesity', in J. Dixon and D. Broom (eds) *The Seven Deadly Sins of Obesity: How the Modern World Makes Us Fat*, University of New South Wales Press, Sydney, Australia, pp148–171

Grantham-McGregor, S., L. Fernald and K. Sethuraman (1999) 'Effects of health and nutrition on cognitive and behavioural development in children in the first three years of life: Part 2: Infections and micronutrient deficiencies: Iodine, iron and zinc', *Food and Nutrition Bulletin*, vol 2, pp76–99

Grillo, I. (2009) 'Mexico's growing obesity problem', *Tucson Sentinel*, 5 September, www.tucsonsentinel.com/nationworld/report/090509_mexico_obese, last accessed 29 December 2010

Gupta, R. and A. Seth (2007) 'A review of resource conserving technologies for sustainable management of the wheat-cropping systems of the Indo-Gangetic Plains (IGP)', *Crop Production*, vol 26, no 3, pp436–447

Hall, A. (2010) 'Overweight people should pay "fat tax" to cover healthcare costs, German MP says', *The Telegraph*, 22 July, www.telegraph.co.uk/news/worldnews/europe/germany/7904990/Overweight-people-should-pay-fat-tax-to-cover-healthcare-costs-German-MP-says.html, last accessed 29 December 2010

Harvie, A. and T. Wise (2009) *Sweetening the Pot: Implicit Subsidies to Corn Sweeteners and the US Obesity Epidemic*, Global Development and Environmental Institute, Tufts University, Policy Brief No 09-01, February, www.ase.tufts.edu/gdae/Pubs/rp/PB09-01SweeteningPotFeb09.pdf, last accessed 15 August 2010

Hayne, C., P. Morgan and M. Ford (2004) 'Regulating environments to reduce obesity', *Journal of Public Health Policy*, vol 25, no 3–4, pp391–407

Hoddinott, J. and B. Kinsey (2001) 'Child growth in the time of drought', *Oxford Bulletin of Economics and Statistics*, vol 63, no 4, pp409–436

Horgen, K. and K. Brownell (2002) 'Comparison of price change and health message interventions in promoting healthy food choices', *Health Psychology*, vol 21, no 5, pp505–512

Horton, S. and J. Ross (2003) 'The economics of iron deficiency', *Food Policy*, vol 28, pp51–75

Institute of Medicine (2004) *Advertising, Marketing and the Media: Improving Messages*, Institute of Medicine, Washington, DC, www.iom.edu/~/media/Files/Report%20Files/2004/Preventing-Childhood-Obesity-Health-in-the-Balance/factsheetmarketingfinaBitticks.ashx, last accessed 17 December 2010

John, T. and P. Eyzaguirre (2007) 'Biofortification, biodiversity and diet: A search for complementary applications against poverty and malnutrition', *Food Policy*, vol 32, pp1–24

Johns, T. and B. Sthapit (2004) 'Biocultural diversity in the sustainability of developing country food systems', *Food and Nutrition Bulletin*, vol 25, pp143–155

Lake, A., T. Townshend and S. Alvanides (2010) *Obesogenic Environments: Complexities, Perceptions, and Objective Measures*, Wiley and Sons, Ames, IA

Laraia, B., A. M. Siega-Riz, J. Kaufman and S. J. Jones (2004) 'Proximity of supermarkets is positively associated with diet quality index for pregnancy', *Preventative Medicine*, vol 39, pp869–875

Life Magazine (1963) 'Red China's greatest weakness', *Life*, 20 December, vol 55, no 25, pR2

MacDonald, J. and P. Nelson (1991) 'Do the poor still pay more? Food price variations in large metropolitan areas', *Journal of Urban Economics*, vol 30, pp344–359

Maillot, M., N. Darmon, F. Vieux and A. Drewnowski (2007) 'Low energy density and high nutritional quality are each associated with higher diet costs in French adults', *American Journal of Clinical Nutrition*, vol 86, no 3, pp690–696

Major, P. (1997) *The Death of KPD: Communism and Anti-Communism in West Germany, 1945–1956*, Oxford University Press, New York, NY

Martorell, R., K. L. Khan and D. Schroeder (1994) 'Reversibility of stunting: Epidemiological findings in children from developing countries', *European Journal of Clinical Nutrition*, vol 48, ppS45–S57

McMurray, R., J. Harrell, S. Bangdiwala and S. Deng (1999) 'Cardiovascular disease risk factors and obesity of rural and urban elementary school children', *Journal of Rural Health*, vol 15, pp365–374

Monivais, P., J. Mclain and A. Drewnowski (2010) 'The rising disparity in the price of healthful foods: 2004–2008', *Food Policy*, vol 35, no 6, pp514–520

Morland, K., S. Wing and A. Diez Roux (2002a) 'The contextual effect of the local food environment on resident's diet: The atherosclerosis risk in communities study', *American Journal of Public Health*, vol 92, pp1761–1767

Morland, K., S. Wing, A. Diez Roux and C. Poole (2002b) 'Neighborhood characteristics associated with the location of food stores and food service places', *American Journal of Preventative Medicine*, vol 22, pp23–29

Morland, K., A. Diez Roux and S. Wing (2006) 'Supermarkets, other food stores and obesity: The atherosclerosis risk in communities of study', *American Journal of Preventative Medicine*, vol 30, no 4, pp333–339

Morris, C. and D. Sands (2006) 'The breeders dilemma – yield or nutrition?', *Nature Biotechnology*, vol 24, no 9, pp1078–1080

Morton, L., E. A. Bitto, M. Oakland and M. Sand (2005) 'Solving the problems of Iowa food deserts: Food insecurity and civic structure', *Rural Sociology*, vol 70, pp94–112

Mudry, J. (2009) *Measured Meals: Nutrition in America*, SUNY University Press, Albany, NY

Murlin, J. (1913) 'Scientific standards for the governmental regulation of foods', *Popular Science*, vol 83, pp344–354

Nestle, M. (2007) *Food Politics: How the Food Industry Influences Nutrition and Health*, University of California Press, Berkeley, CA

Nestle, M. (2010) 'Food politics blog', 16 April, www.foodpolitics.com/2010/04/can-pepsico-help-alleviate-world-hunger, last accessed 15 August 2010

Nichols, B. (1994) 'Atwater and USDA nutritional research and service', *The Journal of Nutrition*, vol 124, pp1718S–1727S

OECD (Organisation for Economic Co-operation and Development) (2010) *Health at a Glance: Europe 2010*, OECD Publishing, www.ec.europa.eu/health/reports/docs/health_glance_en.pdf, last accessed 29 December 2010

Pingali, P. (2006) 'Westernization of Asian diets and the transformation of food systems: Implications for research and policy', *Food Policy*, vol 32, pp281–298

Pollan, M. (2007) 'You are what you grow', *The New York Times Magazine*, 22 April, www.michaelpollan.com/articles-archive/you-are-what-you-grow, last accessed 14 August 2010

Pollan, M. (2008) *In Defense of Food: An Eater's Manifesto*, Penguin, New York, NY

Popkin, B. (1998) 'Key economic issues', *Food Nutrition Bulletin*, vol 19, no 2, pp117–121

Popkin, B. (2007) 'The world is fat', *Scientific American*, vol 297, no 3, pp60–66

Putman, J., J. Allshouse and L. Kantor (2002) 'US per capita food supply trends', *Food Review*, vol 25, no 3, pp2–15, www.ers.usda.gov/publications/FoodReview/DEC2002/frvol25i3a.pdf

Robinson, T., M. Saphir, H. Kraemer, A. Varady and K. Haydel (2001) 'Effects of reducing television viewing on children's requests for toys', *Developmental and Behavioral Pediatrics*, vol 229, no 3, pp179–184

Robinson, T. D. Brozekowski, D. Matheson and H. Kraemer (2007) 'Effects of fast food branding on young children's taste preferences', *Archive of Pediatrics and Adolescent Medicine*, vol 161, no 8, pp792–797

Rose, D. and R. Richards (2004) 'Food store access and household fruit and vegetable use among participants in the US Food Stamp Program', *Public Health and Nutrition*, vol 7, pp1081–1088

Ruel, M. (2003) 'Operationalizing dietary diversity: A review of measurement issues and research priorities', *The Journal of Nutrition*, vol 133, no 11, pp3911S–3926S

Sanghvi, T. G. (1996) *Economic Rationale for Investing in Micronutrient Programs: A Policy Brief Based on New Analyses*, Office of Nutrition, Bureau for Research and Development, United States Agency for International Development, Washington, DC

Schafft, K., E. Jensen and C. Hinrichs (2009) 'Food deserts and overweight schoolchildren: Evidence from Pennsylvania', *Rural Sociology*, vol 74, no 2, pp153–177

Schoonover, H. and M. Muller (2006) *Food without Thought: How US Farm Policy Contributes to Obesity*, Institute for Agriculture and Trade Policy, Environment and Agriculture Program, March, www.iatp.org/iatp/factsheets.cfm?accountID=258&refID=89968, last accessed 15 August 2010

Scrinis, G. (2008) 'On the ideology of nutritionism', *Gastronomica*, vol 8, no 1, pp38–48

Seshadri, S. (2001) 'Prevalence of micronutrient deficiency particularly of iron, zinc and folic acid in pregnant women in South East Asia', *British Journal of Nutrition*, vol 85, ppS87–S92

Sloane, D., A. Diamant, L. V. Lewis, A. K. Yancey, G. Flynn, L. Nascimento, W. J. McCarthy, J. Guinyard and M. R. Cousineau (2003) 'Improving the nutritional resource environment for healthy living through community-based participatory research', *Journal of General Internal Medicine*, vol 18, pp568–575

Starmer, E., A. Witteman and T. Wise (2006) *Feeding the Factory Farm: Implicit Subsidies to the Broiler Chicken Industry*, Global Development and Environment Institute, Tufts University, Working Paper No 06-03, www.ase.tufts.edu/gdae/Pubs/wp/06-03BroilerGains.pdf, last accessed 15 August 2010

Stein, A. (2006) *Micronutrient Malnutrition and the Impact of Modern Plant Breeding on Public Health in India: How Cost-Effective is Biofortification?*, Cuvillier Verlag, Göttingen

Taras, H. J. Sallis, T. Patterson, P. Nader and J. Nelson (1989) 'Television's influence on children's diet and physical activity', *Journal of Developmental and Behavioral Pediatrics*, vol 10, pp176–180

Tillotson, J. (2004) 'America's obesity: Conflicting public policies, industrial economic development, and unintended human consequences', *Annual Review of Nutrition*, vol 24, pp617–643

Tripp, R. (2001) 'Can biotechnology reach the poor? The adequacy of information and seed delivery', *Food Policy*, vol 26, pp249–264

UNICEF, WTO and WFP (United Nations Children's Fund, World Trade Organization and World Food Programme) (2007) *Preventing and Controlling Micronutrient Deficiencies in Populations Affected by an Emergency*, Joint Statement, Geneva, Switzerland, www.helid.desastres.net/en/d/Js13449e/1.html, last accessed 9 August 2010

Unnevehr, L., C. Pray and R. Paarlberg (2008) 'Addressing micronutrient deficiencies: Alternative interventions and technologies', *AgBioForum*, vol 10, no 3, Article 1

USDA ERS (2002) *USDA ERS Food Marketing System*, www.ers.usda.gov/publications/aer811/aer811.pdf

Wallinga, D., H. Schoonover and M. Muller (2009) 'Considering the contribution of US agricultural policies to the obesity epidemic: Overview and opportunities',

Journal of Hunger and Environmental Nutrition, vol 4, no 1, pp3–19

Warner, M. (2006) 'A sweetener with a rap', *New York Times*, 2 July, www.nytimes.com/2006/07/02/business/yourmoney/02syrup.html?_r=1& pagewanted=1, last accessed 15 August 2010

Weis, T. (2007) *The Global Food Economy: The Battle for the Future of Farming*, Zed Books, New York, NY

Welsh, R. and R. Graham (1999) 'A new paradigm for world agriculture', *Field Crops Research*, vol 60, pp1–10

Welsh, R. and R. Graham (2002) 'Breeding crops for enhanced micronutrient content', *Plant and Soil*, vol 245, no 1, pp205–214

Wiecha J., K. Peterson, D. Ludwig, J. Kim, A. Sobol and S. Gortmaker (2006) 'When children eat what they watch: Impact of television viewing on dietary intake in youth', *Archives of Pediatric and Adolescent Medicine*, vol 160, no 4, pp436–442

Zenk, S., A. Schulz, B. Israel, S. James, S. Bao and M. Wilson (2005) 'Neighborhood racial composition, neighborhood poverty, and the spatial accessibility of supermarkets in Metropolitan Detroit', *American Journal of Public Health*, vol 95, pp66–67

5

Cheap Meat

At one point it stretched over 202,340ha (500,000 acres). Today, that figure is down to 71,000ha (175,000 acres). Located on the Big Island of Hawaii, Parker Ranch is now the fifth largest cow–calf ranch in the US. Living some 4100km (2500 miles) from mainland US, one might think Hawaiians – and the millions of tourists who visit the islands annually – eat their fill of grass-fed local beef. 'It's too lean and tough; we don't actually eat the island meat.' That's what I was told by a waiter in Kona, a town located about an hour from the Parker Ranch. Granted, this is a bit of an overstatement. For example, I learned from Parker Ranch employees that they turn animals that don't 'breed up' (produce 'improved' progeny) into grass-fed beef. Yet, most of the ranch's animals – some 10 million pounds annually – are mainland bound, where they are 'finished' (Holt, 2009, p209), a typical fate for most of the Big Island's beef. So there I was, seated in a restaurant just a short drive from one of the largest ranches in the US, on an island in the middle of the Pacific Ocean, staring at a menu with cuts of beef that all came from well over 4000km away. Maybe some of the beef started just down the road at the Parker Ranch, before being shipped to the mainland to be finished off with corn, only to then return to the Big Island and the tables around me. In what possible world can a food system that does this be classified as 'efficient', I thought to myself? 'Ours', I realized, and then ordered the fish of the day.

Animal rights proponents – those whose ethical proclivities demand that all forms of animal agriculture be dismantled – will not find this chapter terribly satisfying. I believe we can affordably make room for animal products at our dinner tables. I think we can, if you will, have our cake and eat meat too. But to reach this point, something significant must change. Of all commodities discussed in this book, few are less affordable than cheap meat.

This chapter was both the easiest and hardest to write. The difficulty lies in the fact that eating large quantities of meat has become a cultural imperative throughout much of the world, having become a sign of affluence and modernity and a 'right' of consumer choice. The interests favouring the status quo as it pertains to meat production are also immensely powerful. After all, the costs discussed in this chapter not only threaten the livestock industry – a multi-billion dollar a year sector of the global economy. Collectively, cattle, pigs and poultry consume roughly half the world's wheat, 90 per cent of the world's corn, 93 per cent of the world's soybeans, and close to all the world's barley not used for brewing and distilling (Tudge, 2010, p15); the grain industry and agro-processors, therefore, also have a similar interest in keeping meat 'cheap'. Finally, having been around animal agriculture my whole life, I can't help but think of the farmers who rely upon livestock to help put food on the table, clothes on their children and money in their bank accounts.

Weighing against all of this are the facts. Quite simply: *we cannot continue down the current path*. On present trends, the global consumption of meat is set to double by 2050, which means that livestock will be consuming basic food to feed 4 billion people (which, by the way, was the world's population during the early 1970s). This would place the world's *effective* population in 2050 at 13.5 billion, rather than at 9.5 billion as most models predict (Tudge, 2010, pp15–16). I frequently hear arguments about how eating meat is a matter of individual choice. If everyone in the world ends up eating meat at rates comparable to that found in the US – which is at around 125kg (276 pounds) per person per year – who or what has the right to stop them? Simple: the laws of physics. If everyone in the world were to suddenly consume meat at a rate equal to the average US consumer, total grain harvests could only support a global population of about 2.6 billion people (or 38.8 per cent of the existing population) (Roberts, 2008, p211). In lights of these facts, to deny the need for a policy change is essentially to deny the existence of certain laws of nature!

Cheap Meat and Nutritional Health

A review in *Nutritional Journal* examines the state of the peer-reviewed literature on the nutritional value of grass-fed beef. As the authors explain: 'Research spanning three decades suggests that grass-only diets can significantly alter the fatty acid composition and improve the overall antioxidant content of beef' (Daley et al, 2010, p1). This altered fatty acid composition took the form of fewer 'bad fats', which were found in higher quantities in grain-fed beef, and more 'good fats', found in grass-fed protein. This fits with a growing body of research showing that lifelong pastured livestock raised on grass-based diets yield meat that is lower in total fat and calories, higher in good fats such as Omega 3s and conjugated lineolic acid (CLA), higher in precursors for vitamin A and E, and with greater concentrations of antioxidants such as vitamin C and beta-carotene (see, for example, Duckett et al, 1993; Couvreur et al, 2006; Duckett et al, 2009). Jo Robinson, an investigative journalist and a *New York Times* best-selling author, explains that 'If you eat a typical amount of beef (66.5 pounds a year), switching to lean grass-fed beef will save you 17,733 calories a year', which, if the remainder of one's diet remains the same, translates into about 6 pounds of weight loss a year.[1]

Having their movement restricted while being fed a ration of corn and/or silage, hay and distillers grains – plus, in many countries, a liberal dose of antibiotics – allows cattle to gain prodigious amounts of weight in a short period of time. This process also gives meat its distinct 'marbled' look: an emerging global aesthetic standard. The greater the amount of this intramuscular fat, the higher a cut's rating.[2] It is a commentary about our times when 'fatter' equals 'better' when evaluating cuts of meat.

It is interesting to note that this was not always the case. As corn became more abundant (and cheap) during the 1920s, the cattle industry worked to enhance the desirability of heavily marbled meat. With the help of key players such as Alvin Sanders (editor and owner of the *Breeder's Gazette*) and William Jardine (US secretary of agriculture), these efforts culminated in the US Department of Agriculture (USDA) initiating a grading programme in 1927 based on visible marbling, which is still in place today (Gewertz and Errington, 2010, p172).

Save for individuals associated with either the Crown or the Church, our ancestors ate nowhere near the amount of meat consumed in countries such as the US, the UK and Demark. The Danes, by the way, love their meat, having the world's highest per capita meat consumption in the world at 146kg (322 pounds).[3] But in addition to eating less meat, what meat our ancestors did consume was, on average, significantly leaner than that available at your nearest fast-food restaurant, grocery store or steakhouse. A venison (deer) steak, for example, derives 18 per cent of its calories from fat (the rest from protein). A 'choice' sirloin cut (which is still less marbled than 'prime'), in contrast, derives 84 per cent of its calories from fat (Gewertz and Errington, 2010, p172)

While levels of omega-6 fatty acids are roughly the same in the meat of corn-fed and grass-fed cattle, levels of omega-3 are higher in the fully pastured cow. The ratio of omega-6 to omega-3 in grass-fed beef is roughly 1.56:1, while in grain-fed beef it averages about 7.65:1. A healthy diet is believed to supply these fats in the range of 1:1 to 4:1. Diets in the West, however, tend to have ratios in the range of 11:1 to 30:1, which is hypothesized to be a significant factor in the rising rate of inflammatory disorders in the US (Daley et al, 2010, p5). Ratios in the 'healthy' range have been associated with inflammation suppression in patients with rheumatoid arthritis, decreased colorectal cancer cells and reductions in the risk of breast cancer among women (Young, 2010, p91). The modern chicken suffers from similar omega-3 deficiencies. Industrial chickens contain one fifth of the amount of omega-3 fatty acids compared to wild birds, giving modern broilers an omega-6 to omega-3 ratio of close to 10:1. The increasingly unhealthy ratio of omega-6 to omega-3 fats in the average Western diet, thanks in large part to the changing fatty acid profile of modern meat, has been linked to the rise of dementia in countries such as the US and England (Crawford et al, 2009, pp208–210; Young, 2010, p91).

In another study, 0.5 million individuals throughout the US – aged 50 to 71 – were tracked for ten years. To increase the sample size of respondents eating minimal amounts of meat, populations of Seventh-Day Adventists in the US and vegetarians in Europe were targeted and surveyed. The study found an increased risk in overall mortality among persons consuming high levels of red meat compared to those who ate meat sparingly (if at all). High levels of meat consumption were linked not only with higher levels of mortality *per se*, but also with cause-specific mortality due to cancer and cardiovascular disease (Sinha et al, 2009). High levels of red meat consumption were also linked to an increased risk of colon cancer (see also Larsson and Wolk, 2006). A study involving a cohort of approximately 150,000 individuals – aged 50 to 74 years – yielded similar results, finding that one's risk of colon cancer goes up with processed meat consumption. Colon cancer was also positively correlated with a high ratio of red meat to poultry and fish consumption (Chao et al, 2005).

Costs to General Human Health

Evolution didn't prepare cattle for the three to six months they spend in a feedlot. A diet rich in corn makes the cow's stomach, which normally has a neutral pH, abnormally acidic. Left untreated – this is where the antibiotics come into play – acidosis can kill the animal. But usually they end up really sick, which is a problem from a productivity standpoint (to say nothing about animal welfare). Sick cows do not eat a whole lot, which means they will not grow a whole lot either.

A handful of countries have banned growth promotion uses of antibiotics, such as Demark and member countries of the European Union (EU). For most of the world, however, antibiotic use for livestock continues to rise. An estimated 70 per cent of all antibiotics used in the US go towards healthy livestock (Union of Concerned Scientists, 2008). Non-therapeutic use of antibiotics accounts for 90 per cent of all antibiotics used in the US livestock industry in the form of low-level feed additives (Donham and Thelin, 2006, p345).

The costs of using farm antibiotics can no longer be denied. The links between antibiotics in animal agriculture and the emergence of antibiotic-resistant bacteria (such as dangerous *E. coli* strains that account for millions of bladder infections each year) and resistant types of salmonella and other microbes are clear. Asked by a reporter for the *New York Times* about these links, Gail Hansen, a veterinarian and senior officer of the Pew Charitable Trusts (which has recently been campaigning for new restrictions on farm antibiotics), responded by saying: 'Those who say there is no evidence of risk are discounting 40 years of science. To wait until there's nothing we can do about it doesn't seem like the wisest course' (cited in Eckholm, 2010).

As microbes evolve on the farm, it makes sense that sooner or later a mutation is going to occur, giving birth to a strain immune to antibiotics. As these drug-resistant strains of microbes evolve on farms, they make their way into peoples' homes through the meat they buy at the grocery store. Once in our homes the microbes could easily find their way into our bodies either from our handling of the uncooked meat or by eating it (in cases where the meat is insufficiently cooked). Livestock have been identified as the most likely sources of drug-resistant strains of such microbes as salmonella, which annually strike tens of thousands of people annually in the US alone, and campylobacter, which is believed to cause more than 2 million illnesses, 13,000 hospitalizations and over 100 deaths each year in the US (Scott et al, 2009).[4] Genetic studies of drug-resistant *E. coli* strains found on poultry and beef from grocery stores and strains in sick patients have proven to be virtually identical (Eckholm, 2010).

Research also suggests that we may be ingesting antibiotics through vegetables, perhaps even organic vegetables. A team of researchers at the University of Minnesota planted green onions and cabbage in manure-treated soil. After six weeks, cutting about 2.5cm above the soil surface, the tops of the crops were removed and analysed. They were found to have absorbed chlor-tetracycline, a widely used antibiotic among pig producers. In the second experiment, corn, green onions and cabbage were planted in soil treated with liquid hog manure. An analysis showed low levels of antibiotics in the harvested plants. While highly processed commodities such as corn would probably have these antibiotics removed before being consumed, minimally processed foods such as spinach and lettuce would not. Risks associated with low-level exposure to antibiotics include the development of antibiotic resistance among human pathogens, allergic reactions (especially among children) and toxicity due to interaction effects from the ingestion of multiple antibiotics (Kumar et al, 2005).

As much as 90 per cent of the antibiotics administered to animals end up excreted either as urine or manure (Kumar et al, 2005). This drug-laced excrement may be used instead of chemical fertilizers on an organic farm. While there are restrictions on the use of raw manure in US organic farming operations due to concerns over bacteria, no rules exist for treating manure containing antibiotics or hormones. While high-temperature composting of manure is designed to kill pathogens, as required by USDA organic standards, growers are not expected to test soils for the presence of drugs after having been treated (Cimitile, 2009).

As of early 2011, federal regulators in the US appear to be on the verge of issuing the strongest guidelines to date on the use of antibiotics in agriculture for non-therapeutic purposes. Some claim that this move by the Food and Drug Administration (FDA), the agency responsible for overseeing the use of antibiotics in the US, 'would end farm uses of the drugs simply to promote faster animal growth' (Eckholm, 2010). I see the effect of the guidelines, if they ever see the light of day, a little differently. Even if the guidelines are strict, they are still *guidelines*. Compliance will be voluntary. What is the incentive for producers to reduce their use of antibiotics? Granted, there are also calls for tighter oversight by veterinarians to monitor the use – and overuse – of antibiotics. But again, what is the incentive for veterinarians to do this? It seems to me, veterinarians employed by feedlots will continue to feel the same pressures as their employers. Absent *regulations* (e.g. bans), I cannot see how guidelines will change business as usual in concentrated animal feeding operations (CAFOs) concerning their use of antibiotics.

Now let me play devil's advocate: should we even be tinkering with 'higher-level' troubles such as farm antibiotics while leaving deeper problems unaddressed? Sure; antibiotics in agriculture is a problem. But let's ask ourselves why we need antibiotics in the first place. A worldwide ban on non-therapeutic uses of antibiotics in farming changes few of the costs discussed in this book. This point is important because such piecemeal 'solutions' to cheap meat risk doing as much harm as good. Shortly into the non-therapeutic antibiotic ban in Denmark, reports surfaced of heightened post-weaning diarrhoea problems and outbreaks of ileitis (inflammation of a portion of the small intestine) in younger pigs. Mortality rates also began rising in some Danish nurseries (Vansickle, 2005). The Danish hog industry has responded by making *deeper structural and managerial changes* to how they raise pigs, such as by extending the weaning period, providing more space for each animal, and increased veterinary scrutiny (Eckholm, 2010). Granted, this has increased the *retail* price of pork in Denmark by roughly 5 cents per pound (Eckholm, 2010). But I wouldn't say this has increased the pork's *real* cost. The costs were always there. They're just less socialized now.

In many cases, antibiotics are used to produce cheap meat because the animals need them. You might say antibiotics in conventional CAFO environments actually improve an animal's welfare (that's not a ringing endorsement for antibiotics, but a sad commentary about the welfare of a CAFO animal). This is not to argue in favour of antibiotics, but to argue in favour of changing the *conditions* for animals so they no longer *need* antibiotics. Take organic livestock production. Proponents of cheap food are not shy about pointing to the organic dairy industry and their higher culling and disease rates when arguing against any bans on non-therapeutic drug use in animal agriculture. These higher rates are the result of farmers being unable to treat sick cows with antibiotics.

Yet, a CAFO cow and a CAFO organic cow is not a valid comparison. I would be surprised if culling rates *didn't* increase if antibiotics were banned in countries such as the US if nothing else changed. Take organic dairies. It is difficult to compare small extensive organic dairies to large total-confinement operations employing a very different nutritional and management strategy. Organic restrictions on the use of insecticides and herbicides in the production of animal feed tend to encourage many organic dairies to employ grazing to an extent not seen in total confinement operations. And more grazing equals less intensive feeding, less confinement, less need for antibiotics and, ultimately, lower culling and disease rates than what's reported in large CAFO operations (Sato et al, 2005, p106). Evidence of this 'efficiency' is offered by the US-based Organic Center, which recently released a study where convention and organic dairy systems were compared (Benbrook et al, 2010). According to the study's authors, compared to milk cows on high-production dairy farms, lactating cows on organic dairy farms live 1.5 to 2 years longer, milk through 4 or 4.5 lactations (in contrast to less than 2), milk through shorter lactations averaging 313 to 337 days (versus 410 days), lose only 10 to 16 per cent of successful conceptions as a result of embryonic loss or spontaneous abortions (compared to 27 per cent), and require only 1.8 to 2.3 breeding attempts per calf carried to term (compared to 3.5 attempts).

In 2005, China experienced a massive outbreak of the zoonotic pig pathogen *Streptococcus suis*. Over 200 cases of human disease were associated with the outbreak, resulting in approximately 40 deaths (WHO, 2005b). Although rarely diagnosed in Europe and the US, *Streptococcus suis* is a relatively common cause of bacterial meningitis among people residing in Southeast Asia. According to the World Health Organization (WHO), the most important risk factor in acquiring the infection is contact with pigs or uncooked pig products. Farmers, veterinary personnel, slaughterhouse workers and butchers are therefore particularly at risk of infection (WHO, 2005a). The strain of *Streptococcus suis* that ravaged China in 2005 was arguably the most virulent ever and was linked by the WHO as well as China's Ministry of Commerce to confinement conditions (WHO, 2005a; Greger, 2010, p162).

There are dozens of emerging zoonotic disease threats that pose grave risks to human health (Greger, 2010, p164). Some of the more famous include the H5N1 virus – also known as Severe Acute Respiratory Syndrome (SARS) – which infected thousands while killing hundreds, and the H1N1 virus (aka swine flu), which infected tens of millions while killing over 10,000 (CDC, 2010). For millennia, small farms have brought humans into close contact with animals, thus giving us a long history of animal-to-human transfer of infectious agents (McMichael and Butler, 2010, p181). So what has changed that makes the viruses of today seemingly that much more virulent? According to famed virologist (and the world's foremost expert on avian influenza) Robert Webster, 'farming practices have changed'. While our ancestors ate wild birds and backyard poultry, today:

> ... we put millions of chickens into a chicken factory next door to a pig factory, and this virus has the opportunity to get into one of these chicken factories and make billions and billions of these mutations continuously. And so what we've changed is the way we raise animals and our interaction with those animals. And so the virus is changing in those animals and now finding its way back out of those animals into the wild birds. (Council on Foreign Relations, 2005)[5]

To put all of this into a dollars-and-cents perspective, the costs associated with the 2001 foot-and-mouth disease outbreak in the UK were estimated to be more than US$15 billion (Sones, 2006). The WHO (2007, p39) places the costs of SARS to Asian countries somewhere between US$20 billion and US$60 billion in 2003 alone, after taking into account things such as loss of life (human and animal), healthcare costs, losses in productivity due to illness, and decreases in tourism and global consumer confidence towards certain Asian food commodities. And the antibiotic crisis? The cost of microbial resistance to antibiotics in the US is estimated to be US$4 billion per year (Wang, 2010).

CAFOs are not the only environments responsible for viral outbreaks. Current practices associated with how we shuttle billions of animals around the globe are equally culpable in all of this. To quote the United Nations Food and Agriculture Organization (FAO, 2002):

> Modern animal transport systems are ideally suited for spreading disease. The animals commonly originate from different herds or flocks and they are confined together for long periods in a poorly ventilated stressful environment, all of which will favour the transmission within the group of infectious disease should sick animals be present. If not destined for slaughter, the animals will be introduced into new herds or flocks, where they will be subject to social and dietary stress and an exchange of microorganisms with the resident population.

In the US, over 50 million live cattle, sheep and pigs, plus an unknown number of the more than 9 billion chickens, turkeys and other birds currently found in our food system, are traded annually across state lines (Greger, 2007, p301). The value of the live export trade has been increasing annually by roughly 4 per cent, or more than US$10 billion (Phillips, 2008, p138). While a sizeable portion of this trade is between neighbouring countries (such as the US and Canada or across European boarders), sea trade in live animals across great distances is not uncommon. This is especially the case for animals originating from New Zealand and Australia. The Australian live export trade is the largest in the world, worth approximately US$990 million (Dunlevy, 2010). Australia annually exports roughly 1 million cattle, 6 million sheep and 100,000 goats to over 40 overseas destinations. Ports found in the Middle East (where the animals are immediately slaughtered) and Southeast Asia (where they are 'finished') make up the majority of these destinations. Australia's live sheep export industry is worth approximately US$340 (Dunlevy, 2010), with 95 per cent of these animals going to the Middle East (Phillips, 2008, p138).

One study assessed the role that transportation has on increasing the prevalence of *E. coli* and salmonella in cattle. Two samples – a hide swab and a faecal sample – were collected from 200 steers and heifers from a large (65,000-head capacity) feedlot prior to and after shipping to a commercial packing facility. While the levels of *E. coli* on hides and faeces decreased modestly during transport – 18 to 4.5 per cent and 9.5 to 5.5 per cent, respectively – levels of salmonella on hides and faeces increases dramatically – 6 to 89 per cent and 18 to 46 per cent, respectively (Barham et al, 2002, p280). While cleaning trailers with disinfectants falls within most operations' biosecurity protocols, one survey out of Kansas found that only 16 per cent of respondents washed their transport vehicles between loads and less than 5 per cent did so using disinfectants (Greger, 2007, p301).

The long-distance confinement associated with animal transport has also been shown to facilitate the spread of animal pathogens, with the potential to affect human health (Greger, 2007, p302). One of the more tragic examples of this – for humans and animals involved – occurred in 1998. Location: the peninsular region of Malaysia. The outbreak caught public health officials off guard because it involved a pathogen, the 'Nipah' virus (named after the village Sungai Nipah where the first cases were reported), unknown to scientists. Of the 257 humans infected, 105 died (Ryan and Glarum, 2008, p103).

The virus was traced to an industrial pig farm. In humans, the virus causes rapidly progressive encephalitis. In pigs, it leads to severe respiratory illness followed by swift neurological degeneration. Infected pigs first develop violent coughs. After this come the neurological symptoms. Pigs first begin to act aggressively – smashing their heads against walls is common. This is quickly followed by the life-ending spasms. Humans, in contrast, develop high fevers and headaches as their brains swell. If the brain swelling is not controlled, seizures, a coma and, finally, death follow. A few months after the initial outbreak, health officials began receiving reports from other geographic locations of a number of human cases of encephalitis with mortality. It was eventually established that the outbreak was the work of the Nipah virus. It had spread across the entire country in just three months. The contagion's rapid dissemination is widely attributed to the movement of infected pigs. Prior to the outbreak, Malaysia's domestic pig population was roughly 2.4 million, with a total annual value of US$400 million. By the end of the outbreak, 1.1 million of these animals were destroyed at an estimated loss of US$217 million (Greger, 2007, p302; Ryan and Glarum, 2008, p103).

Cheap meat creates vulnerabilities by opening up vectors of infection that could be avoided if meat's real costs were folded into the system. Diane Carmen, writing for the *Denver Post*, wrote back in 2002 that 'if 19 million pounds of meat distributed to half of this country had been contaminated with a deadly strain of *E. coli* bacteria by terrorists, we'd go nuts. But when it's done by a Fortune 100 corporation, we continue to buy it and feed it to our kids.'[6] The US Department of Justice has expressed grave concern over the use of the foot-and-mouth disease in acts of agro-terrorism. Twenty times more infectious than smallpox, a foot-and-mouth disease attack against the US livestock industry could cost taxpayers as much as US$60 billion (DOJ, 2006, p1).

While you might not have heard of the term 'agro-terrorism', it has become one of the hottest buzzwords in centres of governmental power throughout the world. My concern is not that we're talking about agro-terrorism; we certainly need to do everything we can to protect the health and well-being of humans as well as the animals that help feed us. It's *how* we are dealing with the subject that bothers me. Rather than changing the conditions that introduce these vulnerabilities into the system, we have chosen instead to militarize the commodity chain. We're further 'locking down' feedlots, packing plants, egg-laying facilities and dairy parlours in an attempt to minimize intentional as well as unintentional contamination. But this action is not without costs. The increasing gulf between humans and livestock has been convincingly shown to produce ambivalence and ignorance among consumers towards agriculture, in general, and meat production and consumption, in particular (Carolan, 2011).

As stories abound of children failing to identify that milk comes from cows and that ham comes from pigs, with few having ever (at least in affluent countries) set foot on a farm or laid hand on a steer, pig or chicken, do we really want to push animal agriculture further into the shadows?

Costs of Living Near or Working in CAFOs

Health, as defined by the WHO – a definition that dates back to 1948 – is 'a state of complete physical, mental, and social well-being and not merely the absence of disease or infirmity'.[7] In this section, attention centres on a specific population: those living near or working in what are colloquially called 'factory farms'. Adopting a broad definition of health, like that taken by the WHO, discussion is broken up around the subjects of physical health, community well-being and social injustice as it applies to cheap meat agriculture.

Physical health

A highly cited review of the health effects of CAFOs was published in 2007. At the time of its publication, its authors calculated that 70 papers had been published documenting the adverse health effects of the confinement environment on swine producers (Donham et al, 2007, p317). Work in animal agricultural industries ranks among the most hazardous of all occupations (Mitloehner and Schenker, 2007, p309; see also Box 5.1). The majority of worker injuries can be attributed to accidents with machinery and animals (Miller et al, 2004). Yet chronic diseases that result from environmental conditions are proving to also pose a considerable risk (Mitloehner and Schenker, 2007, p309).

Over 25 per cent of CAFO workers suffer from respiratory diseases, ranging from bronchitis to mucous membrane irritation, asthma and acute respiratory distress syndrome (Donham et al, 2007, p318). Exposure to high concentrations of bio-aerosols (airborne particles that are biological in origin), which are common environmental contaminants in livestock buildings, has been linked to organic dust toxic syndrome – a disease that affects more than 30 per cent of swine workers (Donham et al, 2007, p318). While the risk of illness increases with exposure, short-term exposure can bring disease. Acute respiratory symptoms have been documented early in the work history of some individuals sufficiently severe to warrant immediate and permanent removal from the workplace (Dosman et al, 2004, p698). The greatest risk, however, is among those working full time for extended periods (Mitloehner and Calvo, 2008, p167). Roughly 60 per cent of swine confinement workers who have worked for over five years in a CAFO environment experience, on average, at least one respiratory symptom (Donham, 2010, p107). And among workers with underlying conditions – such as asthma, bronchitis and/or heart disease – the disease potential of the CAFO environment is greatly amplified (Essen and Auvermann, 2006).

An influential report by the University of Iowa–Iowa State University CAFO Study Group documents the dangerous concentrations of hydrogen sulphide,

Box 5.1 The real cost of poultry

Peter Applebome's (1989) report for the *New York Times* in 1989 about the dangerous working conditions found in poultry processing plants in the US was probably the first time most readers were exposed to the issue. Since then books such as *Fast Food Nation* (Schlosser, 2001) and *The Omnivore's Dilemma* (Pollan, 2006) have brought the issue to mainstream consciousness. In light of all this publicity, many simply assume that these problems have been dealt with, a sad but past chapter in the history of meat. Think again.

The *Charlotte Observer* pulled few punches when it published its six-day special report entitled *The Cruelest Cuts*, from 10 to 15 February 2008.[8] The series, whose tag line is 'the human cost of bringing poultry to your table', with interactive graphics and online videos, documents the injuries, abuses and racism that still plague workers in poultry processing plants. The series concentrates on House of Raeford – a major meat processing firm – which has been cited for 130 serious workplace safety violations since 2000. Yet, as the *Charlotte Observer* notes, if the company's records are accurate, they have one of the country's safest chicken and turkey plants. The catch is that injury reporting is totally voluntary. Moreover, the government rewards companies that report low injury rates by inspecting them less often. And given that regulators rarely bother to confirm the accuracy of a firm's injury logs, there is tremendous incentive to under-report workplace trauma. When reporters for the *Charlotte Observer* compared company injury logs to reports from workers in surrounding neighbourhoods, a dozen injuries showed up that were not reported in company documents. When an Occupational Safety and Health Administration (OSHA) employee was asked by the paper about federal oversight of these plants, reporters were told that the OSHA is leaving businesses to regulate themselves. After the series was published, this official was placed on paid administrative leave.

ammonia, inhalable particulate matter and endotoxins in livestock buildings (University of Iowa–Iowa State University, 2002); though it is worth mentioning that air emissions – gases and particulate matter – can differ substantially from farm to farm in accordance with management practices, regional locale and animal species. Dust and gas concentrations tend to increase in winter months when houses are sealed up tight to conserve heat (Mitloehner and Calvo, 2008, p171). Dust concentrations also increase whenever there is significant animal movement, such as when livestock are moved, handled and fed (Reynolds et al, 1996, p36). Public health scientists have placed the maximum recommended exposure of dust in livestock buildings at 2.5mg per cubic metre, well below the 15mg per cubic metre rate set by the Occupational Health and Safety Administration (Donham, 2010, p107). It is not unusual to find in livestock buildings dust concentrations at 10mg to 15mg per cubic metre during cold weather or when animals are being shuttled from one space to another (Mitloehner and Calvo, 2008, p175; Donham, 2010, p107).

Workers are not the only ones who incur a cost because of CAFOs. Those living in the vicinity of concentrated livestock operations also pay, sometimes quite dearly, for our cheap meat (e.g. Wing and Wolf, 2000; Thu, 2002; Avery et al, 2004; Carolan, 2008). Neighbours of large-scale CAFOs have been shown to have higher levels of respiratory and digestive disturbances (Wing and Wolf, 2000; Bullers, 2005; Radon et al, 2007). As a population they also have abnormally high rates of psychological disorders, such as anxiety, depression and sleep disturbances (Schiffman et al, 1995; Carolan, 2008). Residents living near CAFOs, even if exposed to concentrations below government-established occupational limits, often have *overall levels* of exposure that far exceed that of a 40-hour-a-week worker. Indeed, populations most vulnerable – namely, the elderly, the ill and the young – not only risk a greater chance of illness due to this physiological vulnerability to environmental pollutants; being elderly, young or ill may also keep these individuals at home for extended periods of time, resulting in possibly *unlimited* exposure levels from breathing in contaminates 24 hours a day, 7 days a week, 52 weeks a year.

Children living on and near hog farms have abnormally high rates of asthma (Chrischilles et al, 2004). These rates have been shown to increase proportionally with the size of the CAFO operation (Donham et al, 2007). Officials at middle schools in North Carolina that were within 4.8km (3 miles) of one or more large hog-feeding facilities reported rates of asthma among the student population well above state averages (Mirabelli et al, 2006). Post-traumatic stress disorder (PTSD) has also been reported among residents living near CAFOs. CAFO-induced PTSD has been linked to the chronic stress that these facilities sometimes cause neighbours in the form of a reduced sense of quality of life and decreased socio-economic vitality of the surrounding community (Donham et al, 2007, p318).

Community well-being

Numerous studies point to a variety of negative social impacts attributable to the presence of one or multiple CAFOs in a rural community. The noxious odours can diminish a household's ability to entertain family and friends, which can have a significant impact upon their sense of community belonging and personal identity (Carolan, 2008). Another study calculated an average reduction in property prices of US$144 per hectare within 3.2km of a CAFO (Seipel et al, 1998). Research also points to the social conflict – specifically between growers and their neighbours – that can emerge when CAFOs are sited within a community (Durrenberger and Thu, 1998; Constance and Tuinstra, 2005).

Wright and colleagues (2001) studied six counties in Minnesota containing multiple large-scale livestock facilities. While the findings pointed to mixed impacts of CAFOs for residents in these counties, the 'losers' appeared to outnumber the 'winners'. Winners were farmers who owned the large-scale animal facilities (although, as discussed in Chapter 9, not all operators of CAFOs are owners and, thus, potential 'winners'). Those negatively affected included neighbours of CAFOs who saw their ability to enjoy their property (and their property's value) diminish; younger and mid-sized farmers who were unable to expand because the CAFOs had restricted their access to markets; and older producers who expressed anxiety over the future of farming and rural life, more generally. These findings are consistent with the CAFO sociological literature as a whole where community impacts have been documented (Lobao and Stofferahn, 2008).

Social injustice

Studies point to how a disproportionate share of swine CAFOs are located in low-income and often non-white communities (Wing and Wolf, 2000; Ladd and Edwards, 2002; Wilson et al, 2002) or near low-income and often non-white schools (Mirabelli et al, 2006). Poverty represents a socio-economic vulnerability that is a significant risk factor to CAFO-related disease and illness. Low-income communities are particularly affected by CAFOs. Malnutrition, poor housing, unprotected sources of drinking water, and poor (or lack of) healthcare make the poor particularly susceptible to livestock pollutants. Lower levels of formal schooling, lack of access to legal remedies and marginal political representation all further hinder these communities' abilities to protect their health and community well-being from being negatively affected by CAFOs (Wing et al, 2008, p1367).

Costs to the Environment

Large-scale animal facilities have long been known to threaten groundwater and to foul air. The USDA (2006) estimates that more than 335 million tonnes of 'dry matter' waste – the waste remaining after liquids are removed – is produced annually on farms in the US alone. An 'all in' figure – wet and dry matter waste – has been calculated for the Chinese livestock industry. The cumulative weight of both solid and liquid animal waste in China is approximately 4 billion tonnes annually, an amount 4.1-fold higher than what the country produces in industrial waste (Xing et al, 2010, p393) (of the roughly 900,000 pigs in the world, 500,000 reside in China – see Lawrence and Stott, 2010, p149). One 2500-head dairy farm produces as much excrement as a city with roughly 411,000 residents (EPA, 2004, p7). While human sewage, at least in affluent countries, is treated to kill pathogens, animal waste is not. Proponents of cheap meat have told me not to worry, that well-managed lagoons kill off a good share of viruses and bacteria – roughly 85 to 90 per cent of the former and 45 to 50 per cent of the latter (Kirby, 2010, p70). Yet what that tells me is that 'treated' animal waste is still teeming with viruses and bacteria. And of those bacteria that do remain, because of the steady diet of non-therapeutic drugs fed to animals while in the feedlot, there is a real chance they possess antibiotic resistance.

A study published in *American Journal of Public Health* concludes that over half of Iowa's 5600 manure storage structures leak at rates above legal limits (Osterberg and Wallinga, 2004). In the absence of government regulation, studies have found that producers will apply manure to soil at levels that would harm its agronomic health (Innes, 2000; Keplinger and Hauck, 2006). This is not surprising as almost all hog producers, at least in the US, have more manure than they know what to do with. Ribaudo and colleagues (2004) found that 82 per cent of large hog operations in the US lack sufficient land to meet agronomically sound nitrogen application rates, while 96 per cent do not have enough land to sustainably apply phosphorus. Were the US government to require more sustainable manure application rates, it is estimated that these regulations would increase operating costs for most large-scale hog producers by between 4 and 5.5 per cent. The one exception to this would be large CAFOs in the Midwest. Due to their proximity to significant stretches of cropland, meeting such standards would probably only increase their operating costs by 1 per cent (Starmer and Wise, 2007b). By turning a blind eye to CAFO manure-disposal practices, communities, the public sector and taxpayers are subsidizing the livestock industry by paying for costs – namely, environmental clean-up and negative human health impacts – that ought to be part of the retail price of meat.

With affluence comes effluent. Projections expect global livestock population to double by 2050. That's a lot of manure. Yet I believe even graver environmental threats loom because of our addiction to cheap meat – after all, there are some technological options (such as anaerobic digesters) that could mitigate some of the risks that go along with having prodigious amounts of animal effluent. The link between cheap meat and climate change seems, at least to me, like a dirty little secret that the livestock industry and supporting organizations are either ignoring or denying. The generally accepted estimate asserts that animal agriculture is responsible for approximately 18 per cent of total greenhouse gas emissions annually (Stehfest et al, 2009, p83) – though other estimates, such as that from the World Watch Institute (Goodland and Anhang, 2009, p11), place the figure as high as 51 per cent. 18 per cent also happens to be the global equivalent of total annual greenhouse gas emissions from deforestation. World leaders are beginning to collectively link deforestation with climate change. International collaborations are therefore forming to mitigate the latter by way of slowing the former, as evidenced by the mobilization behind the United Nations collaborative programme on Reducing Emissions from Deforestation and Degradation (REDD) (Porritt, 2010, p275). In contrast, there is no international support among world leaders to do anything about the 18 per cent that the livestock industry is responsible for.

I remember sitting through a presentation attended almost exclusively by agricultural scientists and members of the livestock feeder industry. Up came a chart of emissions attributable to animal agriculture. At the very top of the chart, I believe along with ammonia (which contributes significantly to acid rain), was nitrous oxide. Roughly 65 per cent of all human-related nitrous oxide (N_2O) emissions can be traced to livestock (most coming from all that manure I was discussing earlier) (United Nations, 2006). Nitrous oxide, because greenhouse gases are not equal, has roughly 300 times the global warming potential (GWP) of carbon dioxide (CO_2). Tracking the graph to the right, I watched the nitrous oxide line rocket upwards as the amount of manure – and, thus, N_2O – over the next 40 years is projected to double. Questions were asked about the graph, but not one touched on that remarkable line at the top, plotting nitrous oxide emissions (I return shortly to subject of N_2O). That's what I meant earlier when I likened cheap meat's relationship to climate change to a 'dirty little secret'. Cheap meat proponents know the link is there. They may quibble over its magnitude; but they know about cheap meat's greenhouse gas footprint.

Governments are equally culpable in keeping the links between cheap meat and climate change from view of the general public. I recently came across a report prepared for the US Congress entitled *Nitrous Oxide from Agricultural Sources*. Emissions from livestock are only mentioned once in the entire report, and even then only methane is discussed (that's right, in a US Congressional report on N_2O in agriculture, livestock emissions seem to have been ignored). Tucked away in a *footnote*, the sentence reads: 'Also in the agricultural sector, animal digestive systems and manure management account for a large portion of US methane emissions' (Bracmort, 2010, p1).

When emissions from land use and land-use change are included in the calculation, the livestock sector accounts for 9 per cent of CO_2 deriving from human-related activities (United Nations, 2006). It has been calculated that producing 1kg of cheap beef generates as much CO_2 as driving 250km in an average European car or using a 100W bulb continuously for 20 days (Ogino et al, 2007, p424). Animal agriculture is also responsible for roughly 37 per cent of all human-induced methane emissions, which has a GWP 23 times that of carbon dioxide (United Nations, 2006).

A study published in *Nature Geoscience* seeks to explain atmospheric concentrations of nitrous oxide (Davidson, 2009). Increasing use of nitrogen fertilizer applications is often pointed to as the main source of atmospheric N_2O. Such was the case, for example, in the aforementioned Congressional report (Bracmort, 2010), where apparently livestock have no N_2O footprint. Yet, it seems unlikely that fertilizer alone can explain the historical trends of atmospheric concentrations of nitrous oxide. Analysing atmospheric concentrations of N_2O, industrial sources of nitrous oxide, fertilizer applications and manure production since 1860, the paper's conclusions force us to revisit conventional wisdom about the source of this greenhouse gas.

According to the study, 2 per cent of manure nitrogen and 2.5 per cent of fertilizer nitrogen were converted to nitrous oxide between 1860 and 2005. Nevertheless, the production of the former so far exceeded that of the latter that manure's *overall* N_2O atmospheric footprint is greater than nitrogen's (see Figure 5.1). Nitrogen fertilizers will undoubtedly increase N_2O emissions in future decades, due in no small part to increases in global per capita meat consumption. The author of the study, however, notes that fertilizer use to support animal agriculture will generate nearly twice as much N_2O as would its use for crops destined for direct human consumption. This is because, to quote the author, 'N$_2$O is first produced when the fertilizer is applied to the cropland for growing the animal feed grain and then is produced a second time when the manure-N, which has been re-concentrated by livestock consuming the feed, is recycled onto the soil or otherwise treated or disposed of' (Davidson, 2009, p662).

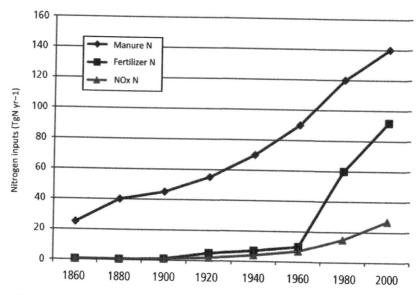

Figure 5.1 *Historical reconstructions of atmospheric N$_2$O and its inferred anthropogenic sources*

Source: adapted from Davidson (2009, p661)

Red meat is, on average, approximately 150 per cent more greenhouse gas intensive than chicken or fish (Weber and Matthews, 2008, p3508). Ruminants – cows, sheep and goats – are terribly inefficient at converting forage into useful products when compared to monogastrics, such as pigs and poultry. That said, it is one thing to be inefficient at converting grain – something we *can* eat – into animal protein. It is quite another, however, to be inefficient at converting cellulose – something we *can't* eat – into something edible to humans. We need to be mindful that there are a variety of animal rearing systems. Not every hamburger, chicken breast, pork tenderloin or leg of lamb comes from a CAFO. Two Holsteins (a breed of cattle) can have a completely different ecological footprint depending upon how they were raised. I have no reason to doubt that red meat is, *on average*, 150 per cent more greenhouse gas intensive than chicken or fish because the vast majority of red meat is 'cheap' and cheap meat, as I've been saying repeatedly, costs the environment dearly. Yet this is not a statement about whether beef (or chicken, milk, etc.) ought to be included in our food system. What concerns me, rather, is *how* it is included – for example, is the animal pasture fed or grain fed? But even here additional contextualization is needed. For example, one fifth of the Amazon rainforest has been destroyed, 80 per cent of which is due to making room for cattle pastures (and some of the remaining 20 per cent is used to raise soybeans that are then fed to cattle) (Gura, 2010, pp57–58). Pasture-fed cattle have the ecological benefit of not being fed grain; but arguably those 'gains' are cancelled out when the cost of rainforest deforestation is added to the equation.

A recent collaboration between Dalhousie University and Iowa State University highlights the importance of context when assessing the ecological impact of meat (Pelletier et al, 2010). Consistent with other studies (Basset-Mens and van der Werf, 2005; Thomassen and de Boer, 2008), the authors determine feed production to be the largest contributor to life-cycle energy use in beef production. From there, however, their findings diverge from other studies (and conventional wisdom) which suggest that pasture-finished beef production is less energy intensive than feedlot-finished beef production. The authors point out that grass-fed beef in Iowa and Minnesota, the two states that were examined, consume large amounts of hay in the winter months when pastures can be covered by as much as 1m of snow. But hay production and transportation have energy costs, which may even exceed those of substitutable feed inputs. Furthermore, the managed pastures typical in Iowa and Minnesota are very different from unmanaged rangeland in that they are dependent upon energy inputs in the form of fertilizer production and application, seeding and periodic renovation. Factoring in all of this, the paper arrives at the following conclusion: 'Our results suggest that pasture-finished beef (19.2kg CO_2-e/kg) from managed grazing systems as currently practiced in the US Upper Midwest is more greenhouse gas intensive than feedlot-finished beef (14.8kg CO_2-e/kg) when viewed on an equal live-weight production basis' (Pelletier et al, 2010, p386). The authors note how their conclusion is consistent with previous research that finds that higher-quality diets and increased growth rates reduce an animal's greenhouse gas footprint (see, for example, Lovett et al, 2005).

Yet isn't this obvious? Grass-fed cattle take roughly twice as long to reach their ideal slaughter weight. Any animal that lives half as long will, by definition, breathe, shit, fart and burp less than a longer-living (grass-fed) contemporary. The term 'greenhouse gas intensive' is also something that needs to be keyed in on. Animal scientists and the livestock industry continually tout how corn-fed beef is steadily becoming more and more efficient as defined by the amount of greenhouse gases generated *per unit produced*. The problem, however, is that the rate by which animal production is increasing is vastly outpacing any gains in per unit efficiency. So while a unit of meat produced today may have a lower greenhouse gas intensity than, say, a unit from ten years ago, we are still emitting into the atmosphere more greenhouse gases *overall* because we are producing so much more meat. Don't get me wrong; a reduction in greenhouse gas intensity is not, in itself, problematic. Yet there is not a shred of empirical evidence to indicate that greenhouse gas efficiencies will *ever* increase fast enough to offset the doubling of global meat consumption between now and 2050. Moreover, as mentioned earlier, there is simply not enough grain in the world to feed all the humans, animals and (if biofuel trends continue) cars that projections say will need to be fed in decades to come. This all makes the point about per unit efficiencies moot. What ought to matter is the *overall* greenhouse gas footprint of the meat system. And on this point, cheap meat can never be defined as 'sustainable'.

Costs to Food Security

Animals will soon eat us out of house and home. The global consumption of meat is expected to double by the middle point of this century. How do we expect to feed these animals that are raised to feed us? In 2010, with a global population of 45 billion chickens, there were as many of these animals being reared for food as there have been of humans who have ever lived on the planet (Butterworth, 2010, p133). Setting aside the fact that global animal population growth has exploded during recent decades, per capita grain production has been unable, since the mid 1980s, to keep up with *human* population growth. Since 1984, per capita grain production has been on the decline (FAO, 2007). Bumper crops, high-yielding hybrid cereals and biotechnology have been unable to keep the world's grain reserves from steadily falling, shrinking from 130 days in the peak year of 1986 to 55 days in 2007 (Schade and Pimentel, 2010, p251).

Faced with these facts, now recognize that you don't get as much out of CAFO livestock – calorie (and often protein) wise – as you put into them. Cheap beef are the most notorious, with a grain to animal protein conversion ratio that can reach as high as 16:1 (Sachs, 1999). Yet, as Table 5.1 illustrates, even more efficient grain converters, like such monogastrics (animals with single-chambered stomachs) as hogs and chickens, end up losing a sizeable amount of energy (calories) and protein in the conversion process.

Table 5.1 *Feed, calories, and protein needed to produced one kg of chicken, pork, and beef*

Grain (calories/protein in grams) ⟶	Animal (calories/protein in grams)
2kg feed grain (6900/200)	1kg Chicken (1090/259)
3kg feed grain (10,350/300)	1kg Pork (1180/187)
7kg feed grain (24,150/700)*	1 kg Beef (1140/226)
16kg feed grain (55,200/1600) **	1 kg Beef (1140/226)

* Low range conversion ratio
** High range conversion ratio
Based upon data from Sachs (1999) and Patnaik (2009)

Cheap meat advocates do not seem bothered by these conversion inefficiencies, whether 16:1, 6:1 and even 2:1. In light of these ratios, I have asked them, how can the livestock industry possibly think a doubling of meat consumption is attainable? Their answer goes something like this (quoting from the *Encyclopedia of Animal Science*):

> Globally, 74 mmt [million metric tonnes] of human-edible protein is fed to livestock, which produce 54 mmt of human food protein. This gives an input/output ration of 1.4:1, or an efficiency of about 70 percent. The biological value of animal protein, a measure reflecting digestibility and essential amino acid content and balance, is on average 1.4 times that of plant protein, suggesting no loss of human food protein value from livestock production. (Bradford, 2005, p245)

The problem with this reasoning is that livestock eat a lot more than just human-edible protein. Typically, feedlot diets, whether for beef or sheep, consist of somewhere between 15 to 20 per cent crude proteins on a dry-matter basis. Or, to put it another way, over 80 per cent of what livestock eat is something other than protein. Corn, for example, is less than 10 per cent protein. CAFO animals eat a lot of corn. Yet, its starchy derivatives would not be included in these protein ratio comparisons that attempt to paint factory-farmed livestock as efficient food producers. After factoring in for total calories, proteins and water (discussed in Chapter 6), the 'output' of CAFOs suddenly does not appear so efficient.

Figure 5.2 depicts the world's total animal population (lest we forget, humans are animals too): approximately 56 billion. It is very difficult to estimate what portion of this figure is fed by our food system. By definition, all 6.7 billion humans eat human-edible food. The majority of the world's cows, pigs and chickens, given their CAFO existence, also consume a significant share of calories that would have otherwise been eaten by humans. There are even camel feedlots. Some of the animals listed in the figure, however, eat material that either cannot be eaten by humans (such as cellulose) or will not be eaten by humans (such as food waste or certain animal parts). Remember also that by mid century, if trends continue, the human population is expected to top 9 billion and the global consumption of meat will double from where it is today. Brute facts of reality being what they are, I cannot see how we will be able to feed all these 'mouths' grains and soybeans.

Costs to Taxpayers

To all taxpayers (vegans included): you are helping to subsidize cheap meat. Burger King recently aired a commercial for a quarter-pound cheeseburger for US$1. A remarkable price, really – a testament to what taxpayer funds and unenforced anti-trust acts can accomplish. Sarcasm aside, those quarter-pound cheeseburgers cost a lot more than US$1. I know of one estimate (Dunne, 1994)

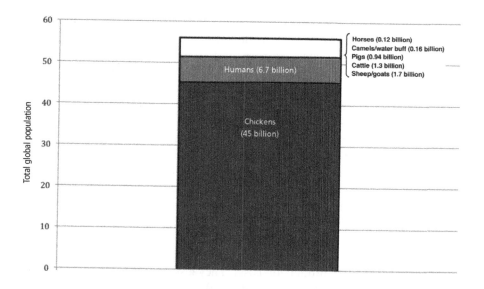

Figure 5.2 *Breakdown of the world's 55.92 billion animals that are part of the food system*

Source: based on data from Gerrits et al (2005, p283); FAO (2008, p53); Butterworth (2010, p133); and http://cattle-today.com

that places the real cost of a hamburger at over US$200! Most of the cost, however, has been socialized, pushed onto taxpayers, the environment and future generations. How can Burger King sell its quarter-pound cheeseburgers for US$1 and still presumably make a profit? Let's explore this question.

Timothy Wise and Elanor Starmer, from the Global Development and Environment Institute at Tufts University, have done some first-rate sleuthing to highlight links between commodity subsidies and the livestock industry – an industry, I should mention, that prides itself on being subsidy free. Between 1997 and 2005, animal agriculture saved approximately US$3.9 billion annually because they are able to purchase corn and soybeans at prices below the cost of production. These savings reduced operating costs by 5 to 15 per cent for producers who fed corn and soybeans to their animals. According to Starmer and Wise (2007a), total savings to industrial hog, broiler, egg, dairy and cattle operations totalled almost US$35 billion over the nine-year period. Broiler producers saved US$1.250 billion, hog producers US$0.945 billion, dairy producers US$0.733 billion, beef feeders US$0.5 billion, and egg producers US$0.433 billion. In terms of total operating cost savings, these figures translate into a 15 per cent reduction for hog producers, 13 per cent for both broiler and egg producers, 6 per cent for dairy producers, and 5 per cent for beef feeders (Starmer and Wise, 2007a).

Another report authored by Starmer and Wise (2007b) turns attention to the swine industry. They note that roughly 60 per cent of operating costs (95 per cent of which goes to buying corn and soybeans) among industrial hog operations goes towards buying feed. Their ability to purchase corn and soybeans on the market at a price below cost of production – thanks to subsidies – gave large-scale hog producers a 10 per cent feed discount before 1996 and a 26 per cent discount after this date. Starmer and Wise calculate that a 26 per cent feed discount translated into a 15 per cent reduction in firms' operating costs between 1997 and 2005, or US$945 million annually. For Smithfield Farms, the largest hog-producing company in the US with a 31 per cent market share, this comes out to total savings (between 1997 and 2005) of US$2.54 billion and annual savings of US$284 million.

Starmer and Wise conclude that any advantages that CAFOs are said to have over smaller operations may have less to do with efficiency gains accruing to larger operations than with governmental policies that reduce corn and soybean prices below production costs. They also note that lax environmental regulations offset anywhere from 2.4 to 10.7 per cent of a CAFO's operating costs. When combined – discounted feed *and* lax environmental regulation – the present policy environment has reduced the operating costs of large hog producers by a total of between 17.4 and 25.7 per cent. Folding these costs back into our food system would eliminate the apparent cost advantage that CAFOs are said to have over mid-sized diversified hog farmers.

To support this conclusion, the authors point to a USDA study finding that economies of scale are most evident between small (1 to 499 head) and medium (500 to 1999 head) hog producers, levelling off when operations expand beyond 2000 head. According this research, there is little difference in production costs between large (2000 to 4999 head) and so-called industrial (5000 + head) operations (McBride and Key, 2003). It appears that organizational and managerial abilities might be stretched beyond optimal limits in these larger operations, thus creating a type of inefficiency. Moreover, mid-sized hog producers are far more likely to grow their own feed (corn and soybeans). Mid-sized diversified farms growing and milling their own feed would probably be able to compete with large-scale industrial operations if the latter paid the full cost. Factor also into the equation previously externalized pollution and health costs, and mid-sized hog farmers should easily surpass CAFOs in efficiency.

Figure 5.3 illustrates the differences in operating costs between large CAFO hog operations and mid-sized hog farms. The current practice of socializing some of the costs of meat to society *in toto* favours CAFOs. As those costs are internalized one by one back into the system from where they came, however, large-scale pork production becomes increasingly less attractive, less efficient and, ultimately, less affordable. By merely adding the costs of full-priced feed and enforcing environmental regulations already on the books, CAFOs are placed at a distinct price disadvantage to mid-sized diversified hog producers. If we were to do this, the operating cost of CAFOs and mid-sized producers is estimated to be US$50.11 to $45.85 per hundredweight, respectively. If we were to include additional costs (as indicated in Figure 5.3 by the dotted lines shooting upward), such as environmental and occupational regulations and costs to animal well-being, non-industrial animal agriculture would further outperform large CAFO facilities.

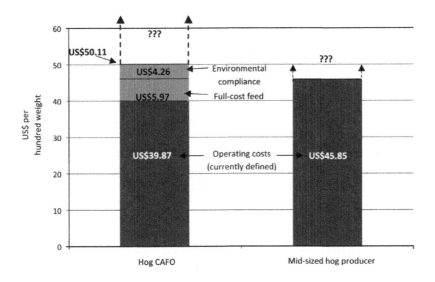

Figure 5.3 *Operating costs of industrial hog production versus non-industrial hog production*

Source: adapted from Starmer and Wise (2007b)

Costs to Animal Welfare

Some readers will undoubtedly take offence to my placing animal welfare issues at the end of the chapter. Philosophically, I appreciate the attention animal welfare activists give to the issue of animal pain and suffering. The pragmatist in me, however, recognizes how long it has taken just to establish minimal welfare standards in our treatment of livestock. Some of my own research points to the ineffectiveness of ethical arguments alone for altering attitudes and behaviours, whether towards the environment (Carolan, 2007) or food and animals, in particular (Carolan, 2011). This is partially why I've chosen to frame my argument in terms of 'costs' rather than, say, 'rights'. The concept of 'cost' also has some flexibility to it. While its everyday usage might speak to something quantifiable, like the per hundredweight operation costs discussed earlier, it can just as easily speak to an irreducible quality, like the incalculable costs associated with animal suffering. I certainly don't think one can put a price tag on an animal's misery (no more than we can on human suffering). It is worth pointing out, however, that these two understandings of 'cost' – one calculable, one incalculable – have a fair amount in common. There is a well-established negative correlation between animal pain and animal productivity. To put it plainly (and rather instrumentally): a stressed animal doesn't 'produce' as much (see, for example, Grandin, 2008).

Earlier I talked about the costs of long-distance live animal transportation to human health. Millions of animals are packed daily into small spaces, whether a trailer or an ocean vessel. The capacity of sea vessels designed specifically for livestock transport range from 900 to 6000 head of cattle and 8000 to 85,000 head of sheep (USDA, 2002). Once packed, their trip can extend up to 5000km. Confinement for extended periods of time creates environments ideally suited for the spreading of pathogenic bacteria. Yet all of this says nothing about the costs of long-distance live animal transport to animal welfare.

An estimated 80,000 pigs die annually in the US during transport, which equates to an US$8 million annual loss to the pork industry (excluding carcass disposal fees) (Greger, 2007, p305). While cattle mortality is relatively rare during hauling, transportation-associated injuries – broken bones, bruising, blood spots and lacerations – lead to carcass devaluation. Bruising alone causes more than US$100 million in annual economic losses to US beef producers (Greger, 2007, p305). Stressed animals also lose weight. 'Carcass shrinkage', as it is called in the livestock industry, is positively correlated with live transport distance (Costa, 2009).

Plans to make livestock transport more humane throughout the EU were recently scrapped by the European Commission. The regulations under consideration would have reduced animals' journey times and lowered stock densities during transport. Surprisingly, farming groups *and* animal welfare activists cheered the commission's decision to drop the proposals. Both groups agreed that existing animal welfare laws must first be properly enforced before stiffer regulations are implemented. To quote Rob Livesey, chairperson of the National Farmers Union in Scotland, 'Solving a problem of poor enforcement by increased red tape is a seriously flawed approach to regulation' (cited in Masters, 2009).

Some have suggested substituting a long-distance live animal transport system for a carcass only (aka dead animal) trade (Greger, 2010, p167). This solution, however, is not without its drawbacks. For example, dead animals cannot move under their own power and risk spoiling. Refrigeration costs in terms of greenhouse gas emissions alone would be significant if dead frozen animals were shipped around the globe. Killing animals before transport would also drastically decentralize the slaughtering and packing process. From an economic standpoint, this may result in inefficiencies due to losses of economies of scale. From an animal welfare standpoint, slaughtering animals where they are fed makes the monitoring of animal decency standards difficult as inspectors would be spread dangerously thin.

Recently, mobile abattoirs (slaughtering facilities on wheels) have been developed in a handful of countries to alleviate animal welfare problems due to long-distance transportation to slaughterhouses (Carlsson et al, 2007). Their future, however, remains uncertain. In the UK, for example, the Food Standards Agency (FSA) is proposing to charge these small mobile abattoirs the full cost of meat hygiene inspections.[9] Many expect that such a move would kill off small- and medium-sized abattoirs. This is because only large 'factory' slaughterhouses have sufficient economies of scale to justify the full-price inspections.

There are also costs to ignoring the needs of non-human animals. Temple Grandin's (2010, p39) widely cited 'core standards' are informed by the belief that animals have certain basic needs. A failure to fulfil these needs places animals under unnecessary stress. The following are some of these standards (see also National Research Council, 2010, pp242–243):

- Animals must be given opportunity to care for, interact with and nurture their young.
- They must be able to build nests during farrowing.
- They must have sufficient space to move, exercise and socialize with herd mates.
- They must have a dry area where they can lay down at the same time without soiling their bellies.
- They require clean air (which includes non-excessive ammonia levels).

There are often sound evolutionary reasons why these 'needs' exist. Take the case of chickens. Most chickens – those not bred or genetically engineered at least for life in a cage – are clearly 'built' for life outdoors. Their feathers provide waterproofing, heat and the ability to fly (at least a little). Spend any time around free-range or backyard chickens and you will bear witness to a wide range of adaptive behaviours that are not fulfilled by birds in intensive indoor poultry factories (Butterworth, 2010, p146). Furthermore, providing animals the environments that they 'need' not only increases their overall physiological welfare, but can make them less susceptible to illness. For example, the ultraviolet radiation in sunlight has been shown to destroy many of the viruses that threaten animal health. Thirty minutes of direct sunlight is all it takes to inactivate many viruses, whereas in the shade or indoors, viruses can survive – which means multiply and potentially mutate – for weeks (Greger, 2010, p166).

More specific attempts to place an economic value on animal welfare are offered in Chapter 7. Caring about the cost of food does not preclude one from caring about the welfare of animals. Animal suffering is not an inconsequential cost. Even looking at it from an amoral production standpoint (for those undeterred by talk of needing to care for animals), a stress-free animal will always out-produce (and outlast) one in pain.

Notes

1   See www.eatwild.com/healthbenefits.htm, last accessed 10 April 2010.
2   For example, the USDA's grading system, which rewards marbling, has eight

different grades. The highest is 'prime' followed by 'choice', 'select', 'standard', 'commercial', 'utility', 'cutter' and 'canner'.

3   See www.guardian.co.uk/environment/datablog/2009/sep/02/meat-consumption-per-capita-climate-change, based upon FAO data.
4   See www.about-campylobacter.com.
5   The last point about wild birds addresses claims that SARS originated from non-domesticated birds.
6   Article reproduced at www.organicconsumers.org/Toxic/ecoli0702.cfm, last accessed 15 October 2010.
7   See www.who.int/about/definition/en/print.html.
8   See www.charlotteobserver.com/poultry.
9   See www.meatinfo.co.uk/news/fullstory.php/aid/11869/Abattoir_bosses_concerned_over_ l meat_charging_proposals.html.

References

Applebome, P. (1989) 'Worker injuries rise in poultry industry as business booms', *New York Times* 6 November, http://query.nytimes.com/gst/fullpage.html?res= 950DE2D7133DF935A35752C1A96F948260, last accessed 3 December 2010

Avery, R., S. Wing, S. Marshall, and S. Schiffman (2004) 'Perceived odor from industrial hog operations and suppression of mucosal immune function in nearby residents', *Archives of Environmental Health*, vol 59, pp101–108

Barham, A., B. Barham, A. Johnson, D. Allen, J. Blanton and M. Miller (2002) 'Effects of the transportation of beef cattle from the feedyard to the packing plant on prevalence levels of *Escherichia coli* O157 and *Salmonella* spp', *Journal of Food Protection*, vol 65, pp280–283

Basset-Mens, C. and H. van der Werf (2005) 'Scenario-based environmental assessment of farming systems: The case of pig production in France', *Agriculture, Ecosystems and Environment*, vol 105, pp127–144

Benbrook C, C. Carman, E. Clark, C. Daley, W. Fulwider, M. Hansen, C. Leifert, K. Martens, L. Paine, L. Petkewitz, G. Jodarski, F. Thicke, J. Velez and G. Wegner (2010) *A Dairy Farm's Footprint: Evaluating the Impacts of Conventional and Organic Farming Systems*, The Organic Center, www.organic-center.org/ reportfiles/COFEFFinal_Nov_2.pdf, last accessed 3 December 2010

Bracmort, K. (2010) *Nitrous Oxide from Agricultural Sources: Potential Role of Greenhouse Gas Emission Reduction and Ozone Recovery*, Congressional Research Service, CRS Report for US Congress, 3 May, www.nationalaglawcenter.org/assets/crs/R40874.pdf, last accessed 20 October 2010

Bradford, E. (2005) 'Contributions to society: Conversion of feed to food', in W. Pond and A. Bell (eds) *Encyclopedia of Animal Science*, Marcel Dekker, New York, NY, pp245–250

Bullers, S. (2005) 'Environmental stressors, perceived control, and health: The case of residents near large-scale hog farms in eastern North Carolina', *Human Ecology*, vol 33, pp1–16

Butterworth, A. (2010) 'Cheap as chicken', in J. D'Silva and J. Webster (eds) *The Meat Crisis: Developing More Sustainable Production and Consumption*, Earthscan, London, pp133–148

Carlsson, F., P. Fryblom and C. Lagerkvist (2007) 'Consumer willingness to pay for farm animal welfare: Mobile abattoirs versus transportation to slaughter', *European Review of Agricultural Economics*, vol 34, no 3, pp321–344

Carolan, M. (2007) 'Introducing the concept of tactile space: Creating lasting social and environmental commitments', *Geoforum*, vol 38, no 6, pp1264–1275

Carolan, M. (2008) 'When good smells go bad: A sociohistorical understanding of agricultural odor pollution', *Environment and Planning A*, vol 40, no 5, pp1235–1249

Carolan, M. (2011) *Embodied Food Politics*, Ashgate, Farnham, UK

CDC (Centers for Disease Control) (2010) *2009 H1N1 Flu: Situation Update*, CDC, www.cdc.gov/h1n1flu/update.htm, last accessed 14 October 2010

Chao, A., M. Thun, C. Connell, M. McCullough, E. Jacobs, D. Flanders, C. Rodriguez, R. Sinha and E. Calle (2005) 'Meat consumption and risk of colorectal cancer', *JAMA*, vol 293, no 2, pp172–182

Chrischilles, E., R. Ahrens, A. Kuehl, K. Kelly, P. Thorne, L. Burmeister and J. Merchant (2004) 'Asthma prevalence and morbidity among rural Iowa schoolchildren', *The Journal of Allergy and Clinical Immunology*, vol 113, no 1, pp66–71

Cimitile, M. (2009) 'Worried about antibiotics in your beef? Vegetables may be no better', *Scientific American*, 6 January, www.scientificamerican.com/article. cfm?id=vegetables-contain-antibiotics&print=true, last accessed 13 October 2010

Constance, D. and R. Tuinstra (2005) 'Corporate chickens and community conflict in east Texas: Growers' and neighbors' views on the impacts of the industrial broiler production', *Culture and Agriculture*, vol 27, no 1, pp45–60

Costa, L. (2009) 'Short-term stress: The case of transport and slaughter', *Italian Journal of Animal Science*, vol 8, no 1, pp241–252

Council on Foreign Relations (2005) *Council on Foreign Relations Conference on the Global Threat of Pandemic Influenza, Session 1: Avian Flu – Where Do We Stand?*, Transcript, www.cfr.org/publication/9230/council_on_foreign_ relations_conference_on_the_global_threat_of_pandemic_influenza_session_1.html, last accessed 14 October 2010

Couvreur, S., C. Hurtaud, C. Lopez, L. Delaby and J. L. Peyraud (2006) 'The linear relationship between the proportion of fresh grass in the cow diet, milk fatty acid composition, and butter properties', *Journal of Dairy Science*, vol 89, no 6, pp1956–1969

Crawford, M., R. Bazinet and A. Sinclair (2009) 'Fat intake and CNS functioning: Ageing and disease', *Annals of Nutrition and Metabolism*, vol 55, no 1–3, pp202–228

Daley, C., A. Abbott, P. S. Doyle, G. A. Nader and S. Larson (2010) 'A review of fatty acid profiles and antioxidant content in grass-fed and grain-fed beef', *Nutritional Journal*, vol 9, no 10, pp1–12

Davidson, E. (2009) 'The contribution of manure and fertilizer nitrogen to atmospheric nitrous oxide since 1860', *Nature Geoscience*, vol 2, pp659–662

DOJ (United States Department of Justice) (2006) *Agroterrorism – Why We're Not Ready*, DOJ, Washington, DC, www.ncjrs.gov/pdffiles1/nij/214752.pdf, last accessed 15 October 2010

Donham, K. (2010) 'Community and occupational health concerns in pork production: A review', *Journal of Animal Science*, vol 88, pp102–111

Donham, K. and A. Thelin (2006) *Agricultural Medicine: Occupational and Environmental Health for the Health Professions*, Blackwell, Oxford, UK

Donham, K., S. Wing, D. Osterberg, J. Flora, C. Hodne, K. Thu and P. Thorne (2007) 'Community health and socioeconomic issues surrounding concentrated animal feeding operations', *Environmental Health Perspectives*, vol 115, no 2, pp317–320

Dosman, J., J. Lawson, S. Kirychuk, Y. Cormier, J. Biem and N. Koehncke (2004) 'Occupational asthma in newly employed workers in intensive swine confinement facilities', *European Respiratory Journal*, vol 24, no 4, pp698–702

Duckett, S. K., D. G. Wagner, L. D. Yates and H. G. Dolezal (1993) 'Effects of time on feed on beef nutrient composition, *Journal of Animal Science*, vol 71, no 8, pp2079–2088

Duckett, S. K., J. P. S. Neel, J. P. Fontenot and W. M. Clapham (2009) 'Effects of winter stocker growth rate and finishing system on: III. Tissue proximate, fatty acid, vitamin, and cholesterol count', *Journal of Animal Science*, vol 87, no 9, pp2961–2970

Dunlevy, G. (2010) Parties clash over live animal exports, *The Sydney Morning Herald*, 14 October, http://news.smh.com.au/breaking-news-national/parties-clash-over-live-animal-exports-20101014-16keg.html, last accessed 14 March 2011

Dunne, D. (1994) 'Why a hamburger should cost 200 dollars: The call for prices to reflect ecological factors', *Financial Times*, 12 January

Durrenberger, P. and K. Thu (1998) *Pigs, Profits, and Rural Communities*, SUNY Press, Albany, NY

Eckholm, E. (2010) 'US meat farmers brace for limits on antibiotics', *New York Times*, 14 September, www.nytimes.com/2010/09/15/us/15farm.html, last accessed 13 October 2010

EPA (US Environmental Protection Agency) (2004) *Risk Management Evaluation for Concentrated Animal Feeding Operation*, EPA, Office of Research and Development, www.epa.gov/nrmrl/pubs/600r04042/600r04042.pdf, last accessed 20 October 2010

Essen, S. and B. Auvermann (2006) 'Health effects from breathing air near CAFOs for feeder cattle or hogs', *Journal of Agromedicine*, vol 10, no 4, pp55–64

FAO (Food and Agriculture Organization of the United Nations) (2002) *Improved Animal Health for Poverty Reduction and Sustainable Livelihoods*, FAO, Animal Production and Health Division, Rome, Italy, www.fao.org/DOCREP/005/Y3542E/Y3542E00.HTM, last accessed 14 October 2010

FAO (2007) *World Grain Consumption, 1961–2006*, FAO, Rome, Italy

FAO (2008) *The State of Food Insecurity in the World 2008*, FAO, International Fund for Agricultural Development, World Food Programme, www.fao.org/docrep/011/i0291e/i0291e00/htm, last accessed 13 April 2010

Gerrits, R., J. Lunney, L. Johnson, V. Pursel, R. Kraeling, G. Rohrer and J. Dobrinsky (2005) 'Perspectives for artificial insemination and genomics to improve global swine populations', *Theriogenology*, vol 63, pp283–299

Gewertz, D. and F. Errington (2010) *Cheap Meat: Flap Food Nations in the Pacific Islands*, University of California Press, Berkeley, CA

Goodland, R. and J. Anhang (2009) 'Livestock and climate change', *World Watch Magazine*, November/December, pp10–19

Grandin, T. (2008) *Humane Livestock Handling: Understanding Livestock Behavior and Building Facilities for Healthier Animals*, Storey Publishing, North Adams, MA

Grandin, T. (2010) 'Implementing effective standards and scoring systems for assessing animal welfare on farms and slaughter plants', in T. Grandin (ed) *Improving Animal Welfare: A Practical Approach*, Wallingford, UK, CABI, pp32–49

Greger, M. (2007) 'The long haul: Risks associated with livestock transport', *Biosecurity and Bioterrorism*, vol 5, no 4, pp301–311

Greger, M. (2010) 'Industrial agriculture's role in the emergence and spread of disease', in J. D'Silva and J. Webster (eds) *The Meat Crisis: Developing More Sustainable Production and Consumption*, Earthscan, London, pp161–172

Gura, S. (2010) 'Industrial livestock production and biodiversity', in J. D'Silva and J. Webster (eds) *The Meat Crisis: Developing More Sustainable Production and Consumption*, Earthscan, London, pp57–79

Holt, K. (2009) 'Parker Ranch: Maintaining the legacy', *Angus*, October, pp209–210, www.angusjournal.com/ArticlePDF/Parker%20Ranch%2010.09.pdf, last accessed 9 October 2010

Innes, R. (2000) 'The economics of livestock waste and its regulation', *American Journal of Agricultural Economics*, vol 82, pp97–117

Keplinger, K. and L. Hauck (2006) 'The economics of manure utilization: Model and application', *Journal of Agricultural and Resource Economics*, vol 31, no 2, pp414–440

Kirby, D. (2010) *Animal Factory: The Looming Threat of Industrial, Pig, Dairy, and Poultry Farms to Humans and the Environment*, St Martin's Press, New York, NY

Kumar, K., S. Gupta, S. Baidoo, Y. Chander and C. Rosen (2005) 'Antibiotic uptake by plants from soil fertilized with animal manure', *Journal of Environmental Quality*, vol 34, pp2082–2085

Ladd, A. and B. Edwards (2002) 'Corporate swine and capitalist pigs: A decade of environmental injustice and protest in North Carolina', *Social Justice* vol 29, no 3, pp26–46

Larsson, S. and A. Wolk (2006) 'Meat consumption and risk of colorectal cancer: A meta-analysis of prospective studies', *International Journal of Cancer*, vol 119, no 11, pp2657–2664

Lawrence, A. and A. Stott (2010) 'Sustainable pig production: Finding solutions and making choices', in J. D'Silva and J. Webster (eds) *The Meat Crisis: Developing More Sustainable Production and Consumption*, Earthscan, London, pp149–157

Lobao, L. and C. Stofferahn (2008) 'The community effects of industrialized farming: Social science research and challenges to corporate farming laws, *Agriculture and Human Values*, vol 25, no 2, pp219–240

Lovett, D., L. Stack, S. Lovell, J. Callan, B. Flynn, M. Hawkins and F. O'Mara (2005) 'Manipulating enteric methane emissions and animal performance of late lactation dairy cows through concentrate supplementation at pasture', *Journal of Dairy Science*, vol 88, pp2836–2842

Masters, D. (2009) 'Europe ditches proposals for humane livestock transport', *Fairhome*, 23 September, www.fairhome.co.uk/2009/09/23/europe-ditches-proposals-for-humane-livestock-transport, last accessed 25 October 2010

McBride, W. and N. Key (2003) *Economic and Structural Relationships in US Hog Production*, United States Department of Agriculture, Economic Research Service, Washington, DC, www.ers.usda.gov/publications/aer818, last accessed 24 October 2010

McMichael, A. and A. Butler (2010) 'Environmentally sustainable and equitable meat consumption in a climate change world', in J. D'Silva and J. Webster (eds) *The Meat Crisis: Developing More Sustainable Production and Consumption*, Earthscan, London, pp173–189

Miller, R., J. Webster and S. Mariger (2004) 'Nonfatal injury rates of Utah agricultural producers', *Journal of Agricultural Safety and Health*, vol 10, pp285–293

Mirabelli M., S. Wing, S. Marshall and T. Wilcosky (2006) 'Asthma symptoms among adolescents who attend public schools that are located near confined swine feeding operations', *Pediatrics*, vol 118, no 1, pp66–75

Mitloehner, F. and M. Calvo (2008) 'Worker health and safety in concentrated animal feeding operations', *Journal of Agricultural Safety and Health*, vol 14, no 2, pp163–187

Mitloehner, F. and M. Schenker (2007) 'Environmental exposure and health effects from concentrated animal feeding operations', *Epidemiology*, vol 18, no 3, pp309–311

National Research Council (2010) *Toward Sustainable Agricultural Systems in the 21st Century*, National Academy Press, Washington, DC

Ogino, A., H. Orito, K. Shimada and H. Hirooka (2007) 'Evaluating environmental impacts of the Japanese beef cow-calf system by the life cycle assessment method', *Animal Science Journal*, vol 78, pp424–432

Osterberg, D. and D. Wallinga (2004) 'Addressing externalities from swine production to reduce public health and environmental impacts', *American Journal of Public Health*, vol 94, no 10, pp1703–1709

Patnaik, U. (2009) 'Origins of the food crisis in India and developing countries', *Monthly Review*, July/August, https://www.monthlyreview.org/090727patnaik.php, last accessed 21 October 2010

Pelletier, N., R. Pirog and R. Rasmussen (2010) 'Comparative life cycle environmental impacts of three beef production strategies in the upper Midwestern United States', *Agricultural Systems*, vol 103, pp380–389

Phillips, C. (2008) 'The welfare of livestock during sea transport', in M. Appleby, V. Cussen and L. Lambert (eds) *Long Distance Transport and the Welfare of Farm Animals*, CABI, Wallingford, UK, pp137–156

Pollan, M. (2006) *The Omnivore's Dilemma: A Natural History of Four Meals*, Penguin Press, New York, NY

Porritt, J. (2010) 'Confronting policy dilemmas', in J. D'Silva and J. Webster (eds) *The Meat Crisis: Developing More Sustainable Production and Consumption*, Earthscan, London, pp275–286

Radon, K., A. Schulze, V. Ehrenstein, R. van Strien, G. Praml and D. Nowak (2007) 'Environmental exposure to confined animal feeding operations and respiratory health of neighboring residents', *Epidemiology*, vol 18, pp300–308

Reynolds, S., K. Donham, P. Whitten, J. Merchant, L. Burmeister and W. Popendorf (1996) 'Longitudinal evaluation of dose–response relationships for environmental exposures and pulmonary function in swine production workers', *American Journal of Industrial Medicine*, vol 29, pp33–40

Ribaudo, M., A. Cattaneo and J. Agapoff (2004) 'Cost of meeting manure nutrient application standards in hog production: The roles of EQIP and fertilizer offsets', *Review of Agricultural Economics*, vol 26, no 4, pp430–444

Roberts, P. (2008) *The End of Food*, Houghton Mifflin Company, New York, NY

Ryan, J. and J. Glarum (2008) *Biosecurity and Bioterrorism: Containing and Preventing Biological Threats*, Elsevier, Burlington, MA

Sachs, J. (1999) 'Food at the center of global crisis', The World Food Prize, 2009, *Norman E. Borlaug International Symposium, Food, Agriculture, and National Security in a Globalized World*, Des Moines, IA, 14–16 October, http://208.109.245.191/assets/Symposium/2009/transcripts/2009-Borlaug-Dialogue-Sachs.pdf, last accessed 8 March 2011

Sato, K., P. Bartlett, R. Erskine and J. Kaneene (2005) 'A comparison of production and management between Wisconsin organic and conventional dairy herds', *Livestock Production Science*, vol 93, pp105–115

Schade, C. and D. Pimentel (2010) 'Population crash: Prospects for famine in the twenty-first century', *Environment, Development and Sustainability*, vol 12, no 2, pp245–262

Schiffman, S., E. Miller, M. Suggs and B. Graham (1995) 'The effect of environmental odors emanating from commercial swine operations on the mood of nearby residents', *Brain Research Bulletin*, vol 37, pp369–375

Schlosser, E. (2001) *Fast Food Nation: The Dark Side of an All-American Meal*, Houghton Mifflin, New York, NY

Scott, L., P. McGee, C. Walsh, S. Fanning, T. Sweeney, J. Blanco, M. Karczmarczyk, B. Earley, N. Leonard and J. J. Sheridan (2009) 'Detection of numerous verotoxigenic *E. coli* serotypes, with multiple antibiotic resistance from cattle faeces and soil', *Veterinary Microbiology*, vol 134, no 3–4, pp288–293

Seipel, M., M. Hamed, J. Sanford Rikoon and A. M. Kleiner (1998) 'The impact of large-scale hog confinement facility sitings on rural property values', *Conference Proceedings: Agricultural Systems and the Environment*, Des Moines, Iowa, July, pp415–418

Sinha, R., A. Cross, B. Graubard, M. Leitzmann and A. Schatzkin (2009) 'Meat intake and mortality: A prospective stud of over half a million people', *Archives of Internal Medicine*, vol 169, no 6, pp562–571

Sones, K. (2006) 'Global trade in livestock: Benefits and risks to developing countries', *New Agriculturalist*, May, www.new-ag.info/focus/focusItem.php?a=1157, last accessed 14 October 2010

Starmer, E. and T. Wise (2007a) *Feeding at the Trough: Industrial Livestock Firms Saved $35 Billion from Low Feed Prices*, Policy Brief No 07-03, Global Development and Environment Institute, Tufts University, Medford, MA

Starmer, E. and T. Wise (2007b) *Living High on the Hog: Factory Farms, Federal Policy, and the Structural Transformation of Swine Production*, Working Paper No 07-04, Global Development and Environment Institute, Tufts University, Medford, MA

Stehfest, E., L. Bouwman, D. Vurren, M. Elzen, B. Eickhout and P. Kabat (2009) 'Climate benefits of a changing diet', *Climate Change*, vol 95, pp83–102

Thu, K. (2002) 'Public health concerns for neighbors of large scale swine production', *Journal of Agricultural Safety and Health*, vol 8, no 2, pp175–184

Thomassen, M. and I. de Boer (2008) 'Life cycle assessment of conventional and organic milk production in the Netherlands', *Agricultural Systems*, vol 96, pp95–107

Tudge, C. (2010). 'How to raise livestock – and how not to', in J. D'Silva and J. Webster (eds) *The Meat Crisis: Developing More Sustainable Production and Consumption*, Earthscan, London, pp9–21

Union of Concerned Scientists (2008) *Preservation of Antibiotics for Medical Treatment Act*, Union of Concerned Scientists, Cambridge, MA, www.ucsusa.org/food_and_agriculture/solutions/wise_antibiotics/pamta.html, last accessed 13 October 2010

United Nations (2006) 'Rearing cattle produces more greenhouse gases than driving cars', United Nations, 29 November, www.un.org/apps/news/story.asp?newsID=20772&CR1=warning#, last accessed 20 October 2010

University of Iowa–Iowa State University (2002) *Iowa Concentration Animal Feeding Operation Air Quality Study*, University of Iowa, Environmental Health Sciences Research Center, Iowa City, Iowa, www.public-health.uiowa.edu/ehsrc/cafostudy.htm, last accessed 18 October 2010

USDA (United States Department of Agriculture) (2002) *Report to Congress on Use of Perishable Commodities and Live Animals in Food Aid Programs*, USDA, Foreign Agricultural Service, 10 September, www.fas.usda.gov/info/speeches/cr091002.pdf, last accessed 25 October 2010

USDA (2006) *FY-2005, Annual Report Manure and Byproduct Utilization National Program 206*, USDA, Washington, DC, www.ars.usda.gov/research/programs/

programs.htm?np_code=206&docid=13337, last accessed 20 October 2010

Vansickle, J. (2005) 'Denmark's ban adds to pig health problems', *National Hog Farmer*, 15 July, http://nationalhogfarmer.com/mag/farming_denmarks_ban_adds, last accessed 13 October 2010

Wang, B. (2010) 'Free antibiotics hurt patients', *The Daily Targum*, 26 September, www.dailytargum.com/opinions/free-antibiotics-hurt-patients-1.2343092, last accessed 14 October 2010

Weber, C. and S. Matthews (2008) 'Food miles and the relative climate impacts of food choices in the United States', *Environmental Science and Technology*, vol 42, no 10, pp3508–3513

WHO (World Health Organization) (2005a) *Streptococcus suis, Fact Sheet*, WHO, 2 August, www.wpro.who.int/media_centre/fact_sheets/fs_20050802.htm, last accessed 13 October 2010

WHO (2005b) *Outbreak Associated with Streptococcus suis in Pigs in China: Update*, WHO, 16 August, www.wpro.who.int/media_centre/news/news_20050816.htm, last accessed 13 October 2010

WHO (2007) *The World Health Report 2007: A Safer Future – A Global Public Health Security in the Twenty-First Century*, WHO, Geneva, Switzerland

Wilson, S., F. Howell, S. Wing and M. Sobsey (2002) 'Environmental injustice and the Mississippi hog industry', *Environmental Health Perspectives*, vol 110, suppl 2, pp195–201

Wing, S. and S. Wolf (2000) 'Intensive livestock operations, health, and quality of life among eastern North Carolina residents', *Environmental Health Perspective*, vol 108, pp233–238

Wing, S., R. Horton, S. Marshall, K. Thu, M. Tajik, L. Schinasi and S. Schiffman (2008) 'Air pollution and odor in communities near industrial swine operations', *Environmental Health Perspectives*, vol 116, no 10, pp1362–1368

Wright, W., C. Flora, K. Kremer, W. Goudy, C. Hinrichs, P. Lasley, A. Maney, M. Kronma, H. Brown, K. Pigg, B. Duncan, J. Coleman and D. Morse (2001) *Technical Work Paper on Social and Community Impacts*, Prepared for the Generic Environmental Impact Statement on Animal Agriculture and the Minnesota Environmental Quality Board

Xing, Y., Z. Li, Y. Fan and H. Hou (2010) 'Biohydrogen production from dairy manures with acidification pretreatment by anaerobic fermentation', *Environmental Science and Pollution Research*, vol 17, no 2, pp392–399

Young, R. (2010) 'Does organic farming offer a solution?', in J. D'Silva and J. Webster (eds) *The Meat Crisis: Developing More Sustainable Production and Consumption*, Earthscan, London, pp80–96

6

Cheap Food and the Environment

Upon learning that I'm writing a book entitled *The Real Cost of Cheap Food*, most immediately assumed that the 'cost' I will be discussing is to the environment, as it is no secret that today's food system places tremendous stress on ecosystems. Soil erosion, deforestation, aquifer depletion, 'dead zones', pesticides leaching into underground wells, industrial livestock facilities fouling our air and water ... the price that the environment has paid in the name of 'efficiency' is quite remarkable. Ultimately, however, it we who are (or will be) actually *paying* for this system of accounting that seems to live by the proverbial principle of seeing no evil, hearing no evil and speaking no evil. Granted, on a planet with some 6.7 billion people (and roughly 2 billion more in 40 years), global food security is *going* to have an environmental impact, no matter how lightly we tread. And if global consumption of meat (especially beef) continues to increase, and if we continue to demand foods out of season, then those costs will increase accordingly. This chapter will unpack some of this complexity by showing the location of certain ecological costing 'hotspots' within our current food system.

We all know the saying about never biting the hand that feeds us. Similarly, we cannot bite the hand that feeds what feeds us either: the ecological productive base. The ideology of cheap food has helped to give shape to an industry that is undermining its (and, thus, *our*) very existence. What we are currently doing in the name of food security is actually having just the opposite effect. We have created a system that is inherently unsustainable. And there is nothing secure about that.

The Food Miles Surprise

Cheap food has rightly been criticized for being energy intensive. The food system – from field to fork – consumes a tremendous amount of non-renewable energy resources. In the US, it is estimated that each resident requires approximately 2000 litres a year in oil equivalents to eat as they do, with animal products accounting for half of this figure. This constitutes about 19 per cent of the total energy used annually in the country (Pimentel et al, 2008, p459). Many assume that the global food system's massive carbon (and, more generally, greenhouse gas) footprint is due mainly to the sheer distance that our food travels. After all, the often-cited average distance between field and fork, at least in the US, is 1500 miles (2400km) (Black, 2008). It has been calculated that every calorie of food produced expends about 10 calories of fossil fuel (Frey and Barrett, 2007). Surely a significant amount of these calories are consumed during transport – right? Intuitively, the food miles critique makes sense. It is practically a truism: food from further away must travel greater distances and therefore, by definition, consumes more energy than food produced locally.

The externalization of ecological cost is not the sole providence of any one food system model. As long as we hold onto our newly formed tastes for out-of-season and non-native foods, we will continually be presented with uncomfortable choices. Do we want, for example, to incur the ecological cost associated with food miles or would we rather produce something locally, even if the latter comes with levels of energy consumption far higher than what it would take to grow that commodity in a country with agro-ecological conditions better suited for its production? Pick your poison.

Given the earlier mentioned development strategy of encouraging poorer countries to focus on high-value commodities such as fruits for exports, I do not expect the supply of these foods to dry up any time soon. Indeed, I am hesitant to say that we should even be striving for such an end in the absence of policies designed to protect poor farmers in the developing world who rely upon these exports for survival. Fruits and vegetables not only account for about one quarter of exports in some developing countries, but represent, in many cases, the agricultural segment with the greatest growth potential (Desrochers and Shimizu, 2008, pp5–6).

Looking more closely at food miles analyses, one quickly learns that not all miles (or kilometres) are equal. A highly cited study by the UK Department for Environment, Food and Rural Affairs (Defra) (Smith et al, 2005) uses two measurements for food transport: vehicle kilometres (the distance travelled by vehicles carrying food regardless of the amount transported) and tonnes per kilometre (distance multiplied by load). Of vehicle kilometres, 82 per cent were generated within the UK, half of which were car miles incurred driving the commodity from store to home. As for tonne kilometres, although sea transport is a well-known contributor to food miles, the method of transport was shown to be highly energy efficient at the per unit level.

Let us say that one large container ship can transport 1 million mandarin oranges and that this ship travels from Japan (where one quarter of the world's crop is grown) to San Francisco. The trip is roughly 8200km (5100 miles). You could say that each tangerine has a mileage of 5100 miles after the trip, in which case the entire ship racked up a staggering 5.1×10^{12} food miles. Alternatively, you could (dividing rather than multiplying) argue that the entire shipment travelled 5100 miles, in which case each piece of fruit has associated with it a mere 0.0000051 food miles. The tonnes per kilometre unit involves this latter form of arithmetic. When calculated in this manner, the results can be counter-intuitive. Another study finds that for products such as milk, apples and lamb, total CO_2 emissions per unit were less when the items were produced in New Zealand and shipped to the UK than when they were produced in the UK (Saunders et al, 2006, pp27–28).

The earlier-mentioned Defra study (Smith et al, 2005) also compares emissions from energy used for UK and Spanish tomatoes, including in their analysis the production and transportation stages. It was estimated that UK tomato production emits 2394kg of CO_2 per tonne compared to 630kg per tonne for Spanish tomatoes. Production in the UK was so much more carbon intensive because of energy requirements associated with the use of heated greenhouses, as opposed to Spain, where tomatoes are raised in unheated plastic-sheeted greenhouses. Related to the use of heated greenhouses is the issue of seasonality. Out-of-season cold storage – one of the many things that makes the year-round consumption of many fruits and vegetable possible – for UK apples consumed 2069 megajoules per tonne and had a production carbon footprint of 85.5kg of CO_2 per tonne.[1] Interestingly, while energy consumption is comparable to what would be required to ship New Zealand apples to the UK (2030 megajoules per tonne), the carbon footprint of UK-produced apples far exceeds that of New Zealand apples (60.1kg of CO_2 per tonne) due to the latter country's more favourable apple-growing conditions (see also Box 6.1) (Saunders et al, 2006, pvii).

Box 6.1 Fair miles instead of food miles

The Africa Research Institute (a British-based think-tank) recently released a provocative report on Kenya's horticulture industry (Muuru, 2009). The report contains a small section written by Stephen Mbithi, chief executive for Fresh Produce Exporters' Association of Kenya (FPEAK) (Muuru, 2009, p7). Dr Mbithi tells of when FPEAK contacted a number of leading supermarkets in the UK to express concern about their plans to label air-freighted foods. The supermarkets made it clear to Dr Mbithi that the labels were the product of immense pressure from European farm lobby groups. The belief among these groups was that by presenting air-freighted produce as environmentally unsustainable, they hope to encourage the consumption of domestically raised food.

The Africa Research Institute report insists just the opposite: Brits would be eating more sustainably if they were to consume air-freighted food from Kenya. On the sustainability of Kenyan horticulture, Dr Mbithi notes that the country's 'temperatures are 20 to 25 degrees Celsius all year round; we use no hothouses unlike in Europe, and with two acres there's no space to drive a tractor so we have to use manual means' (Manson, 2009). African horticultural imports represent 0.1 per cent of all of Britain's carbon emissions (Muuru, 2009, p7). Kenyan Horticultural imports to the US represent even less: 0.05 per cent of carbon emissions. As mentioned in Chapter 2 of this book, more than 60 per cent of Kenya's produce comes from the country's smallholder farmers. Total flights from Nairobi are 0.5 per cent of those from Heathrow (a major London airport). The UK's government buildings alone emit more carbon than all of Kenya. Those in the UK who wish to cut down on greenhouse emissions, contends Dr Mbithi, are looking in the wrong places (Manson, 2009).

What about consumer behaviour? As participants of the food system, our behaviours too ought to be included in footprint and mileage analyses. A study from the UK estimates that the cost of bringing food from the store to home imposes an additional UK£1.28 billion a year of externalized (environmental and health) costs (Pretty et al, 2005). Another study, looking at different distribution schemes, concludes that 'if a customer drives a round-trip distance of more than 7.4km in order to purchase their organic vegetables, their carbon emissions are likely to be greater than the emissions from the system of cold storage, packing, transport to a regional hub and final transport to customer's doorstep used by large-scale [non-organic] vegetable box suppliers' (Coley et al, 2009, p154). And what about when the food gets home? What is the footprint of food preparation? A study examining CO_2 emissions of organically grown potatoes in Sweden found a significant footprint in the kitchen. Since potatoes are never eaten raw and take a while to cook, their energy requirements at the 'fork' end were quite high (Mattson and Wallén, 2003).

I have two concerns about this newly formed popular obsession over food miles. First, 'food miles', at least as used in popular debates, remains terribly ambiguous. Are we talking about, for example, vehicle kilometres or tonnes per kilometre? And, second, I am not convinced that mileage matters more than methods of production, storage and preparation. In fact, research indicates that greenhouse gas emissions associated with food come primarily from the production phase, contributing 83 per cent of the average US household's CO_2 footprint for food consumption. *What* is being produced, it turns out, matters more than how (and how far) that commodity travels after leaving the farm gate. One analysis found that transportation represents only 11 per cent of life-cycle greenhouse emissions (Weber and Matthews, 2008, p3508). The authors also note that food groups vary considerably in terms of greenhouse gas intensity, where, for example, red meat is about 150 per cent more intensive than either chicken or fish. The authors therefore argue that a change in diet could actually be a more effective means of lowering one's greenhouse gas footprint than merely buying local. They note: 'Shifting less than one day per week's worth of calories from red meat and dairy products to chicken, fish, eggs, or a vegetable-based diet achieves more GHG [greenhouse gas] reduction than buying all locally sourced food' (Weber and Matthews, 2008, p3508).

Agriculture, by substituting energy for both labour and internal controls (e.g. pest ecology, nutrient recycling), is growing more than food and fibre. It is also growing its greenhouse gas footprint. Changing *how* we grow food – and perhaps even *what* we grow – will go a long way towards reducing the ecological cost of the things we eat.

From Soil to Sewage: The Life-Cycle Analysis

The types of studies discussed in the previous section are known as life-cycle analyses (LCAs). The rationale behind the LCA is simple: commodities affect the environment at each 'link' of the commodity chain. This ecological footprint stretches from the very beginning, such as when raw materials are mined or grown in a field, to the very end, whether involving incineration, landfill, sewage treatment plant or recycling faculty.

Traditional approaches to environmental impacts tend to focus on a single plant or 'link' in the commodity chain and on one or two pollutants. Doing this, however, can lead to 'footprint shifting', where ecological efficiencies might be achieved at one stage of the life cycle only to increase environmental impacts in another. Clearly, sustainability can only be achieved by reducing the ecological footprint of the *entire* life cycle of a commodity. Examining the entire life cycle of a product also allows 'hotspots' to be identified – a stage where the ecological footprint is particularly large and where reductions can be achieved at minimal cost and effort (Basset-Mens and van der Werf, 2007).

The LCA can (and has) been used to assess the overall ecological footprint for a variety of food items, from whole foods such as farmed salmon (Pelletier et al, 2009) to multi-ingredient, ready-to-eat foods such as meat pies (Defra, 2008a) (see Box 6.2). Doing this gives a more complete picture of the environmental costs associated with our food system than when attention focuses on only one stage (e.g. transportation). Moreover, as indicated previously, LCAs show us that hotspots can even vary within the *same* commodity, depending upon, for example, what the season is. The farm gate footprint of commodity X will vary widely depending upon whether or not it is grown in season.

LCAs can also be used to assess *methods* of production. A Defra-funded report from 2006 found that organic wheat production utilized about 27 per cent less energy compared with non-organic wheat, but little difference was recorded in the case of potatoes. While organic practices result in large reductions in energy used by avoiding synthetic fertilizers, these savings are (to various degrees) offset by lower organic yields and higher inputs into fieldwork, such as more cultivations (Williams et al, 2006). Although not in the report, the lead author of the study, Adrian Williams, conveyed to me in a personal communication that some organic vegetable operations use flame throwers for weed control, which avoids herbicides while minimizing the manual work associated with pulling weeds. For these operations, this practice would further offset any footprint-reducing 'gains' made by avoiding petrochemical inputs. Livestock production methods were also assessed in the report. Thus, for example, while free-range (non-organic) poultry saves energy by relying less

Box 6.2 The cottage pie study

Jeremy Wiltshire, Gary Tucker and Savvas Tassou performed a life-cycle analysis (LCA) on a UK food staple: the cottage pie (Defra, 2008a) (for those unfamiliar with this dish, it is a meat pie with a crust made from mashed potatoes). The modelling involved in this analysis must be appreciated as the ready-to-eat cottage pie consists of over 20 different ingredients, although the main ingredients, which together make up over 70 per cent of the pie, are mashed potatoes and cooked beef. The farm was shown to be an emission hotspot, contributing 65 per cent to the life-cycle greenhouse gas emissions. The manufacturing, retail and consumer-use stages made up most of the remaining 40 per cent, contributing 12, 12 and 9 per cent, respectively ('disposal' constituted the remaining 2 per cent).

Beef production far and away exceeded all other stages when it came to the pie's greenhouse gas footprint. One of the study's more unexpected findings – as least for me – was how the method of household preparation influenced the overall ecological footprint of the meal. The researchers found that if a microwave is used instead of an electric fan oven to reheat the cottage pie, the greenhouse gas footprint of the consumer-use stage drops significantly, from 9 to 2 per cent of the total greenhouse gas emissions.

upon heated and lighted housing 24 hours a day, these savings are partially offset by increases in feed demands. The additional energy demands resulting from being mobile increase energy use by 20 per cent for meat production and 15 per cent for egg production (though it could be argued that these energy 'costs' are well worth the price of increased animal welfare).

Table 6.1 *Average fuel efficiency and energy intensity by vehicle type*

Vehicle class	Ave. fuel efficiency (km/litre)	Energy intensity by volume (ml/pallet-km)	Energy intensity by weight (ml/ton-km)
Medium rigid	3.87	33.0	83.8
Large rigid	2.91	31.8	37.1
City freezer	3.14	21.4	36.4
32 ton freezer	3.35	19.1	26.4
38 ton freezer	2.79	18.0	26.0

Developed utilizing data from DEFRA (2008b, p9)

Similarly, LCAs have been performed to identity hotspots at points 'above' the farm gate. An exemplary study with this scope was funded in 2008 by Defra.

Led out of Brunel University, UK, this LCA examines the greenhouse gas impacts of food retailing (Defra, 2008b). The study found, for example, that high fuel-efficiency retail supply trucks do not automatically signify an efficient distribution operation, especially if those trucks are not filled to their capacity. Instead, the report argues that a better measure of energy efficiency is energy intensity – an expression of fuel consumption on a pallet kilometre basis. As Table 6.1 illustrates, distribution vehicles that are most fuel efficient do not perform nearly as well once full efficiency is *combined* with payload capacity. Whether an expression of volume or weight, large trucks carry more per drop (or petro) than the smaller more fuel-efficient models.

The study also examined the energy intensity for a sample of 640 stores. The average energy intensity was 1540 kilowatt hours (kWh) per square metre of retail space. More interesting, however, was that the average electrical energy intensity was inversely correlated with sales area (see Figure 6.1). To put it simply, as store size increased, the energy intensity of the retail outlet decreased. One possible explanation is that smaller stores have to make the most of what limited space they have. It is also important to keep in mind that though smaller groceries appear to be more energy intensive, their *overall* energy footprint is lower than that of their big-box competitors. Finally, as much of this energy is used to keep our out-of-season, out-of-region and highly processed foods chilled, these energy intensities actually say more about our dietary patterns than they do about anything inherent to the scale of retail space.

Virtual Water and Real Problems

Rain-fed agriculture makes up approximately 80 per cent of all global cultivated areas and produces 60 per cent of the world's food, while irrigated agriculture produces the remaining 40 per cent on just 18 per cent of the world's farmland (Khan and Hanjra, 2009, p131). Despite a globally orchestrated attempt to boost crop yields, the productivity of irrigated fields, most notably in the developing world, has been in decline, further threatening the food security of millions (Hussain and Hanjra, 2003, p435). Nearly 15 million hectares – about one tenth of the world's irrigated land – in developing countries have experienced significant yield declines due to soil salinity and water logging (roughly 75 per cent of all irrigated land is located in developing countries). Moreover, salinization is destroying between 1 and 2 million additional hectares of productive farmland annually (Korten, 2001, p36).

Losses due to these problems risk offsetting a significant proportion of the gains in agricultural productivity achieved through the Green Revolution (Khan and Hanjra, 2009, p131). One estimate, for example, shows that the degradation of irrigated lands used to produce rice and wheat in the Punjab region that straddles India and Pakistan reduced the gains made by breeding and infrastructural and educational investments by approximately 33 per cent (Murgai et al, 2001, pp214–215; Khan and Hanjra, 2009, p131). During the last 50 years, global cropland has decreased by 13 per cent and pasture by 4 per

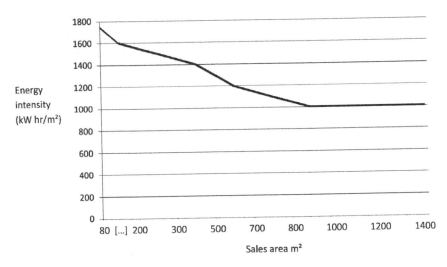

Figure 6.1 *Average electrical energy intensity and sales area for stores between 240 and 1400 square metres (m²)*

Source: developed from Defra (2008b, p17)

cent (Hanjra and Qureshi, 2010, p367). The Food and Agriculture Organization of the United Nations (FAO) predicts that world agricultural production growth will fall by 1.5 per cent per year until 2030 and then decrease an additional 0.9 per cent until 2050 (versus 2.3 per cent growth per year since 1961) (FAO, 2003).

In addition to decreasing amounts (and quality) of arable cropland, we are also witnessing increasing demands upon water resources. Irrigated agriculture currently accounts for between 70 and 80 per cent of global freshwater withdrawals (Khan and Hanjra, 2009, p131; Hanjra and Qureshi, 2010, p366). The United Nations recently concluded that water scarcity – not a lack of arable land – will be the number one constraint on food security for the next few decades (UNDP, 2007). Global demand for water has increased threefold since the 1950s (Hanjra and Qureshi, 2010, p366). A recent study shows that the per capita water requirement for food in China has increased by almost 400 per cent since 1961. As indicated in Figure 6.2, this change is due primarily to increases in the consumption of animal products (Liu and Savenije, 2008, p887). How will additional food-related water demands within China be satisfied knowing that there are already more than 300 cities within the country experiencing water shortages of various intensities (Khan et al, 2009, p350)?

China has an annual per capita water requirement for food of approximately 900 cubic metres per year – over half of which, as illustrated in Figure 6.2, is attributable to animal products. In the US, this figure is approximately 1800 cubic metres per year (two-thirds of which is due to the use and consumption of livestock) (Liu and Savenije, 2008, p893). Due to increases in

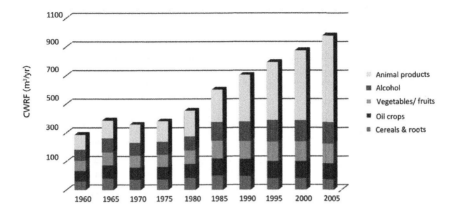

Figure 6.2 *Annual per capita water requirements for food (CWRF) in China*

Source: adapted from Liu and Savenjie (2008, p891)

population and incomes, global cereal and meat consumption is projected to increase by well over 50 per cent by 2050 (de Fraiture et al, 2007). At the same time, if the past is any guide, freshwater resources will continue to shrink. What does this all mean for the future of cheap food and, more importantly, global food security?

Domestic food security looks to be a chimera for many countries that are water poor. While estimates vary on the minimum amount of water needed, on a per capita basis, to achieve living standards equivalent to those found in developed countries, the majority place this figure at around 2000 cubic metres (Qadir et al, 2007, p3). A per capita freshwater supply of below 1000 cubic metres places countries in the category of 'severe water stress'. Table 6.2 shows some of the countries that fall well below, or perilously close to, the 1000 cubic metre threshold, and how, by 2030, all are expected to be facing severe water stress.

Enter virtual water. The transportation of water is notoriously energy intensive (one of the reasons why bottled water has been heavily criticized). From a variety of standpoints – infrastructural, economic and, increasingly, political – the trading of water for agricultural purposes doesn't make sense. So rather than import water to raise crops, countries are importing cereals and meats. Virtual water represents the water embedded within these imported agricultural commodities. It can refer either to the water actually used or to the water that would have been used had the product been produced domestically. Many have latched onto the virtual water concept to describe how water-poor countries can enhance their food security by importing water-intensive food and feed crops (Allan, 1998; Wichelns, 2001; Kumar and Singh, 2005). An analysis in 2007 lists the top virtual water exporter as the US, while Sri Lanka comes first among virtual water-importing countries (see Table 6.3) (Hoekstra and

Chapagain, 2007).

Table 6.2 *Renewable water resources (RWR) per capita for selected countries, 2005 and 2030*

Country	2005 (m³/yr)	2030 (m³/yr)
Kuwait	7	5
United Arab Emirates	48	37
Saudi Arabia	94	56
Libya	94	56
Singapore	137	122
Jordan	157	104
Yemen	191	81
Israel	254	190
Oman	331	191
Algeria	435	324
Tunisia	458	372
Rwanda	604	387
Egypt	779	534
Morocco	919	682
Kenya	919	734
Lebanon	1170	938
Somalia	1257	553
Pakistan	1382	820
Syria	1410	915
Ethiopia	1483	865

Adapted from Qadir et al (2007, p4)

There is very little evidence that a country's virtual water trade is determined by its water situation. The fact is that virtual water often flows out of water-poor, land-rich countries into land-poor, water-rich countries (Kumar and Singh, 2005, p759). Many countries often import food not because they lack water, but because they lack sufficient arable land that can be put to cultivation. Clearly, some countries – like the Kuwaits of the world – have no choice but to trade in virtual water. Yet, not all countries have the financial resources of Kuwait, the United Arab Emirates or Saudi Arabia, either. Better to be water poor and economically rich than poor on both accounts. For those countries that fall into the latter category, *virtual* aptly describes the type of water and food security that they will be facing in the future.

The standard method for calculating the virtual water content – or water footprint – of a crop is to divide total water used (the sum of crop transpiration and soil evaporation during the crop growing period) by crop yield. The virtual water content of animal products takes into consideration the total volume of water used for growing and processing feed, for drinking and for cleaning housing facilities. Assessing the virtual water content for commonly consumed

Table 6.3 *Top virtual water exporting and importing countries (annual net volume)*

Top exporting (billions of cubic metres)	Top importing (billions of cubic metres)
Unites States (758.3)	Sri Lanka (428.5)
Canada (272.5)	Japan (297.4)
Thailand (233.3)	Netherlands (147.7)
Argentina (226.3)	Korea (112.6)
India (161.1)	China (101.9)
Australia (145.6)	Indonesia (101.7)

Data from Hoekstra and Chapagain (2007)

foods reveals a wide discrepancy between the most and least 'thirsty' agricultural commodities. Table 6.4 shows the virtual water and water to calorie ratio (an expression of kilocalories produced per cubic metre of water) of a variety of popular agricultural commodities. While the table reflects data calculated for Chinese goods, it is instructive for a general discussion on the subject. Easily the thirstiest commodity is beef, requiring 12.56 cubic metres of water for every kilogram of live weight gain. At the other end of the continuum are potatoes and other starchy roots, needing a relatively thrifty 0.23 cubic metres of water for every kilogram produced. As measured by the calories to water ratio, beef again fares poorly, producing 161 kilocalories for every cubic metre of water, while calorically dense sugar tops the list, producing 3423 kilocalories per cubic metre of water.

Cheap food is also premised upon cheap – and, thus, inefficiently utilized – water. Irrigation efficiency refers to the ratio of water evaporated to what saturates the soil. The efficiency of traditional gravity irrigation, a method widely used in the developing world, is about 40 per cent. Sprinkler systems have an efficiency range of between 60 and 70 per cent, and drip irrigation systems are between 80 and 90 per cent efficient (Seckler, 1996, p1).

Let's unpack the term 'irrigation efficiency'. What does it mean in practical food policy terms? As prescribed by the ideology of cheap food, it usually refers to the minimal amount of water that can be applied to achieve maximum yield. But there are other ways to think about irrigation efficiency. We could instead strive to maximize crop production per unit of applied water – a goal that would not result in a crop reaching maximum yield. The reason for this is because the relationship between yields and applied water is not entirely linear. The relationship is linear up to approximately 50 per cent of 'full irrigation' (the amount of water needed to achieve maximum yield); after this, however, the yield curve takes on a curvilinear shape due to increased surface evaporation, runoff and deep percolation, until turning downward from anaerobic root zone conditions, disease and the leaching of nutrients (English et al, 2002, p268).

This general relationship is illustrated in Figure 6.3. Usually, farmers irrigate for maximum yield – the point right before the curve turns downward. But if the price of water were to one day became more fully internalized within the food system, 'maximum yield' would cease to be synonymous with 'maximum profits'. Under such a scenario it might make more sense – and, literally, more *cents* – to shoot for, say, 80 per cent yields if those yields could be had with 50 per cent of the water that would otherwise be applied to the field. This approach seeks to maximize 'crop per drop' (see, for example, Molden, 2007), rather than maximize water application rates (which can occur at the expense of crop yields). It has been estimated that if California corn and cotton growers irrigated for maximum profit versus maximum yield, enough water would be saved to sufficiently supply a city of 600,000 residents (English et al, 2002, p268).

Table 6.4 *Virtual water content and calories to water ratio for selected commodities*

Commodity	Virtual water content (m³/kg)	Calories to water (kcal/ m³)
Rice	1.31	2770
Wheat	0.98	2701
Corn	0.84	3403
Potatoes & other starchy roots	0.23	3107
Sugar cane	1.02	3423
Soybeans	3.20	1035
Beef	12.56	161
Pork	4.46	785
Poultry	2.39	715
Mutton	4.50	446
Fish	5.00	99
Eggs	3.55	410

Adapted from Liu and Savenije (2008, p889)

How are we incentivizing the inefficient use of water in agriculture? A major culprit is subsidies. Let me give a couple of examples.

Over 90 per cent of India's non-titled freshwater is used for agriculture, most through irrigation (Myers and Kent, 2001, p134) (firms have laid claim to a sizeable – largely free – chunk of India's freshwater, even in the country's most thirsty regions; see Box 6.3). Landholders in India are free to extract water that runs below their property. With few restrictions on the amount used by farmers, overuse is creating a crisis in the rural countryside where groundwater supplies 85 per cent of their water. A major culprit is an indirect water subsidy:

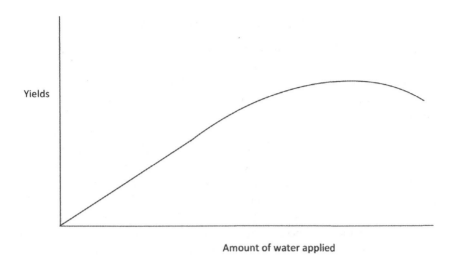

Figure 6.3 *General relationships between yields and total amount of water applied*

Source: English et al (2003)

subsidized electricity. The electricity that powers water pumps is not metered and a flat tariff is charged based upon the horsepower of the pump. With a marginal cost of power of zero, farmers have little incentive to use electricity – and, thus, water – more efficiently (Government of India Planning Commission, 2007, p8). It has been estimated that farmers in Punjab, a major agricultural region of the country, could reduce water use by 15 per cent without any negative effect to crop yields (Myers and Kent, 2001, p134).

Saudi Arabia spent US$40 billion during the 1980s to develop its agricultural sector, much of it going to subsidizing water. If you only look at the calories produced, this investment was a tremendous success. From 1980 to 1992, the country went from producing virtually no wheat to annually harvesting more than 4 million tonnes. The water cost for each of those calories, however, was tremendous. More than 75 per cent of the water pumped to the surface to irrigate wheat fields in Sandi Arabia is lost to evaporation (Myers and Kent, 2001, p135).

Then there is the US, a country with some of the most generous water subsidies in the world. The cost recovery for Bureau of Reclamation irrigation projects averages at around 10 to 20 per cent total costs. Irrigation subsidies for the western half of the US alone exceed US$500 million a year. A study by the US Department of the Interior in 1988 estimated the total value of the irrigation subsidy from 1902 to 1986 to be just under US$10 billion (or approximately 17.5 billion in today's dollars) (National Research Council, 1989, p55). The value of the water subsidy during 2002 to 2003 to California growers is

Box 6.3 You can't irrigate with Coke

While Coca-Cola improved its water efficiency by 6 per cent between 2003 and 2004, its global average is still 2.72 litres per 1 litre of final product (*The Economist*, 2005). A reporter for the *Toronto Star* recently went to Kala Dera, India, and reported on the local conflict arising over the 36 million litres that a *single* Coke plant uses annually (Westhead, 2010). Another plant, located in the drought-prone state of Rajasthan, extracts 500,000 litres of water daily. Still another, located in the highly impoverished village of Plachimada in the state of Kerala, extracts between 500,000 and 1.5 million litres daily. Coca-Cola gets its water in India for a nominal fee, plus a small fee for discharging the plant's wastewater. The cost is around 14 paise per 1000 litres (1 Indian rupee consists of 100 paise and US$1 is equal to between 40 and 50 rupees). The price of water, even after state-of-the-art treatment using reverse osmosis and membranes, is thus far cheaper in India than in the US, the home of bottled-drink giants Coca-Cola and PepsiCo (Aiyer, 2007, p642). The use of exorbitant amounts of cheap (even free) water so that middle- and upper-class Indians can enjoy their favourite carbonated beverage is occurring at the expense of India's smallholder farmers who struggle to obtain sufficient fresh water for their crops (see, for example, Nash, 2007).

estimated to have been over US$100 million (Peterson, 2009, p117). In many cases, subsidized water in the US is used to raise subsidized crops such as corn, rice, wheat, oats and soybeans – the double subsidy.

Food Waste

When was the last time you threw out food? Chances are, very recently. According to one estimate, approximately 50 per cent of all food in the US is wasted (Stuart, 2009, pxvi), costing the American economy at least US$100 billion annually (Jones, 2005, p2). US per capita food waste has increased by 50 per cent since 1974 to more than 1400 kilocalories per person per day, or 150 trillion kilocalories per year (Hall et al, 2009, p1). The energy embedded in wasted food in the US represents approximately 2 per cent of the country's total annual energy consumption (Cuéllar and Weber, 2010, p6464). A Swedish study of food waste in the home ranks vegetables as the second most wasted food group, after meat and dairy (Sonesson et al, 2005). For every eight portions of vegetables prepared in the Swedish homes studied, almost four never reached the cooking stage, due primarily to spoilage. Looking at household food waste in the UK, it is estimated that approximately 7 million tonnes of food – roughly one third of the food purchased by consumers – was discarded. At a retail cost of around UK£10.2 billion (US$19.5 billion), this waste has a CO_2 equivalent of 18 million tonnes –

an amount equal to the annual emissions of one fifth of the UK's total car fleet (Desrochers and Shimizu, 2008, p9). During 2002 to 2003, the total amount of food wasted in the UK, from industrial, commercial and agricultural sources, was 94.6 million tonnes (Garnett, 2006, p8). A rural US county in upstate New York was found to be wasting 10,205 tonnes of food annually. The authors of this study calculated the waste to be more than 8.8 billion kilocalories, enough to feed the county's residents for 1.5 months (Griffin et al, 2009, p67).

There also appear to be waste 'hotspots' along the supply chain, though this varies considerably by product type. Waste levels for certain imported fruits and vegetables have been shown to be as high as 40 per cent in the air-freighted foods sector (Garnett, 2006, p68). I am unaware of any food miles analyses that take into consideration risk of waste due to transportation. Showing 'more miles' to be correlated with 'more waste', at least with regard to certain foods, would add a new layer of complexity to the food miles discussion.

The food waste generated in the US alone constitutes sufficient nourishment to pull approximately 200 million people out of hunger. If you were to include all the feed that went into producing the meat and dairy thrown away annually by consumers, retailers and food services in the US and UK, that number increases to 1.5 billion people (Stuart, 2009, p82). We also need to dispel the belief that because food is biodegradable, sending it to the landfill is an environmentally neutral act, the logic being that it will break down in a very short time. The anaerobic (oxygen-deficient) environment encountered by buried food waste, however, means that organic waste does not decay in landfills. And rotting food, if oxygen happens to be present, emits methane, a greenhouse gas. Remember also that the further along the commodity chain food goes before being discarded, the more virtual energy, water and greenhouse gas emissions it contains. One study calculates that food waste today in the US accounts for more than one quarter of the country's total freshwater consumption and 300 million barrels of oil per year (Hall et al, 2009, p1).

In spite of all this, food waste continues to be largely ignored by the media, politicians, environmentalists and food activists. It is really a remarkable omission. Not once while talking to people about the environmental costs of our food system do I get any sense of outrage over food waste. I hear plenty about the inefficiencies of feeding grain to livestock and the system's dependence on fossil fuels. Occasionally, I even have someone admit to throwing a lot of food away themselves. But that is where it ends. It's a personal problem, I think many believe, not a failure of systemic proportions.

While food waste is a major problem, it is also something that we can fix – an action that would yield significant economic, environmental and food savings. To quote food waste historian Tristram Stuart (2009, pxix): 'In a hypothetical scenario, if we planted trees on the land currently used to grow unnecessarily surplus and wasted food, we could theoretically offset a maximum of 50–100 percent of the world's greenhouse gas emissions.' Another analysis reports that each calorie of food waste saved in the US avoids the use of 7 calories over the entire life cycle (Heller and Keoleian, 2003, p1007).

There are a lot of factors underlying the wastefulness of our current food system. One is aesthetics. Marks & Spencer (a major British retailer), for example, forbids its sandwich suppliers from using four slices of bread from every loaf used – namely, the two crust pieces and the first slice from either end (Stuart, 2009, p45). A recent article in the *Brisbane Times* notes how in the state of Queensland, Australia, over 100,000 tonnes of bananas are disposed of each year for not meeting cosmetic retail standards (Hurst, 2010). Estimates place the amount of bananas wasted globally due to not looking 'right' at between 20 to 40 per cent of the annual harvest (Stuart, 2009, p118). Then there's the story of the fruit seller in Bristol, UK, who was told he could not legally sell his kiwi because they were 2 grams less than the 62 gram minimum prescribed by EU law (which, fortunately, has since been changed) (*Daily Mail Reporter*, 2008).

Tristram Stuart (2009, pp103–111) tells a story of a large British carrot farmer who supplies the supermarket chain Asda. Annually, 25 to 30 per cent of his crop fails to meet the chain's aesthetic standards. 'Asda insists', the farmer explains, 'that all carrots should be straight so the consumer can peel the full length in one easy stroke' (Stuart, 2009, p104). Any – and only – oversized carrots go to food processors, which account for about one third of those unwanted by Asda. Small carrots require too much labour to handle, so food processors usually avoid them. What's left, the remaining two-thirds of that originally rejected by Asda, is fed to livestock.

Another factor contributing to food waste is market concentration in the supermarket sector. Supermarket chains, as discussed in Chapter 9, wield tremendous power over those 'earlier' in the agri-food chain. While supermarkets can select from a range of processors and farmers with whom to work, the inverse is less the case. In many instances, processors and farmers supply only one supermarket chain, giving the latter tremendous leverage when it comes to calling the shots. This allows firms such as Birds Eye, for example, to contractually forbid growers from selling their peas to anyone else, even those rejected by the company for not meeting certain standards. This leaves growers with no choice but to either feed their surplus to livestock or to recycle the vegetables back into the soil (Stuart, 2009, p117).

Market concentration also gives supermarket chains the ability to redistribute risk in ways that encourage overproduction and waste. For instance, in order to retain their lucrative contracts with supermarkets, food processors are hesitant to question what are known as 'forecast orders'. Forecast orders are submitted to suppliers in advance, alerting the food manufacturer to what the supermarket chain *might* ultimately buy in the near future. This gives supermarkets tremendous flexibility in what they can *actually* buy the day the purchase is made. It is not unusual, then, for chains to buy significantly less than what was forecasted. Better to forecast order too much, and place the costs of overproduction on the supplier, than not enough and suffer the costs yourself through loss of sales. And, as in the case of Birds Eye, the suppliers often have clauses in their contracts that state they can't sell their surplus to anyone else (Stuart, 2009, pp47–48; Dunmall, 2010).

Supermarkets, however, do over-purchase. When they do, they attempt to push the cost of waste onto consumers. Enter the buy-one-get-one (BOGO) promotional offers. These schemes might be a great deal if you actually use the 'free' one. Most don't, which is why, at least in the UK, BOGO offers have come under intense scrutiny during recent years (see, for example, Wainwright, 2009). In the US, conversely, BOGO schemes remain popular – or at least they do at the grocery stores I frequent.

While some supermarkets are starting to sell the slightly off-sized discards, such as the UK's Sainsbury's with their Basics packaging – with tag-lines such as 'All shapes, all sizes' and 'Lemons: no lookers, great juicers' – most still do not, especially in the US. Why is this? Supermarkets will tell you that consumers just won't buy them. Place two of the same vegetables side by side, one meeting conventional aesthetic standards and the other slightly misshaped (but otherwise perfectly fine), and the consumer overwhelmingly votes for the former by buying it. But what if the slightly 'off' vegetables where sold for less? Consumers would more likely buy them (Stuart, 2009, p108). Would these purchases be made at the expense of the full-priced produce with wider profit margins? Probably. This gets us back to an earlier point: it's better from the supermarket's standpoint to shift the cost of misshaped produce to suppliers than to pay the cost themselves.

So what does food waste have to do with cheap food and what are the consequences of cutting down on this waste. Even cheaper food? We are willing to waste cheap food precisely because its actual cost is not adequately contained within the food system and passed onto consumers. Instead, we are selfishly charging others – that is, society, the environment and future generations – for the food we eat (and waste).

What about Phosphorus?

What *about* phosphorus? That's the answer I get when I ask students about phosphorus. They all seem to have at least a rudimentary understanding of food's links to phenomena such as soil erosion, climate change and groundwater pollution. Even nitrogen – the media darling of the NPK (nitrogen, phosphorus and potassium) trio – gets regular mention in classrooms, newspapers and books in discussions about the environmental costs of our food system. But phosphorus? As one student once put it to me: 'No one ever talks about phosphorus, so how could it be a problem?' I did not have a really good answer for her, merely responding: 'I expect that will be changing in the years ahead.'

One of the main roles of phosphorus in living organisms is in the transfer of energy, making the element essential (and irreplaceable) for plant growth. Phosphorus's critical role in life explains why it is a limiting nutrient in ecosystems. In other words, the carrying capacity of natural ecosystems appears to be conditioned upon available phosphorus (Dery and Anderson, 2007). Agriculture would therefore not exist without this macronutrient. But it is running out. That's the problem. And that's why I think we will hear a lot more about phosphorus in the not too distant future.

The life cycle of phosphorus is a lesson in geological timescales. Land ecosystems recycle phosphorus an average of 46 times before the element finds its way into the open sea. Once there, marine organisms can recycle it 800 times before it passes into sediments. Tens of millions of years later, with the aid of tectonic uplift, it may find itself again on dry land to start this journey anew (Vaccari, 2009). Harvesting disrupts this cycle by removing phosphorus from one environment (the field) and transporting it somewhere else (increasingly, anywhere in the world).

Until the 1800s, crop production relied exclusively on natural levels of soil phosphorus and locally available organic matter such as manure and human excrement. Eventually, other forms of the element were sought as soils could not be rapidly replenished. For example, by the early 19th century, England began to import significant quantities of bones from its European neighbours. By the mid to late 19th century, local organic matter started to be replaced entirely by phosphorus material from other countries, most notably guano (excrement of seabirds, bats and seals) and phosphate rock (Cordell et al, 2009, p293). To get a sense of just how prized these sources of phosphorus were, the US passed the Guano Islands Act in 1856. This act gave US citizens who 'discovered' (whether indigenous peoples knew about these sources was immaterial) sources of guano the right to take possession of the land and the entirety of the deposits if not within the jurisdiction of other governments. The act authorized the US President to use military force to protect those interests. Finally, the act stipulated that the extracted guano could only be used by US citizens.

The world trade in guano declined as quickly as it expanded. By the end of the 19th century, guano was being replaced with artificial phosphorus. The term 'artificial', however, is a misnomer, as the material is ultimately derived from phosphate rock. And believing for much of the last 100 years that this resource existed in unlimited quantities, the worldwide use of artificial phosphorus grew rapidly throughout the 20th century.[2]

Some phosphorus facts...

- Approximately 90 per cent of the world's phosphorus goes into fertilizer (the rest goes into products such as food additives, glues, flame retardants and detergents) (Christen, 2007, p2078).
- Around 220 million tonnes of phosphate rock is processed globally every year (White and Cordell, 2008, p1).
- Global demand for phosphorus is expected to increase somewhere between 50 and 100 per cent by 2050 as a result of changing diets (more meat), increased food demand and growth in the biofuel industry (Cordell et al, 2009, p293).

Given phosphorus's link to food, phosphate rock reserves are a strategic resource. Major rock deposits are found in Morocco, China and the US (see Figure 6.4). China has recently slashed exports to secure its domestic supply. US phosphate rock reserves increased from 18 million metric tonnes in 1960 to

35 million metric tonnes in 1970, peaking at 54 million metric tonnes in 1980.[3] It is widely believed that the US will exhaust its stores in roughly 30 years (Rosemarin, 2004; Cordell et al, 2009, p293). About 40 per cent of global reserves are controlled by a single country, Morocco – the Saudi Arabia of phosphorus (Vaccari, 2009). But much of this reserve is located on the western Sahara, which Morocco has occupied since 1975 in violation of resolutions by the United Nations Security Council and a decision by the International Court of Justice. Trading with Morocco for phosphate rock, as most countries do, is

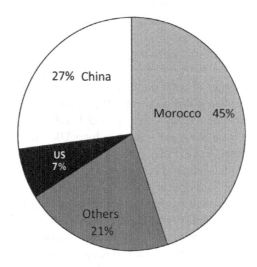

Figure 6.4 *Global resource base (phosphorus)*

Source: adapted from http://seekingalpha.com/article/182522-taking-stock-of-phosphorus-and-biofuels

therefore done in violation of international law.

Worldwide phosphate rock mine production capacity is expected to increase in the short term as governments seek to secure their own phosphate rock reserves. New mines opened in Australia and Peru in 2010, and 2011 promises additional mine openings in Namibia and Saudi Arabia, while expansions to existing operations are planned in Brazil, Canada, China, Egypt, Finland, Morocco, Russia and Tunisia (US Geological Survey, 2010). France, Spain and India (in that order) are the world's three largest importers of phosphorus. India possesses no domestic supply of phosphate rock. While the country has 260 megatonnes of low-grade phosphate rock deposits, they are unsuitable for fertilizer (Rosemarin, 2004).

The International Geological Correlation Programme estimated in 1987 that there could be as much as 163,000 million metric tonnes of phosphate rock remaining on the planet. This represents more than 13,000 million metric tonnes of phosphorus, which at current rates would be enough to last hundreds of years. The problem with this estimate, however, is that it includes types of rocks that are either impractical to mine or that lie in environmentally sensitive areas. When looking at phosphate rock that is actually recoverable, most estimates indicate that the phosphate clock will strike midnight – when reserves are largely exhausted – some time in the next 75 to 100 years (Christen, 2007; Vaccari, 2009). But that's at current rates of consumption. Global supplies of phosphate rock are expected to peak in approximately 30 years, probably some time between 2030 and 2040 (White and Cordell, 2008). And as global demand for phosphorus increases anywhere between 50 and 100 per cent by 2050, we are likely to run out shortly after hitting the state of peak phosphorus.

An early casualty of our growing hunger for phosphate rock is the small island nation of Nauru. Nauru Island, covering just 21 square kilometres, has (or, more accurately, *had*) only one economic resource: rock phosphate. Its deposits ran out, however, during the 1980s. According to a recent issue of the US Central Intelligence Agency's *The World Factbook*: 'intensive phosphate mining during the past 90 years – mainly by a UK, Australia and NZ [New Zealand] consortium – has left the central 90 percent of Nauru a wasteland and threatens limited remaining land resources' (CIA, 2008, p409).

For much of the last century, phosphorus has been applied with near abandon. This has resulted in the eutrophication (over-fertilization) of many watersheds. Excess phosphorus in waterways feeds algal blooms, which can starve fish of oxygen and produce 'dead zones'. According to the International Fertilizer Association, phosphate rock releases carbon dioxide when reacted with sulphuric acid in the phosphate fertilizer manufacturing process.[4] The production of phosphate rock-based fertilizers also produces radioactive by-products and heavy metal pollutants. Each tonne of phosphate processed from phosphate rock generates 5 tonnes of phosphogypsum, which is essentially radioactive gypsum (Cordell et al, 2009, p298).

Cheap food is thus predicated on cheap phosphorus. And as should now be clear, we cannot keep treating phosphorus 'cheaply' any longer. Our addiction to this element was partially spurred by cheap oil. Cheap oil made cheap phosphorus possible by allowing for the cheap extraction, production, transportation and application of this element. It is therefore somewhat counter-intuitive that the demise of cheap oil has actually *increased* global demand for phosphorus. Recent oil price shocks, growing anxiety over climate change, and domestic energy security concerns have stimulated a worldwide push in biofuel crop production, which has dramatically increased demand for phosphate fertilizers (White and Cordell, 2008, p1).

Fortunately, unlike fossil fuels, phosphorus can be recycled. Nevertheless, every estimate of phosphate rock reserves that I have seen indicates a fast approaching natural phosphorus limit. Supplies *will* run out if we are not careful. The clock is ticking.

We can begin by reducing fertilizer usage through more efficient agricultural practices, such as terracing, no-till farming and precision techniques to squeeze the most out of what phosphate rock we have left. We might also think about leaving the inedible biomass in the field, to be returned to the soil. This is especially something to think about as we move towards cellulose-based biofuels, which could strip fields of their biomass. We will also need to make better use of animal waste, utilizing our plentiful reserves of manure and animal bones as phosphorus sources. We might also want to rethink current dietary practices, for both ourselves and our cars. A significant portion of future phosphorus demand will be the result of increasing global meat and biofuel consumption. Changing this demand will have a noticeable impact upon our phosphorus needs.

Half the phosphorus we excrete is in our urine (urine is also a rich source of nitrogen). Recovering phosphorus from human waste will also be necessary from not only a food production standpoint, but also to restore some balance in ecosystem concentrations of phosphorus. For example, most wastewater facilities in Europe and North America already remove phosphorus to prevent eutrophication in receiving waters. Sometimes this sludge is applied to fields. Yet, concerns over contaminants (such as heavy metals, hormones and pharmaceutical residues) in the sludge often cause it to end up in landfills or incinerators (Christen, 2007, p2078).

These concerns can be minimized if sanitation providers and urban planners develop infrastructures that separate human excreta from industrial wastewater streams, as the latter are more likely to carry contaminants not fit for food systems. Moreover, if urine is not mixed with faecal matter in the toilet, the urine (as an essentially sterile compound) can be used safely with simple storage. Two municipalities in Sweden, for example, have mandated that all new toilets be urine diverting. The urine is then piped and stored either under the house or in a communal urine storage tank. Once a year local farmers then collect the urine for use as liquid fertilizer (Cordell et al, 2009, p300).

Notes

1 One megajoule is 1 million joules.
2 As one geologist noted in 1920, referring to phosphate rock deposits in the US: 'Idaho has an unlimited supply of high-grade phosphate in the south-east part of the state' (Stone, 1920, p404).
3 See http://minerals.usgs.gov/minerals/pubs/commodity/phosphate_rock/stat.
4 See www.fertilizer.org/ifa/Home-Page/SUSTAINABILITY/Climate-change/Emissions-from-production.html.

References

Aiyer, A. (2007) 'The allure of the transnational: Notes on some aspects of the political economy of water', *Cultural Anthropology*, vol 22, no 4, pp640–658

Allan, J. A. (1998) 'Virtual water: A strategic resource, global solutions to regional deficits', *Groundwater*, vol 36, no 4, pp545–546

Basset-Mens, C. and H. van der Werf (2007) 'Life cycle assessment of farming systems', in C. J. Cleveland (ed) *Encyclopedia of Earth*, Environmental Information Coalition, National Council for Science and the Environment, Washington, DC, www.eoearth.org/article/Life_cycle_assessment_of_farming_systems, last accessed 19 November 2010

Black, J. (2008) 'What's in a number? How the press got the idea that food travels 1500 miles from farm to plate', *Slate*, 17 September, www.slate.com/id/2200202, last accessed 1 September 2010

Carolan, M. S. (2011) *Embodied Food Politics*, Ashgate, Farnham, UK

Christen, K. (2007) 'Closing the phosphorus loop', *Environmental Science and Technology*, 1 April, p2078

CIA (Central Intelligence Agency) (2008) *The World Factbook*, US Government Printing Office, Washington, DC

Coley, D., M. Howard and M. Winter (2009) 'Local food, food miles, and carbon emissions: A comparison of farm shop and mass distribution approaches', *Food Policy*, vol 34, pp150–155

Cordell, D., J.-O. Drangert and S. White (2009) 'The story of phosphorus: Global food security and food for thought', *Global Environmental Change*, vol 19, pp292–305

Cuéllar, A. and M. Weber (2010) 'Wasted food, wasted energy: The embedded energy in food waste in the United States', *Environmental Science and Technology*, vol 44, no 16, pp6464–6469

Daily Mail Reporter (2008) 'EU forces market trader to pulp thousands of kiwi fruit because they're one millimetre too small', *Daily Mail Reporter*, 27 June, www.dailymail.co.uk/news/article-1029715/EU-forces-market-trader-pulp-thousands-kiwi-fruit-theyre-ONE-MILLIMETRE-small.html, last accessed 3 September 2010

Defra (UK Department for Environment, Food and Rural Affairs) (2008a) *PAS2050 Case Study*, Defra, UK, http://randd.defra.gov.uk/Document.aspx?Document=FO0404_8191_OTH.PDF, last accessed 19 November 2010

Defra (2008b) *SID5 Research Project Final Report*, Defra, http://randd.defra.gov.uk/Document.aspx?Document=FO0405_8189_FRP.pdf, last accessed 19 November 2010

de Fraiture, C., D. Wichelns, J. Rockström, E. Kemp-Benedict, N. Eriyagama, L. Gordon, M. Hanjra, J. Hoogeveen, A. Huber-Lee and L. Karlberg (2007) 'Looking ahead to 2050: Scenarios of alternative investment approaches', in D. Molden (ed) *Water for Food, Water for Life: A Comprehensive Assessment of Water Management in Agriculture*, Earthscan, London, pp91–145, www.iwmi.cgiar.org/assessment/Water%20for%20Food%20Water%20for%20Life/Chapters/Chapter%203%20Scenarios.pdf, last accessed 1 September 2010

Dery, P. and B. Anderson (2007) 'Peak phosphorus', *Energy Bulletin*, 13 August, www.energybulletin.net/node/33164, last accessed 18 November 2010

Desrochers, P. and H. Shimizu (2008) *Yes, We Have No Bananas: A Critique of the 'Food Miles' Perspective*, Policy Primer No 8, Mercatus Center, George Mason University, Washington DC

Dunmall, G. (2010) 'Waste: Uncovering the global food scandal', *Mother Nature Network*, 26 January, www.mnn.com/lifestyle/arts-culture/stories/waste-uncovering-the-global-food-scandal, last accessed 3 September 2010

The Economist (2005) Coca-Cola in hot water: The world's biggest drink firm tries to fend off its green critics, *The Economist* October 6, www.economist.com/node/4492835?story_id=4492835, last accessed 15 March 2011

English, M., K. Solomon and G. Hoffman (2002) 'A paradigm shift in irrigation management', *Journal of Irrigation and Drainage Engineering*, September/October, pp267–277

English, M., K. Soloman, and G. Hoffman (2003) 'A paradigm shift in irrigation management, in Jeffery Russell (ed) *Perspectives in Civil Engineering*, American Society of Civil Engineers, Reston, VA, pp89–99

FAO (Food and Agriculture Organization of the United Nations) (2003) *World Agriculture: Towards 2015/2030 – An FAO Perspective*, FAO, Earthscan, London, ftp://ftp.fao.org/docrep/fao/005/y4252e/y4252e.pdf, last accessed 1 September 2010

Frey, S. and J. Barrett (2007) 'Our health, our environment: The ecological footprint of what we eat', Paper prepared for the International Ecological Footprint Conference, Cardiff, 8–10 May, www.brass.cf.ac.uk/uploads/Frey_A33.pdf, last accessed 1 September 2010

Garnett, T. (2006) *Fruit and Vegetables and UK Greenhouse Gas Emissions: Exploring the Relationship*, Working paper, Food Climate Research Network, UK, www.fcrn.org.uk/fcrnPublications/publications/PDFs/Fruitnveg_paper_2006.pdf, last accessed 3 September 2010

Government of India Planning Commission (2007) *Groundwater Management and Ownership*, Report of the Expert Group, September, New Delhi, http://planningcommission.nic.in/reports/genrep/rep_grndwat.pdf, last accessed 2 September 2010

Griffin, M., J. Sobal and T. Lyson (2009) 'An analysis of a community food waste stream', *Agriculture and Human Values*, vol 26, pp67–81

Hall, K., J. Guo, M. Dore and C. Chow (2009) 'The progressive increase of food waste in America and its environmental impact', *PLoS One*, vol 4, no 1, pp1–6

Hanjra, M. and E. Qureshi (2010) 'Global water crisis and future food security in an era of climate change', *Food Policy*, vol 35, pp365–377

Heller, M. and G. Keoleian (2003) 'Assessing the sustainability of the US food system: A life cycle perspective', *Agricultural Systems*, vol 76, no 3, pp1007–1041

Hoekstra, A. and A. Chapagain (2007) 'Water footprints of nations: Water use by people as a function of their consumption pattern', *Water Resources Management*, vol 21, no 1, pp35–48

Hurst, D. (2010) 'Growers go bananas over waste', *Brisbane Times*, 7 January, www.brisbanetimes.com.au/business/growers-go-bananas-over-waste-20100106-lu7q.html, last accessed 30 July 2010

Hussain, I. and M. Hanjra (2003) 'Does irrigation water matter for rural poverty alleviation? Evidence from South and South-East Asia', *Water Policy*, vol 5, no 5, pp429–442

Jones, T. (2005) 'How much goes where? The corner on food loss', *BioCycle*, 2–3 July

Khan, S. and M. Hanjra (2009) 'Footprints of water and energy inputs in food production – global perspectives', *Food Policy*, vol 34, pp130–140

Khan, S., M. Hanjra and J. Mu (2009) 'Water management and crop production for food security in China: A review', *Agricultural Water Management*, vol 96, no 3, pp349–360

Korten, D. (2001) *When Corporations Ruled the World*, Kumarian Press, Bloomfield, CT

Kumar, D. and O. P. Singh (2005) 'Virtual water in global food and water policy making: Is there a need for rethinking?', *Water Resources Management*, vol 19, pp759–789

Liu, J. and H. Savenije (2008) 'Food consumption patterns and their effects on water requirements in China', *Hydrology and Earth System Science*, vol 12, pp887–898

Manson, K. (2009) 'Kenya's food miles', *Boise Weekly*, 8 September, www.boiseweekly.com/boise/kenyas-food-miles/Content?oid=1168196, last accessed 15 March, 2011

Mattson, B. and E. Wallén (2003) 'Environmental life cycle assessment (LCA) of organic potatoes', *Acta Horticulturae*, vol 619, pp427–435

Molden, D. (ed) (2007) *Water for Food, Water for Life: A Comprehensive Assessment of Water Management in Agriculture*, Earthscan, London

Murgai, R., M. Ali and D. Byerlee (2001) 'Productivity growth and sustainability in post Green-Revolution agriculture: The case of Indian and Pakistani punjabs', *World Bank Research Observer*, vol 16, no 2, pp199–218

Muuru, J. (2009) 'Kenya's flying vegetables: Small farmers and the food miles debate', Africa Research Institute, London, www.africaresearchinstitute.org/files/policy-voices/docs/kenyas-flying-vegetables-small-farmers-and-the-food-miles-debate-0v6s400wzm.pdf, last accessed 15 March 2010

Myers, N. and J. Kent (2001) *Perverse Subsidies: How Tax Dollars Can Undercut the Environment and Economy*, Island Press, Washington, DC

Nash, J. (2007) 'Consuming interests: Water, rum, and Coca-Cola from ritual propitiation to corporate expropriation in Highland Chiapas', *Cultural Anthropology*, vol 22, no 4, pp621–639

National Research Council (1989) *Irrigation Induced Water Quality Problems*, National Academy Press, Washington, DC

Pelletier, N., P. Tyedmers, U. Sonesson, A. Scholz, F. Ziegler, A. Flysjo, S. Kruse, B. Cancino and H. Silverman (2009) 'Not all salmon are created equal: Life cycle assessment of global salmon farming systems', *Environmental Science and Technology*, vol 43, no 23, pp8730–8736

Peterson, E. W. (2009) *A Billion Dollars a Day: The Economics and Politics of Agricultural Subsidies*, Wiley-Blackwell, Malden, MA

Pimentel, D., S. Williamson, C. Alexander, O. Gonzolel-Pegan, C. Kontak and S. Mulkey (2008) 'Reducing energy inputs in the US food system', *Human Ecology*, vol 36, pp459–471

Pretty, J., A. S. Ball, T. Lang and J. Morison (2005) 'Farm costs and food miles: An assessment of the full cost of the UK weekly food basket', *Food Policy*, vol 30, pp1–19

Qadir, M., B. Sharma, A. Bruggeman, R. Choukr-allah and F. Karajeh (2007) 'Non-conventional water resources and opportunities for water augmentation to achieve food security in water scarce countries', *Agricultural Water Management*, vol 87, pp2–22

Rosemarin, A. (2004) 'The precarious geopolitics of phosphorous', *Down to Earth*, June, www.indiaenvironmentportal.org.in/node/542, last accessed 18 November 2010

Saunders, C., A. Barber and G. Taylor (2006) *Food Miles: Comparative Energy/Emissions Performance of New Zealand's Agriculture Industry*, Research Report 285, Agribusiness and Economics Research Unit, Lincoln University, Lincoln, New Zealand, www.jborganics.co.nz/saunders_report.pdf, last accessed 31 August 2010

Seckler, D. (1996) *The New Era of Water Resources Management: From 'Dry' to 'Wet' Water Savings*, Research Report, International Irrigation Management Institute, Colombo, Sri Lanka, www.iwmi.cgiar.org/Publications/IWMI_Research_Reports/PDF/pub001/REPORT01.pdf, last accessed 2 September 2010

Smith, A., P. Watkis, G. Tweddle, A. McKinnon, M. Browne, A. Hunt, C. Treleven, C. Nash and S. Cross (2005) *The Validity of Food Miles as an Indicator of Sustainable Development: Final Report*, Defra, Report No ED50254, issue 7, www.wildchicken.com/grow/defra%20foodmiles%20execsumm.pdf, last accessed 31 August 2010

Sonesson, U., F. Anteson, J. Davis and P.-O. Sjödén (2005) 'Home transport and wastage: Environmentally relevant household activities in the life cycle of food', *Ambio*, vol 34, no 4–5, pp371–375

Stone, R. W. (1920) 'Phosphate rock', in J. Spurr (ed) *Political and Economic Geology and the World's Mineral Resources*, McGraw-Hill, New York, NY, pp402–410

Stuart, T. (2009) *Waste: Uncovering the Global Food Scandal*, Norton, New York, NY

UNDP (United Nations Development Programme) (2007) *Human Development Report 2006 – Beyond Scarcity: Power, Poverty and the Global Water Crisis*, UNDP, New York, http://hdr.undp.org/en/media/HDR06-complete.pdf, last accessed 1 September 2010

US Geological Survey (2010) *Phosphate Rock*, US Geological Service, Washington, DC, http://minerals.usgs.gov/minerals/pubs/commodity/phosphate_rock/mcs-2010-phosp.pdf, last accessed 18 November 2010

Vaccari, D. (2009) 'Phosphorus famine: The threat to our food supply', *Scientific American*, 3 June, pp54–59

Wainwright, M. (2009) 'Supermarket offers and food waste targeted in government's food strategy', *The Guardian*, 10 August, www.guardian.co.uk/environment/2009/aug/10/food-security-climate-change, last accessed 3 September 2010

Weber, C. and H. S. Matthews (2008) 'Food miles and relative climate impacts of food choices in the United States', *Environment, Science and Technology*, vol 42, pp3508–3513

Westhead, R. (2010) 'A new issues percolates throughout India: How much to charge for water?', *The Toronto Star*, August 16, www.thestar.com/news/world/india/article/848255--a-new-issue-percolates-throughout-india-how-much-to-charge-for-water, last accessed 15 March 2011

White, S. and D. Cordell (2008) *Peak Phosphorus: The Sequel to Peak Oil*, Information Sheet Two, Global Phosphorus Research Initiative, http://phosphorusfutures.net/peak-phosphorus, last accessed 18 November 2010

Wichelns, D. (2001) 'The role of "virtual water" in efforts to achieve food security and other national goals, with an example from Egypt', *Agricultural Water Management*, vol 49, pp131–151

Williams, A., E. Audsley and D. Sandars (2006) *Determining the Environmental Burdens and Resource Use in the Production of Agricultural and Horticultural Commodities*, Main report, Defra Research Project IS0205, Cranfield University and Defra, Bedford, UK, www.silsoe.cranfield.ac.uk and www.defra.gov.uk

7

Cheap Food... But at What Price?

I am not a professional economist. Perhaps that is why I am always a little uncomfortable with attempts to reduce 'costs' to 'currency'. There are many different ways of assigning a monetary cost to, say, diminished biodiversity or ecosystem services that result from current food system practices. Yet, does that dollar value constitute the *real* cost of cheap food? We can and should do whatever we can to bring economics into ecology. I just worry that these monetary values will dominate food policy discussions, leaving non-monetized values hidden in the shadows of silent speculative thought.

While philosophically and methodologically problematic, I fully appreciate the practical power of these valuation exercises. Discussions about, say, intrinsic value are simply not enough to challenge a system whose existence rests upon its ability to supply 'cheap' food. Ecological economics allows us to (albeit imperfectly) convert the long shadow cast by cheap food into a form that even cheap food proponents can't ignore: dollars and cents.

When Food Meets Neoclassical Economics

Over the last century we have been moving towards a system of production where capital is substituted for labour, where external controls (inputs) are substituted for internal controls (such as natural pest controls and nutrient recycling), and where industrial monocultures are being managed by an ever-shrinking farm population. We could have gone in the other direction (I'll discuss how in a moment). We could have encouraged a more labour-intensive model of agriculture, of smaller farms managed very intensively, utilizing internal controls instead of petrochemical inputs. Under such an arrangement, input use and production costs are significantly lower, units produced per hectare higher, and the integrity of the requisite ecosystems better preserved (Altieri, 1995). We could have gone in this direction but we didn't, though not without reason.

For one, this alternative model is tremendously labour intensive – an intensity that contrasts starkly with what's required on fully mechanized 'modern' farms. It would also require a high percentage of individuals to remain 'shackled' to agriculture: an unacceptable prospect, as mentioned in Chapter 2, for many models of development. Why tie a sizeable portion of a country's population to agriculture and, thus, the countryside, and risk stalling those more urban sectors of the economy that promise to generate real wealth? Small, diverse farms never had a chance considering economic developmental theory's opinion of them.

The greatest efficiency of today's food system lies not in its technology, economies of scale or managerial techniques. It lies in its accounting practices. Any system capable of pushing most of its costs onto society, the environment and future generations will clearly produce a 'cheap' product. How could it not?

Students often ask me: 'How has the food system been able to get away with this for as long as it has?' 'Because it could', I tell them, smiling. But I am only half joking. If agro-ecosystems were so tightly coupled that any pollution would result in total and immediate ecological collapse, then there wouldn't – indeed, couldn't – be much to say about the costs of food upon the environment. And let's not forget, we all share some culpability in this by allowing a system to persist that socializes most of its costs.

Cost minimization is a core task for any firm (farm or otherwise), typically achieved through savings in labour and/or materials. Determining which costs to target depends on the relative price of both. For roughly the last hundred years, material costs have been dropping faster than wages, especially when compared to wages in fully developed economies (Jackson, 2009, pp92–93). Moreover, countries such as the US have very generous tax codes that provide tax deductions on equipment depreciation, which further undervalues material costs. Farms therefore have a financial incentive to invest in labour-saving technologies. But by investing in technologies and cutting labour, farms have also divested themselves from the ecological base that makes agriculture possible. Moreover, these 'cheap' technologies are *themselves* products of the same sloppy accounting (or they are 'incorrectly priced', in conventional economics parlance) that makes our current food system look so attractive. The reason (or at least one of them) for why material costs have been dropping faster than labour costs is because it is easier to externalize the real cost of technology. Human and environmental health comes at a price. Thanks to democracy, it is far easier to exploit the health of the environment than the health of a labourer. If the real cost of a technology was reflected in its retail price, I doubt material costs would have dropped at the rate they have over the last century, making labour more attractive.

Labour-intensive agriculture, therefore, does not *have* to be more expensive than capital-intensive agriculture. It is, but only because of how we've chosen to socialize certain costs and not others. I once advocated labour-intensive agriculture in the presence of an agriculture economist (I often bounce my ideas off economists because they keep me grounded in the realities of the market). This individual denounced my enthusiasm, asking how labour-intensive agriculture could possibility produce more affordable food without yet more subsidies – namely, for labour. Yet this comment misses why capital-intensive agriculture is 'cheaper' than labour-intensive agriculture to begin with. This 'cheapness' is an artefact of economic policy, which we *can* change. For example, a finger can certainly be pointed at incorrectly priced energy. Agriculture has become capital intensive, which is to say *energy* intensive, because energy is cheap – too cheap. This is admittedly not a problem of agriculture's making. But it does give us an example of how capital-intensive agriculture has been made (to the extent of going to war over cheap oil) to produce cheaper food than a more labour-intensive model.

Many of the goods supplied by the environment are public in nature. A good is said to be public if it is non-rival and non-excludable. Many ecological phenomena that we value, such as biodiversity and clean air, exhibit these characteristics. My use/enjoyment of these 'things' does not affect your use/enjoyment of them. They are also non-excludable in the sense that it is usually difficult to prevent others from benefiting from their existence. Biodiversity, for instance, benefits humanity in so many ways – from food security to the manufacturing of drugs and the enrichment of spirit – that excludability means more than just keeping people out of a particularly bio-logically diverse space. Public goods are therefore under-provided – and under-protected – if left to the market (Pascual and Perrings, 2009, p155). In the absence of socio-organizational constraints that would help to offset economic incentives to exploit the environment (Ostrom, 1999), the quality of these public goods suffer, which itself has social justice and food security implications. For instance, the enormous profitability of the agribusiness, and in particular biotechnology, sector (as discussed in Chapter 9) is due, in part, to its ability to free-ride on farmers responsible for conserving global genetic diversity (Pascual and Perrings, 2009, p153).

Until recently, a highly myopic view of the food 'system' helped to mask many of these environmental costs. Folding those costs back into the system from where they came adds considerably to the bottom line of cheap food. One calculation places the annual external costs of agriculture (up to the farm gate) in Germany at US$2 billion; in the UK that figure is US$3.8 billion and in the US the total is US$34.7 billion. These average out to an additional cost of approximately US$81 to US$343 per hectare of arable land and pasture (Pretty et al, 2001). A more recent analysis estimates the external cost of UK agriculture (up to the farm gate) to be US$2.4 billion (see Table 7.1 for a specific breakdown of these costs) (Pretty et al, 2005).[1]

Neoclassical economic theories do not completely ignore environmental realities. But they come close. It wasn't until the 1960s and early 1970s when some economists came to the (at the time radical) realization that the entire economic system was embedded within a broader, more fundamental, *ecological* system. While the environment existed long before the market (and will continue to exist long after it disappears), the reverse will never be true. Destroy the environmental productive base and the entire house that is the global economy will fall.[2]

Table 7.1 *Negative externalities of UK agriculture (2000 dollars)*

Source of debt	Costs (US$ million/yr)
Pesticides in water	220.6
Nitrate, phosphate, soil and *Cryptosporidium* in water	172.5
Eutrophication of surface water	121.9
Monitoring of water systems and advice	20.2
Methane, nitrous oxide, and ammonia emissions to atmosphere	648.7
Direct and indirect carbon dioxide emissions to atmosphere	158.2
Off-site soil erosion and organic matter losses from soils	90.9
Losses to biodiversity and landscape values	231.5
Negative effects to human health from pesticides	1.8
Negative effects to human health from microorganisms and BSE	666.4
Total	2332.7

Adapted from Pretty et al (2005, p4)

In his introductory remarks at the proceedings of the British Association for the Advancement of Science Festival, Partha Dasgupta (2006, pp1–2) spoke of how many of his professional colleagues believe that the services provided by nature are marginal, at best 2 to 3 per cent of an economy's output. Rather than worry about these capital assets of 'negligible importance', Professor Dasgupta tells of how his colleagues believe that knowledge – because it is durable and can be shared collectively – is capable of circumventing all ecological constraints, giving economies (and, I suppose, food systems) the ability to grow indefinitely.

Regardless of what the actual monetary value of these services are, Dasgupta (2006, p3) points out that natural capital is not like your typical capital assets, such as bridges, roads, buildings and machinery. Like capital assets, ecosystems depreciate when misused or overused. But they also differ from reproducible capital assets in three important ways:

1 Depreciated natural capital is often either slow to recover or irreversible (as in the case of species extinction).
2 It is often quite difficult to substitute one depleted or exhausted ecosystem service for another (such as the pollination services provided by bees).
3 Ecosystems can collapse abruptly, without much prior warning, as famously detailed in Jared Diamond's *Collapse* (Diamond, 2006).

The not-so-subtle reductionism of cheap food policy makes it hard for proponents of the status quo to see the systemic value of nature. I have already discussed (in Chapter 4) how this reductionism has played out in the context of 'food' being reduced to 'calories' and more recently to nutritional components.

It is also visible in *how* food is produced, evident by agricultural techniques (and ideologies) that view soil as little more than structural support for roots and the medium by which the holy trinity of NPK (nitrogen, phosphorus and potassium) and water are taken in by the plant. Following this ideology, plants are reduced to genes, water and nitrogen, and crops become little more than protein delivery systems (see, for example, Purdue University, 1975).

Ecologists have long questioned the 'gains' claimed by the Green Revolution, perhaps in part because their disciplinary view of the world makes reductionist accounts of food production appear myopic in nature. Led by famed ecologist David Pimentel, a team of researchers concluded in a 1973 *Science* article that agriculture was using an equivalent of 80 gallons of gasoline to produce 1 acre of corn. They further noted that while the production of corn per acre increased 2.4 times from 1945 to 1970, the input of fuel rose 3.1 times. In other words, yields in corn energy relative to fuel input *declined* 26 per cent during this period. For each calorie of energy in 1945 used to grow corn, 3.7 calories of corn were produced, whereas in 1970 that ratio dropped to 1:2.8 (Pimentel et al, 1973). At around the same time, Charles Butterworth (1974) noted in *JAMA* that it takes four times as many calories to produce 1 kilocalorie of food in the US as in India (noting also that this calculation does not include the large energy costs involved in the preparation of refined, processed and convenience food items).

Nitrogen (N) and pesticides are two of the more problematic aspects of cheap food from an environmental standpoint. Cropping systems, because they are not closed, leak excess nitrogen and pesticide residue into their ecological base. I am aware of the argument that the Green Revolution has saved marginal lands from agricultural conversion by boosting yields on already cultivated lands (Evenson and Gollin, 2003, p758). But the Green Revolution has also been instrumental in instituting around the world a model of food production premised upon the replacing of internal ecological controls (natural pest suppression techniques) with external inputs (pesticides).

The average corn crop today removes approximately 200kg of N per hectare per year from the soil. This calculates out to somewhere between 2kg to 4kg of N per hectare every day during the crop's peak growing window (which lasts between six and eight weeks). Conversely, the yearly input rate of an average unfertilized field is between 6kg and 8kg of N per hectare, recognizing that half of the N is removed annually in the form of harvest protein (Robertson and Swinton, 2005, p39). Today's crops have been bred to be like high-performance engines. Unlike their ancestors, which could thrive on less N, 1 acre of corn today requires the equivalent of high octane fuel. Increasing plant density also plays a role in this, going from, in the case of corn, less than 14,000 plants per acre before the 1940s to over 30,000 today in the US Corn Belt. All of this makes external sources of N an unavoidable necessity – or so I'm often told. Granted, coming up with that much N through internal sources is a tall order. Yet, in light of existing ecological principles, it is not impossible (Gliessman, 1998, p10; Altieri, 1999, p197; Robertson and Swinton, 2005, p39). Relying more on internal sources of N, where soil tilth, microorganism activity and nutrient levels are built up together, should also help to reduce the level of N leaching into the surrounding watershed, an added ecological benefit in light of the low oxygen 'dead zones' found along the world's coastlines from excess nutrients finding their way into the sea.

The same dilemma exists when it comes to the regulation of pest and pathogen populations: we have created a 'need' that previously didn't exist. Here, too, we seem to have painted ourselves into a corner, which is escapable, but only with a concerted effort. David Pimentel is also responsible for leading early research looking at the effects of insecticide use on crop losses. In a study published in 1978, Pimentel and his colleagues note that despite large increases in insecticide use during the 1960s and 1970s crop losses due to insect pests were actually *increasing*. The article cites two major contributing factors for this: the practice of substituting insecticides for sound bio-environmental pest controls; and higher aesthetic standards (Pimentel et al, 1978). The second point highlights a well-documented change that took place over the 20th century in terms of how we judge 'quality' when it comes to evaluating food (Freidberg, 2009; Carolan, 2011). I have already addressed the issue of aesthetics in the previous chapter when discussing food waste. As to Pimentel's first point: prior to the widespread application of pesticides, insect pests were managed using internal population-level controls such as predation. The more diverse commodity profile of farms a century ago also helped to control pest populations, making it much less likely that a single insect would wreak havoc over the entire countryside. Restoring these internal biological controls can be done, but will require new management practices and new incentives to make those practices stick (Robertson and Swinton, 2005, p39).

Valuing Ecosystem Services and Disservices

During recent years, natural and social scientists (I include economists in this lot) have come to better appreciate the services provided by the environment. Attempts have also been made to put a price tag on those services. Perhaps the most cited figure comes from Costanza and colleagues. In a 1997 issue of *Nature*, they estimated the global economic value of 17 ecosystem services to be about US$33 trillion per year (compared to a gross world product of around US$18 trillion per year at the time). More recently, the value of 12 atmospheric services vital to human well-being and the existence of the biosphere were estimated to be at least between 100 and 1000 times the gross world product, which in 2008 was approximately US$68.5 trillion (Thornes et al, 2010, p243).[3]

Examples of agro-ecological services are highlighted in Table 7.2. Agricultural lands have historically focused on maximizing certain provisioning services, with the production of food and fibre at the top of the list. Yet, we also know that food and fibre production is not possible without supporting, regulating and cultural services. The problem, however, is how do you go about valuing what's known as a non-market good – something we all recognize as having some type of value, just not explicit economic value?

Table 7.2 *Examples of 'services' we value that are provided by agro-ecosystems*

Provisioning Services: Products obtained from ecosystems, including:
- Food and fibre
- Fuel
- Genetic resources
- Ornamental resources
- Fresh water

Regulating Services: Benefits from regulation of ecosystem processes, including:
- Air quality maintenance
- Climate regulation
- Water regulation
- Erosion control
- Water purification
- Biological (e.g. pest) control
- Pollination
- Storm protection

Supporting Services: Services supporting aspects of agriculture that have long had economic value, including
- Nutrient cycling
- Soil formation
- Soil hydrology

Cultural Services: Non-material benefits from ecosystems through spiritual enrichment, cognitive development, reflection, recreation, and aesthetic experiences, including:
- Cultural diversity
- Spiritual/religious values
- Knowledge systems
- Inspiration/aesthetic values
- Recreation
- Cultural heritage

Based on data from *Millennium Ecosystem Assessment* (2005, pp56–59) and Zhang et al (2007, p254)

The most advanced evaluation methods are those that look to the market to price ecosystem services. One is known as the 'travel cost method', which examines the amounts that consumers spend to gain access to certain services. Another is 'hedonic price analysis', which looks at real estate values and attempts to estimate the value that ecosystem services contribute to the total price of the property. Far more controversial and problematic are valuation methodologies for non-market goods. Many of these 'stated preference' (also known as 'willingness to pay') methods rely heavily on surveys that ask hypothetical questions about what the respondent would be willing to pay for a particular service (Robertson and Swinton, 2005, p42). This methodology tends to work best when respondents are aware of the ecosystem services in question (Robertson and Swinton, 2005, p42; Hein et al, 2006, pp213–214). Given the complexity of agro-ecology, consumers may not state a preference for a particular service, but only because they do not understand the functioning of ecosystems (Kroeger and Casey, 2007, p325). Another critique of stated preference methods is the well-documented fact that peoples' stated attitudes and values often do not mesh well with actual behaviours (Carolan, 2010b, pp309–310).

In addition to demand-side metrics, which seek to know what consumers are willing to pay for ecosystem services, there are supply-side methods. These seek to discern what producers would be willing to accept – usually in the form of a government payment – to maintain services that society deems worth protecting. Yet, as currently implemented, supply-side valuation policies are not without their problems. The goal of these policies is often to incentivize particular behaviours so that farmers act on behalf of the 'collective good'. Yet, nothing is done to encourage coordinated behaviour (Zhang et al, 2007, p258). The US, for example, currently has programmes that pay farmers to voluntarily adopt land management practices that encourage natural biological pest controls. As decades of research shows (see, for example, Pretty, 1998; Röling and van de Fliert, 1998; Fakih et al, 2003), however, biological pest controls work best when multiple farms work together and 'scale up' these management practices over a large territory.

Then there is the question of what constitutes a 'desirable' ecosystem service. Even if a satisfactory answer to this question is arrived at, further uncertainty clouds the relationship between those 'desirable' services and the farming activities that generate them (Robertson and Swinton, 2005, p42). Part of this uncertainty lies in the fact that a service's value is context dependent. For example, trees in south-western Australia provide an important ecosystem service by increasing the rate of infiltration of water into soil. In South Africa, conversely, trees can provide an ecosystem *disservice* through the transpiration of water (Zhang et al, 2007, p258). It is not that trees function differently in different places. Whether they provide a service or disservice is conditioned upon the socio-ecological conditions of place, which produces different valuations even though we are talking about the same fundamental ecological processes (Zhang et al, 2007, p259).

Philosophically speaking, perhaps the most problematic question of all is: can one really put a price tag on something that's arguably invaluable? I am reminded of something the Nobel Prize winning economist Amartya Sen (1989, p317) once wrote:

> A formal expression can be extremely precise without being at all a precise representation of the underlying concept to be captured. In fact, if that underlying concept is ambiguous, then the demands for precise representation call for capturing that ambiguity rather than replacing it by some different idea – precise in form but imprecise in representing what is to be represented.

Lest we forget: the term 'value' is just as ambiguous as 'ecosystem service'. Below are some examples of how the concept is defined within the ecological economics literature (adapted from Spangenberg and Settele, 2010, p328):

- *Market value:* exchange value or price of a commodity in the open market.
- *Use value:* value of entities that may have little or no market value but have use value.
- *Intrinsic value:* value attached to ecosystems and life forms for their own sake.
- *Existence value:* value of knowing that species, ecosystems and services exist, even if individual does not make active use of them.
- *Bequest/vicarious value:* a willingness to pay to preserve the environment for the benefit of other people and future generations.
- *Present value:* value today of a future asset, discounted to the present.
- *Option value:* willingness to pay today for future use of an environmental asset.
- *Quasi option value:* value of preserving ecological conditions for future use assuming an increasing knowledge about ecosystem functioning.

When talking about valuing ecosystem services, we need to be honest and acknowledge this exercise for what it is: a tool to protect ecosystems, species and services. Towards this end, we must not forget that value judgements anchor all policies directed at the management of ecosystem services (Carolan, 2006c, 2006d). Ultimately, all such policies hinge on how we answer such questions as: 'What definition of value *ought* to be used?', 'What ecosystem service *should* be protected?' and 'What are *acceptable* trade-offs when valuing one ecosystem service over another?'

Figures 7.1 and 7.2 illustrate two different models for valuing ecosystem services. Figure 7.1 represents the cheap food model; Figure 7.2 characterizes what I will call the affordability model. The only 'service' with significant economic value in the cheap food model is the provision service of food and fibre production. This is not to say that other services are not valued. I know a lot of farmers, for example, who value wetlands on their property for their aesthetics and for the habitat that they provide. Still others I know value heritage seeds, not for market reasons, but for the cultural connection they feel with plant varieties that have been in their family for generations. But market value – the left side of these figures – has a special hold on us. As a friend once told me prior to having his wetland drained (to make room for more corn): 'Biodiversity doesn't put cash in the bank or pay off loans.'

This brings me to Figure 7.2, which illustrates the relocation of certain ecosystem service valuations, through policy (see the left side of the figure). First: to anyone who believes this is a blatant attempt at market interference, I suggest they read Karl Polanyi's (2001) book *The Great Transformation*. One of the many brilliant insights found in this piece of historical economic scholarship is the concept of the 'commodity fiction'. Commodity fictions refer to all those

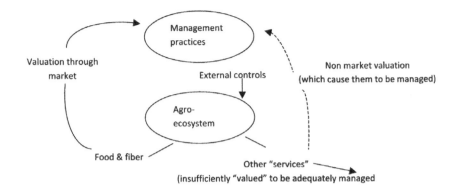

Figure 7.1 *Cheap food's approach to ecosystem services*

Source: Author

aspects of free market capitalism that depend upon the state for their very existence. Polanyi was writing specifically about land, labour and money; but there are many others. In agriculture, for example, we can point to seeds as a commodity fiction. They are very much a commodity, but only because the state and state-sanctioned entities such as the court system – most notably through patent law – bestow upon these artefacts this status (see Carolan, 2010a). Why not grant ecosystem services the same status and, in doing this, help to construct a market that would allow my friend a more balanced 'choice' between keeping his wetland and draining it to make room for more corn?

The hope is that by rewarding farmers to grow services in addition to (or other than) food and fibre (I talk more about this in Chapter 10), internal controls will be substituted for the external controls that wreaked so much environmental havoc over the last century. Under the 'old' paradigm (see Figure 7.1), since most ecosystem services have no market value and because inputs over the last few generations have been relatively cheap, it made (economic) sense for farmers to rely increasingly upon petrochemical inputs. Yet, these management practices have come at a price. Declining crop yield is closely related to decreases in soil quality, whether through salinization, erosion, compaction or nutrient mining. Soil degradation decreases organic matter, diminishes microorganisms' ability to recycle nutrients, and impairs the water-holding capacity of soil (Stocking, 2003, p1357). Until internal controls are better understood we will continue to apply petrochemical inputs to make up for this growing ecological debt. Yet, as more services are valued in the market, and as input costs continue to rise, I expect farm management techniques to change. Under the affordability model (see Figure 7.2), food has the potential to be produced without great cost to the environment, at which point, the healing of the rift we've created separating agricultural systems from ecological systems can begin.

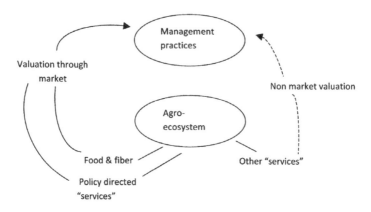

Figure 7.2 *Food production that focuses on affordability*
rather than cheapness

Source: Author

The Cost of Pesticides

About 3 billion kilograms of pesticides are applied each year worldwide, with a purchase price of approximately US$40 billion (Pimentel, 2009, p83). Pesticides have unquestionably enhanced average farm productivity and profitability. Even David Pimentel – who we already know as no shrinking violet when it comes to critiquing pesticides – acknowledges that, in general, 'each dollar invested in pesticide control returns about $4 in protected crops' (Pimentel, 2005, p229). Let's be clear, however: this 4 to 1 ratio is a product of poor accounting. Once we begin talking about the real cost of pesticides, things begin looking far less favourable for these external controls and the cheap food that they help to make possible.

Many economic studies assessing the benefits of pesticides start off with a troubling research question. It is troubling because it tacitly favours the status quo by enquiring what the economic impact would be if pesticides where banned (see, for example, Settle, 1983; Cooper and Dobson, 2007). I hear the same question asked about antibiotics. This has allowed cheap food proponents to argue that if antibiotics were suddenly outlawed – and *nothing else changes* – rates of culling and disease among livestock in total confinement facilities would increase dramatically. Global farm output *would* suffer if pesticides were eliminated. But that is because the conventional system has become almost completely dependent upon these external controls. We have long known about the 'pesticide treadmill' (Van den Bosch, 1978, especially Chapter 3). This is the phenomena where insects evolve to become resistant to pesticides, which leads to more applications and/or higher concentrations and/or new chemicals, which leads to still further resistance, and so on. It is also unrealistic to ask a question that assumes a total and immediate ban of pesticides. The far more likely scenario would involve a transition period, where new technologies replace old and where internal controls are given time to be built back up. As Jules Pretty and Hermann Waibel (2005, p41) point out, any 'benefit–cost ratios derived from 100 per cent chemical reduction scenarios are of no use for drawing policy conclusions'.

There are also methodological problems – namely, their narrow operationalization of 'expertise'. The concept of expertise in agricultural circles is contested and fluid, especially in light of the diversity of management practices employed on farms (Carolan, 2006a, 2006b, 2008). In one now-infamous study (Knutson et al, 1990), 140 social and crop science experts were asked to estimate crop losses in the US in the event of an all-out pesticide ban. Not one respondent had direct working knowledge of organic systems or integrated pest management (IPM) programmes, so, not surprisingly, responses tended to be very pro-pesticide (Buttel, 1993; Pretty and Waibel, 2005).

While pesticides have drawn the ire of ecologists and environmentalists since the early 1960s (see, for example, Carson, 1962), the last ten years have seen a concerted attempt to put an actual price tag on these 'costs'. The global pervasiveness of pesticides has no doubt helped lead to the recent flurry of ecological accounting exercises (see Box 7.1). No longer can their real cost be ignored, as the toll paid by taxpayers, farm labourers, the environment and future generations is beginning to eclipse whatever benefits pesticides promise.

Box 7.1 Coke, Pepsi and pesticides in India

In 2003, the Delhi-based environmental NGO Centre for Science and Environment (CSE) released a stinging report announcing the presence of pesticides – at levels *24 times* greater than allowed under European Union (EU) standards – in a dozen popular beverages manufactured under the Coca-Cola and Pepsi labels. These drinks came from manufacturing and bottling plants located within India.

CSE used the report to underscore not only the problem of water scarcity in India, but also the country's weak pesticide regulations. CSE hoped that nationwide middle-class outrage – a population who assumed their cherished Coke and Pepsi products were safe – could be more successfully channelled to confront the problem of pesticide pollution than could any fragmented peasant movement (Vedwan, 2007, pp659–660). While pesticide residues are present in virtually all groundwater in India and, thus, in many domestic beverages, from Indian-made milk to bottled tea (Bremner and Lakshman, 2006), the tactic of singling out the iconic brands of Coke and Pepsi for criticism sought to bring the issue of pesticides in India to the attention of the international community. Pakistan imported 20,000 tonnes of pesticides in 1991, 40,000 tonnes in 1995, 130,000 tonnes in 2004, and around 200,000 tonnes in 2009 (Shah, 2010).

Costs to public health and well-being

Van Der Hoek and colleagues (1998) estimate that approximately 7.5 per cent of agricultural workers in Sri Lanka experience occupational pesticide poisoning every year. In Costa Rica and Nicaragua, the corresponding figures are 4.5 and 6.3 per cent, respectively (Wesseling et al, 1993; Garming and Waibel, 2009). In a study of farmers in the West African country of Benin, 81 per cent of pineapple farmers and 43 per cent of vegetable farmers reported 'considerable' negative health effects due to pesticide exposure (only 1 per cent of farmers noted no negative effects) (Williamson, 2005, pp171–172). The total number of pesticide poisonings in the US is believed to be around 300,000 a year (see Box 7.2) (Pimentel, 2005, p230). Worldwide, there are more than 26 million cases of non-fatal pesticide poisonings and about 220,000 poisonings resulting in death (Richter, 2002, p3). A survey of epidemiological literature indicates that the incidence of cancer in the US population due to pesticides ranges from about 10,000 to 15,000 cases year (Pimentel, 2005, p231). A recent study on the social costs of pesticides in the Brazilian state of Paraná (situated in the southern region of the country) calculates a total health cost of approximately US$443 million annually, placing a heavy burden on the Brazilian public health system (and Brazilian taxpayers) (Soares and Porto, 2009).

Box 7.2 United Farm Workers of America (UFW) cancer registry study

A seminal study examining the pesticide–cancer link in farm labourers was published in 2001. Supported, in part, by grants from the US Centers for Disease Control and Prevention and the California Cancer Research Program, the study analysed cancer incidence among California Latino farm workers who had been members of the United Farm Workers of America (UFW) union (Mills and Kwong, 2001). Just over 1000 – out of roughly 140,000 – farm workers had been diagnosed with cancer between 1973 and 1997. Compared with the general Latino population, the study found that UFW farm workers were 69 per cent more likely to acquire stomach cancer, whereas the likelihood of developing uterine, cervical and certain types of leukaemia increased by 68, 63 and 59 per cent, respectively.

The authors of the study attributed this cancer rate disparity with the general Latino population to the farm workers' regular exposure to pesticides. Farm workers came into contact with these chemicals while mixing and applying them to crops; during planting, weeding, thinning, irrigating, pruning and harvesting crops; and at home, due to living near treated fields, eating pesticide-contaminated food, and bringing the chemicals home on their clothes. The report also notes that farm workers tended to be diagnosed at a later stage than the general Latino population, indicating a lack of healthcare for most labourers. Furthermore, since many pesticide poisonings go unreported – those affected could move away, never seek treatment or seek treatment in Mexico (Murray, 1994, p47) – the data used in this study probably underestimate the actual trauma rate.

It is widely believed that pesticide toxicity figures based upon government statistics, especially in developing countries, grossly understate the problem. For example, a study of Indonesian farmers found that 21 per cent of those interviewed reported three or more neurobehavioural, respiratory and/or intestinal symptoms indicative of pesticide toxicity. Yet only 9 per cent of the farmers in the study said they had reported being poisoned (Kishi et al, 1995). There are numerous reasons for the under-reporting of pesticide poisonings. In the case of the Indonesian farmers, many of the affected simply ignored the symptoms. Another reason has to do with *whom* epidemiological studies tend to focus on. Often, the sample population comprises males who apply the chemicals. Yet women spend long hours toiling away in freshly sprayed fields. Exposure can also occur by way of a husband's contaminated clothes, which are washed by women and often mixed with other laundry (Kishi, 2005, p26).

In addition to incurring medical expenditures, a loss in earnings from being unable to work is a large cost to farmers who, without illness, are already struggling to make ends meet (Garming and Waibel, 2009). Antle and Capalbo (1994) examined the effects of pesticide use on farmers' health and the effects of farmer ill health on farm productivity, and found that pesticide-related health impairments led to significant reductions in labour productivity and farm output. Pimentel (2005, p235) estimates that in the US alone pesticide use costs farmers and the food system more generally US$520 million due to the need for additional pesticide applications and crop losses. This sizeable expense is the result of the destruction of natural enemies and a built-up resistance among some pests to current pesticides and application rates.

It is unfair to blame the victim in cases of pesticide poisonings, such as by believing that farmers handle chemicals in a risky manner due to ignorance. Most of those exposed know the dangers of pesticides. Nevertheless, 'this knowledge does not necessarily translate into a behavior that mitigates effects, as there are often structural barriers or other reasons that override farmers' concerns about safety when applying pesticides' (Kishi, 2005, p27). Some of these reasons include the following: necessary protective equipment might be too expensive or too uncomfortable when working in oppressive heat and humidity; washing facilities are often inconveniently located relative to fields (especially when one is in the middle of a field); and often agricultural communities in less affluent nations use pesticide-laden irrigation channels for bathing and washing purposes (Murray and Taylor, 2000; Kishi, 2005; Mayer et al, 2010).

The negative health effects of pesticides can be far more significant in children than adults, as children consume more food per unit of weight than adults and thus can consume more pesticides per unit of weight. Other physio-logical traits that increase children's susceptibility to pesticides include their greater skin surface area relative to their size, the fact that they take more breaths per minute, and their not yet developed immune and detoxifying systems (Levine, 2003, p151). The cost of pesticides relative to the meagre income of farmers in developing countries makes this population a high value target for thieves. In order to protect their investment, farmers report storing these chemicals within the home. In Indonesia, 84 per cent of respondents reported storing pesticides at home, 82 per cent reported storing them within reach of their children, and 75 per cent said the chemicals were kept in living or kitchen areas (Kishi et al, 1995).

Costs to the environment

Thousands of domestic animals are accidentally poisoned by pesticides every year. Dogs and cats make up most of these poisonings, as they typically have the ability to freely explore the farm (Berny et al, 2010, p255). Among livestock, 0.5 per cent of all animal illnesses and 0.04 per cent of all animal deaths reported to a veterinary diagnostic laboratory are the result of pesticide toxicosis, which equates to US$21.3 million and US$8.8 million, respectively (Pimentel, 2005, p233). A widely cited study by Carrasco-Tauber (1989) estimates an annual loss of between US$45 and US$120 per hectare due to pesticide resistance in California cotton, or roughly US$348 million annually (and that was back in the 1980s!). David Pimentel (2005, p238) estimates that between 10 and 25 per cent of all pesticides currently used in the US are applied to combat increased resistance. That calculates out to more than US$1.5 billion worth of pesticides annually.

Arguably, the most ignored external environmental cost is the destructive effects of pesticides on our pollinators. Insect pollination is not only a critical ecosystem service, but also an essential component of most agricultural systems. About 84 per cent of the nearly 300 commercial crops are insect pollinated (90 per cent of which is performed by honeybees) (Allsopp et al, 2008, p1). Pollination by European honeybees is estimated to be between 60 to 100 times more valuable than the honey they produce, which explains why farmers dependent upon their services sometimes refer to them as 'flying $50 bills' (Maser, 2009, p69). Unable to discern insect-friend from insect-foe, pesticides have had a major impact upon all insect pollinators (Desneux et al, 2007).

Pimentel (2005, p238) estimates the cost of pesticides due to pollinator poisonings and reduced pollination in the US to be around US$283.6 million annually. Breaking this figure down further, he places the costs for colony losses from pesticides at US$13.3 million a year, honey and wax losses at US$25.3 million a year, loss of potential honey production at US$27 million a year, bee rental for pollination at US$8 million a year, and pollination losses at US$210 million a year. Elsewhere, the annual cost of maintaining feral honeybee pollination ecosystem services in Australia – a more narrowly defined conception of 'cost' compared to Pimentel's – was calculated at between AU$16.4 million and AU$38.8 million (US$12.6 million and US$30.7 million) (Cook et al, 2007). A more 'all-in' estimate of insect pollinators' value is offered by Allsopp and colleagues (2008). Looking at the deciduous fruit industry of the Western Cape of South Africa, they estimate the total annual value of insect pollination services and managed pollination to be US$358.4 million and US$312.1 million, respectively. Between US$28 million and US$122.8 million of this is provided by managed honeybee pollination services, for which only US$1.8 million is currently being paid. This is still better cost internalization than what we see with wild pollinators. They provide us between US$49.1 million and $310.9 million worth of services for free.

Table 7.3 *Estimates of the real costs of pesticides from Leach and Mumford (2008) and Pimentel (2005), in US$*

Costs	UK	US	Germany
(Adapted from Leach and Mumford, 2008)			
Pesticides in sources of drinking water	287,444,066	1,126,337,798	180,522,199
Pollution incidents, fish deaths & monitoring costs	20,360,620	161,720,798	51,355,453
Biodiversity/wildlife losses	29,942,089	207,383,142	6,224,903
Cultural, landscape, tourism, etc.	118,570,678	insufficient data	insufficient data
Bee colony losses	2,395,367	144,597,420	1,556,225
Acute effects of pesticides to human health	2,395,367	167,428,591	28,012,065
Total	461,108,190	1,807,467,750	267,670,848
(Adapted from Pimentel, 2005)			
Public health impacts	not calculated	1,140,000,000	not calculated
Domestic animals deaths and contaminations	not calculated	30,000,000	not calculated
Cost of pesticide resistance and loss of natural enemies	not calculated	2,020,000,000	not calculated
Honeybee and pollination losses	not calculated	334,000,000	not calculated
Crop losses	not calculated	1,391,000,000	not calculated
Bird and fishery losses	not calculated	2,260,000,000	not calculated
Groundwater contamination	not calculated	2,000,000,000	not calculated
Government regulations to prevent damage	not calculated	470,000,000	not calculated
Total	not calculated	9,645,000,000	not calculated

And the total cost is ...

A number of attempts have been made to calculate the real cost of pesticides (see, for example, Pimentel et al, 1992; Bailey et al, 1999; Pretty et al, 2000, 2001; Tegtmeier and Duffy, 2004). Rather than review this literature in its entirety, I will conclude this section by briefly discussing two well-cited estimates, by Leach and Mumford (2008) and Pimentel (2005). As Table 7.3 illustrates, Leach and Mumford's analysis looks at the UK, the US and Germany, while Pimentel's focuses solely on the US. Pimentel's calculation is more inclusive than Leach and Mumford's, which explains his considerably larger cost estimate for the case of the US – a whopping US$9.6 billion annually. Leach and Mumford, however, include in their analysis (at least for the UK) costs that result from lost culture, landscape and tourism (e.g. a pesticide-laced lake can't be fished, which could negatively affect tourism and cultures dependent upon fishing), something missed in Pimentel's calculation.

When tallying up the costs and benefits of pesticides, we would do well to not decide policy based on the column with the largest figure. Earlier I mentioned how each US dollar invested in pesticide control returns about US$4 in protected crops, which calculates out to roughly US$40 billion per year. From a strict (which is another way of saying 'narrowly defined') cost–benefit approach, pesticides handily come out on top. Yet, the moment issues of public health get thrown into the mix, cost–benefit approaches become saturated with thorny ethical questions. As Pimentel (2005, p246) explains: 'assuming that pesticide-induced cancers numbered more than 10,000 cases year and that pesticides returned a net agricultural benefit of $32 billion year, each case of cancer is "worth" $3.2 million in pest control'. Is US$3.2 million in pesticide benefits worth one person falling victim to cancer? Clearly not if you're one of the victims. But even those spared from this disease must ask if similar ends can be achieved without the use of these dangerous chemicals. If we can successfully use agro-ecological (internal) controls in place of pesticides (see, for example, Shennan et al, 2005), then one cancer victim for each US$3.2 million in pesticide benefits seems a terribly high price to pay for our alleged 'cheap' food.

Putting a Price on Meat

What about meat? Would we be stunned if faced with having to pay the real price for our meat? Yes, though the level of shock would vary depending upon how inclusive one was when tallying up the costs.

An example of a more conservative cost estimate comes from researchers at VU University Amsterdam who recently attempted to calculate, in their words, the 'true price of pork' (van Drunen et al, 2010, p1). The unit they attempt to price is 1kg of pork. The authors conclude that:

- The average consumer price (per kilogram) of pork in 2008 was 6.69 Euros. If the externalized costs of climate change were added to the selling price of conventional pork, a conservative estimate would place the total price at 6.87 Euros.
- The external costs for cheap pork upon biodiversity (excluding the effects on climate change) are at least 0.44 Euros per kilogram.
- If internalized, the costs due to pig disease would add 0.32 Euros per kilogram (it is noted that due to being unable to quantify and value global resistance to antibiotics and flu epidemics, this is probably a very conservative estimate).

- Dutch subsidies, if internalized, would add 0.02 Euros per kilogram of meat.
- The total external costs for conventional pork are estimated to be at least 2.06 Euros per kilogram. If internalized, this would increase the per kilogram price of pork by 31 per cent to 8.75 Euros per kilogram.

After supplying this 'conservative estimate' (van Drunen et al, 2010, p17), the authors suggest the introduction of a Pigovian Tax – a tax applied to a market activity that generates a negative externality. They recommend a tax on conventional pork of at least 2.06 Euros per kilogram.

Raj Patel famously places the real price of one hamburger at over US$200 (see also Dune, 1994).[4] Assuming the beef came from a pasture that was once a rainforest – an assumption that admittedly mischaracterizes many of the world's hamburgers – this figure includes, among other things, lost ecological services provided by that rainforest such as the loss of diversity, carbon sequestration and water cycling. Elsewhere, Patel (2009, pp43–44) places the energy cost of the 550 million Big Macs sold in the US every year at US$297 million (producing a greenhouse gas footprint of 2.66 billion pounds of CO_2 equivalent).

Yet, even this estimate is incomplete since it misses the 'costs' paid by the animal (the currency: suffering). Over the last decade, increasing attention has turned towards the thorny question: how do we value animal welfare? For one thing, since animals cannot literally tell us what they value – though, as I've detailed elsewhere (Carolan, 2011; see also Chapter 5), this does not mean they can't 'talk' to us – it is difficult to know which impacts are important for animal welfare as well as how to translate those impacts into monetary figures. For example, consider a move from total confinement to a free-range animal production system. Such a switch would probably improve certain aspects of animal welfare by, among other things, giving the animal greater opportunity to perform natural behaviours. But other aspects of welfare might be compromised as a result of this switch. An animal's health and safety, for example, could be negatively affected due to having less control over predation, thermal and environmental conditions, and the spread of disease (Mench et al, 2009, p31). Through this last point, one could equally make the argument that greater variation in thermal and environmental conditions actually *improves* an animal's welfare as these conditions 'fit' animals' physiology. Take chickens. As mentioned in Chapter 5, chickens are not 'built' for the type of stable environments typically found in total confinement facilities and might therefore actually 'prefer' less predictable (and, thus, less safe) conditions.

Some economists, building upon the works of utilitarian philosophers such as Jeremy Bentham and Peter Singer, have suggested creative ways to think about willingness-to-pay valuation models that consider the well-being of both humans *and* animals (see, for example, Lusk and Norwood, 2009). Matthews and Ladewig (1994), for example, measured pigs' willingness to work – as defined by having the animal press on a nose-plate – for either more food or social contact with other pigs. Assuming that this effort of pressing a nose-plate is analogous to price, they discover that the average demand for food among the pigs studied was inelastic, while demand for social contract was elastic. In other words, the animals were willing to 'pay', through effort, almost anything for additional food, whereas their willingness to pay for social contact was noticeably more limited. Pointing to this research, Lusk and Norwood (2009, p7) propose the following thought experiment: suppose we approach a pig and ask it the amount of corn that it would be willing to give up in exchange for a 1 square foot increase in pen space? An answer to this question could be discerned through the construction of a similar willingness-to-pay experiment as that described above, except instead of (nose-plate) effort, the unit of payment might be feed.

Other studies have focused on poultry. In one experiment, researchers attempted to assess the 'importance' that caged laying hens placed on straw and feather flooring (Gunnarsson et al, 2000). The results suggest a considerable willingness to pay among all layers studied for straw litter substrates, whereas half the birds expressed demand for feathers. In another experiment, individually tested hens pushed a weighted door to enter one of four areas containing either wire floor, sand, wood shavings or peat moss as substrate (de Long et al, 2007). The experiment sought to better understand the substrate preferences of laying hens, particularly with respect to dust bathing and foraging behaviour. The findings indicate that the value of a particular substrate varies according to its suitability for the expression of specific behaviours, though a particularly strong 'demand' was shown for peat moss among birds interesting in dust bathing.

Farmers are understandably concerned that more strict animal welfare standards will cost them in the long run because of a lack of consumers' willingness to pay for more humanely obtained animal products. In some cases, more strict welfare standards will increase production costs as a result of higher input costs such as labour and feed and lower 'output' due to reductions in stocking density (Bornett et al, 2003). Data from the EU suggests that reducing the stocking density of broilers by 20 per cent would increase costs by about 5 per cent, while reducing stocking density by 35 per cent would increase costs by approximately 15 per cent (Moynagh, 2000, pp108–109), costs, at least in the EU, which are borne fully by the producer as increased space for broilers is not associated with improvements in overall productivity as long as the confinement faculties are well ventilated (Mench et al, 2009, p33). In other cases, more humane production models compete quite well against total confinement operations (see Box 7.3). Changing animal husbandry practices to improve animal welfare can also increase productivity, such as through improved feed efficiency or decreased injury (Gonyou and Stricklin, 1998; Wolter et al, 2000). A study in The Netherlands, for example, finds that foot disorders among dairy cows in total confinement facilities results in an annual loss of US$75 per cow due to diminished productivity and costs associated with treatment (Bruijnis et al, 2010).

Box 7.3 Hoop house versus total confinement

The Moultons' farm comprises 200 acres (80ha) of corn, 80 acres (32ha) of beans, 30 acres (12ha) of oats and alfalfa, and 75 acres (30ha) of pasture in eastern Minnesota.[5] In addition, they oversee a 40-cow beef operation and a 120-sow farrow-to-finish operation, sending to market between 2000 and 3000 hogs a year. For many years they managed a standard confinement farrowing unit. That stopped when they started hearing complaints about odour and began worrying about their manure pit contaminating surface and groundwater. In 1995 they erected their first hoop house, which consists of a tarpaulin-covered arched metal frame – resembling a Quonset hut (Nissen hut) – rising up over a clay or earthen floor, slightly sloped away from the raised concrete platform at one end of the structure that supports the feeding and watering units. Hoop farming is generally believed to enhance animal welfare as it offers pigs access to the outdoors, shade, shelter, exercise areas, fresh air and direct sunlight (Gegner, 2005, p8). And as Table 7.4 illustrates, from the standpoint of pocketbook economics, hoop farming enhanced the Moultons' welfare too (see also Kliebenstein et al, 2003). In 1997, the Moultons were named Chisago County (Minnesota) Farm Family of the Year.

Table 7.4 *Comparison between total confinement and hoop house production systems*

Cost	6 hoop buildings (3060 hogs per year)	1000 head confinement (3000 hogs per year)
Building cost	US$60,000	US$180,000
Feed cost per pig	US$43	US$40
Utilities (electric and heat)	US$0	US$1500 (US$0.50 per pig)
Death loss	3.15 percent	3.02 percent
Cost per pig	US$102.60	US$140.50

Developed from www.p2pays.org/ref/21/20975.htm and Kliebenstein (2003)

Studies on willingness to pay for animal welfare suggest that there is a demand for animal welfare-certified food (Bennett, 1997; Rolfe, 1999; Bennett and Blaney, 2003; Lagerkvist et al, 2006; Carlsson et al, 2007; Liljenstolpe, 2008; Nocella et al, 2010). Bennett and Blaney (2003, p88) explain that when UK citizens were asked the maximum that they would be willing to pay in terms of an increase in their weekly household food bill, just 9 per cent expressed a zero willingness to pay. The remainder stated a positive amount between UK£0.10 and £100, with an average of UK£5.50. We also know that the more people learn about animal welfare, the more they are willing to pay for humanely raised products (or the more likely they are to exit meat consumption entirely) (Tonsor and Olynk, 2011).

I am less concerned about what consumers say they are willing to pay for meat that's affordable. Frankly, as long as 'cheap' meat is available, I expect many will continue to buy it. But once we have moved away from cheap meat, do we really *want* consumers to continue eating meat at current (and projected future) levels? The point is to sell animal products at a price that more accurately reflects its real costs, which probably means selling *less* meat precisely because the real cost of cheap meat is outrageously high. I also care about the welfare of producers. I want to ensure that livestock farmers get a fair pay cheque for a good day's work – unlike now, as discussed in Chapter 9, where a fair pay cheque eludes most. We can do this, however, without having to ensure – to play upon an old political slogan – that there's a chicken in every pot and a side of beef in every deep freeze.

While meat consumption in the US has risen slowly over the past 40 years, its impact upon the pocketbook has dropped precipitously since 1970, falling from 4.1 per cent of the average American's annual disposable income to 1.6 per cent in 2008 (Eng, 2010). Yet, during that same period the livestock industry's impact upon the environment has grown exponentially. This is not to say that meat should only be the providence of the wealthy. I am just as much in favour of equalizing meat consumption among the classes as cheap food proponents are. The difference is, however, whereas they want to make the world in the image of the average (meat-crazed) US consumer, I would rather see meat consumption stabilized (and equalized) at a more sustainable level. If we honestly believe meat consumption to be a 'right' – which is an argument I hear occasionally – then perhaps we ought to develop policy instruments that *fairly* enshrine that claim. We could, for example, subsidize a predetermined (affordable) level of animal protein for those least likely to afford it after all its costs are internalized. The issue of 'affordable meat' is taken up further in Chapter 10.

Notes

1   The authors further calculate that a wholesale switch in the UK to organic production could lead to avoided costs of US$1.8 billion, thus cutting agriculture's ecological debt in that country to US$600 million.

2   A number of explanations have been given for why we continue to divorce the economy from the environment: some social-psychological (e.g. prior to the 20th century it was easy to miss the cumulative impact we were having upon the environment; see Boulding, 1966); others socio-structural (e.g. powerful interests kept us from seeing the cumulative impact we were having upon the environment; see Mishan, 1974).

3   To get a sense of what is meant by the term 'service', take watershed services. Among other things, watersheds provide water for drinking, agriculture, the household and industry; allow for hydropower generation, transportation and shipping; store water in soils, wetlands and floodplains, thereby reducing risks associated with flood; and maintain cultural values that support tourism, recreation and traditional ways of life (McNeely, 2009, p141).

4   See interview at www.foodfirst.org/en/node/2963.
5   Case taken from www.p2pays.org/ref/21/20975.htm.

References

Allsopp, M., W. de Lange and R. Veldtman (2008) 'Valuing insect pollination services with cost replacement', *PLoS One*, vol 3, no 9, pp1–8

Altieri, M. (1995) *Agroecology: The Science of Sustainable Agriculture*, Westview Press, Boulder

Altieri, M. (1999) 'The ecological role of biodiversity in ecosystems', *Agriculture, Ecosystems and the Environment*, vol 75, no 1–3, pp19–31

Antle, J. and S. Capalbo (1994) 'Pesticides, productivity, and farmer health: Implications for regulatory policy and agricultural research', *American Journal of Agricultural Economics*, vol 76, no 3, pp598–602

Bailey, A. P., T. Rehman, J. Park, J. D. H. Keatunge and R. B. Trainter (1999) 'Towards a method for the economic evaluation of environmental indicators for UK integrated arable farming systems', *Agriculture, Ecosystems and Environment*, vol 72, pp145–158

Bennett, R. (1997) 'Farm animal welfare and food policy', *Food Policy*, vol 22, no 4, pp281–288

Bennett, R. and R. Blaney (2003) 'Estimating the benefits of farm animal welfare legislation using the contingent valuation method', *Agricultural Economics*, vol 29, no 1, pp85–98

Berny P., F. Caloni, S. Croubels, M. Sachana, V. Vandenbroucke, F. Davanzo and R. Guitart (2010) 'Animal poisoning in Europe, Part 2: Companion animals', *The Veterinary Journal*, vol 183, no 3, pp255–259

Bornett, H., J. Guy and P. Cain (2003) 'Impact of animal welfare on costs and viability of pig production in the UK', *Journal of Agricultural and Environmental Ethics*, vol 16, pp163–186

Boulding, K. (1966) 'The economics of the coming Starship Earth', in H. Jarett (ed) *Environmental Quality in a Growing Economy*, Johns Hopkins Press, Baltimore, MD, pp4–14

Bremner, B. and N. Lakshman (2006) 'Behind the Coke-Pepsi pesticide scare', *Business Week*, 24 August, www.businessweek.com/globalbiz/content/aug2006/gb20060824_932216.htm, last accessed 1 December 2010

Bruijnis, M., H. Hogeveen and E. Stassen (2010) 'Assessing economic consequences of foot disorders in dairy cattle using a dynamic stochastic stimulation model', *Journal of Dairy Science*, vol 93, no 6, pp2419–2432

Buttel, F. (1993) 'Socioeconomic impacts and social implications of reducing pesticide and agricultural chemical use in the United States', in D. Pimentel and H. Lehman (eds) *The Pesticide Question: Environment, Economics and Ethics*, Chapman and Hall, New York, NY, pp151–181

Butterworth, C. (1974) 'The coal-powered heart', *JAMA*, vol 227, no 8, pp934–935

Carlsson F., P. Frykblom and C. Lagerkvist (2007) 'Consumer willingness-to-pay for farm animal welfare: Mobile abattoirs versus transportation to slaughter', *European Review of Agricultural Economics*, vol 34, no 3, pp321–344

Carolan, M. (2006a) 'Sustainable agriculture, science, and the co-production of "expert" knowledge: The value of interactional expertise', *Local Environment*, vol 11, pp421–431

Carolan, M. (2006b) 'Social change and the adoption and adaptation of knowledge claims: Whose truth do you trust in regard to sustainable agriculture?', *Agriculture and Human Values*, vol 23, pp270–285

Carolan, M. (2006c) 'Scientific knowledge and environmental policy: Why science needs values', *Environmental Sciences: The Journal of Integrative Environmental Research*, vol 3, pp229–237

Carolan, M. (2006d) 'The values and vulnerabilities of metaphors within the environmental sciences', *Society and Natural Resources*, vol 19, pp921–930

Carolan, M. (2008) 'Democratizing knowledge: Sustainable and conventional agricultural field days as divergent democratic forms', *Science, Technology and Human Values*, vol 33, no 4, pp508–528

Carolan, M. (2010a) *Decentering Biotechnology: Assemblages Built and Assemblages Masked*, Ashgate, Burlington, VT

Carolan, M. (2010b) 'Sociological ambivalence and climate change', *Local Environment*, vol 15, no 4, pp309–312

Carolan, M. (2011) *Embodied Food Politics*, Ashgate, Burlington, VT

Carrasco-Tauber, C. (1989) *Pesticide Productivity Revisited*, MSc thesis, University of Massachusetts, Amherst, MA

Carson, R. (1962) *Silent Spring*, Houghton Mifflin, New York, NY

Cook, D., M. Thomas, S. Cunningham, D. Anderson and P. De Barro (2007) 'Predicting the economic impact of an invasive species on an ecosystem service', *Ecological Applications*, vol 17, no 6, pp1832–1840

Cooper, J. and H. Dobson (2007) 'The benefits of pesticides to mankind and the environment', *Crop Protection*, vol 26, no 9, pp1337–1348

Costanza, R., R. d'Arge, R. de Groot, S. Farber, M. Grasso, B. Hannon, S. Naeem, K. Limburg, J. Paruelo, R. O'Neill, R. Raskin, P. Sutton and M. van den Belt (1997) 'The value of the world's ecosystem services and natural capital', *Nature*, vol 387, pp253–260

Dasgupta, P. (2006) 'Nature in economics', *Environmental Resource Economics*, vol 39, pp1–7

de Long, I., M. Wolthuis-Fillerup and C. van Reenen (2007) 'Strength of preference for dustbathing and foraging substrates in laying hens', *Applied Animal Behavior Science*, vol 104, no 1–2, pp24–36

Desneux, N., A. Decourtye and J.-M. Delpuech (2007) 'The sublethal effects of pesticides on beneficial arthropods', *Annual Review of Entomology*, vol 52, pp81–106

Diamond, J. (2006) *Collapse: How Societies Choose to Fail or Succeed*, Penguin, New York, NY

Dune, N. (1994) 'Why a hamburger should cost 200 dollars – the call for prices to reflect ecological factors', *Financial Times*, 12 January, pp12–14

Eng, M. (2010) 'Real cost of low-price meat can't be found in the Food Bill', *Chicago Tribune*, 3 October, www.pressherald.com/business/real-cost-of-low-price-meat-cant-be-found-in-the-food-bill_2010-10-03.html, last accessed 29 November 2010

Evenson, R. and D. Gollin (2003) 'Assessing the impact of the Green Revolution, 1960 to 2000', *Science*, vol 300, pp758–762

Fakih, M., T. Rahardjo, and M. Pimbert (2003) *Community Integrated Pest Management in Indonesia: Institutionalising Participation and People Centred Approaches*, International Institute for Environment and Development (IIED) and Institute for Development Studies (IDS), Russell Press, Nottingham

Freidberg, S. (2009) *Fresh: A Perishable History*, Harvard University Press, Cambridge, MA

Garming, H. and H. Waibel (2009) 'Pesticides and farmer health in Nicaragua: A willingness-to-pay approach to evaluation', *The European Journal of Health*

Economics, vol 10, no 2, pp125–133

Gegner, L. (2005) *Hooped Shelters for Hogs*, National Sustainable Agriculture Information Service, http://attra.ncat.org/attra-pub/PDF/hooped.pdf, last accessed 29 November 2010

Gliessman, S. (1998) *Agroecology: Ecological Processes in Sustainable Agriculture*, CRC Press, Boca Raton, FL

Gonyou, H. and W. Stricklin (1998) 'Effects of floor area allowance and group size on the productivity of growing/finishing pigs', *Journal of Animal Science*, vol 76, pp1326–1330

Gunnarsson, S., L. Matthews, T. Foster and W. Temple (2000) 'The demand for straw and feathers as litter substrates by laying hens', *Applied Animal Behavior Science*, vol 65, no 4, pp321–330

Hein, L., K. van Koppenb, R. S. de Groota, and E. C. van Ierland (2006) 'Spatial scales, stakeholders and the valuation of ecosystem services', *Ecological Economics*, vol 57, pp209–228

Jackson, T. (2009) *Prosperity without Growth: Economics for a Finite Planet*, Earthscan, London

Kishi, M. (2005) 'The health impacts of pesticides: What do we now know?', in J. Pretty (ed) *The Pesticide Detox*, Earthscan, London, pp23–38

Kishi, M., N. Hirschhorn, M. Djajadisastra, L. Satterlee, S. Strowman and R. Dilts (1995) 'Relationship of pesticide spraying to signs and symptoms in Indonesian farmers', *Scandinavian Journal of Work, Environment and Health*, vol 21, pp124–133

Kliebenstein, J., B. Larson, M. Honeyman and A. Penner (2003) *A Comparison of Production Costs, Returns and Profitability of Swine Finishing Systems*, Working Paper, March, Department of Economics, Iowa State University, Ames, IA, https://www.econ.iastate.edu/sites/default/files/publications/papers/p3819-2003-03-14.pdf, last accessed 23 November 2010

Knutson, R., C. Taylor, J. Penson and E. Smith (1990) *Economic Impacts of Reduced Chemical Use*, Knutson and Associates, College Station, TX

Kroeger, T., F. Casey (2007) 'An assessment of market-based approaches to providing ecosystem services on agricultural lands', *Ecological Economics*, vol 64, pp321–332

Lagerkvist, C., F. Carlsson and D. Viske (2006) 'Swedish consumer preferences for animal welfare and biotech: A choice experiment', *AgBioForum*, vol 9, no 1, pp51–58

Leach, A. and J. Mumford (2008) 'Pesticide environmental accounting: A method for assessing the external costs of individual pesticide applications', *Environmental Pollution*, vol 151, pp139–147

Levine, M. (2003) *Children for Hire: The Perils of Child Labor in the United States*, Greenwood Publishing, Westport, CT

Liljenstolpe, C. (2008) 'Evaluating animal welfare with choice experiments: An application to Swedish pig production', *Agribusiness*, vol 24, pp67–84

Lusk, J. and B. Norwood (2009) *Speciesism, Altruism, and the Economics of Animal Welfare*, Policy Paper, Oklahoma State University, Tulsa, OK, http://asp.okstate.edu/baileynorwood/FAW/files/PolicyPaper.pdf, last accessed 26 November 2010

Maser, C. (2009) *Earth in Our Care: Ecology, Economy, and Sustainability*, Rutgers University Press, Piscataway, NJ

Matthews, L. and J. Ladewig (1994) 'Environmental requirements of pigs measured by behavioural demand functions', *Animal Behavior*, vol 47, pp713–719

Mayer, B., J. Flocks and P. Monaghan (2010) 'The role of employers and supervisors in promoting pesticide safety behavior among Florida farmworkers', *American Journal of Industrial Medicine*, vol 53, no 8, pp814–824

McNeely, J. (2009) 'Payments for ecosystem services: An international perspective', in K. Ninan (ed) *Conserving and Valuing Ecosystem Services and Biodiversity*, Earthscan, London, pp135–150

Mench, J., H. James and P. Thompson (2009) *The Welfare of Animals in Concentrated Animal Feeding Operations*, UN Food and Agriculture Organization, Animal Production and Health Division, Rome, Italy, www.fao.org/fileadmin/user_upload/animalwelfare/Welfare%20of%20Animals%20in%20Concentrated%20Animal%20Feeding%20Operations1.doc, last accessed 25 November 2010

Millennium Ecosystem Assessment (2005) *Ecosystems and Human Well-being: Synthesis*, Island Press, Washington, DC

Mills, P. and S. Kwong (2001) 'Cancer incidence in the United Farm Workers of America (UFW), 1987–1997', *American Journal of Industrial Medicine*, vol 40, no 5, pp596–603

Mishan, E. (1974) 'What is the optimal level of pollution?', *Journal of Political Economy*, vol 82, no 6, pp1287–1299

Moynagh, J. (2000) 'EU regulation and consumer demand for animal welfare', *AgBioForum*, vol 3, no 2–3) pp107–114, www.agbioforum.org/v3n23/v3n23a06-moynagh.htm, last accessed 15 March 2011

Murray, D. (1994) *Cultivating Crisis: The Human Cost of Pesticides in Latin America*, University of Texas Press, Austin, TX

Murray, D. and P. Taylor (2000) 'Claim no easy victories: Evaluating the pesticide industry's global safe use campaign', *World Development*, vol 28, no 10, pp1735–1749

Nocella, G., L. Hubbard and R. Scarpa (2010) 'Farm animal welfare, consumer willingness to pay, and trust: Results of a cross-national survey', *Applied Economic Perspectives and Policy*, vol 32, no 2, pp275–297

Ostrom, E. (1999) *Governing the Commons: The Evolution of Institutions for Collective Action*, Cambridge University Press, New York, NY

Pascual, U. and C. Perrings (2009) 'Developing mechanisms for in situ biodiversity conservation in agricultural landscapes', in K. Ninan (ed) *Conserving and Valuing Ecosystem Services and Biodiversity*, Earthscan, London, pp151–174

Patel, R. (2009) *The Value of Nothing: How to Reshape Market Society and Redefine Democracy*, Picador, New York, NY

Pimentel, D. (2005) 'Environmental and economic costs of the application of pesticides primarily in the United States', *Environment, Development and Sustainability*, vol 7, pp229–252

Pimentel, D. (2009) 'Pesticides and pest control', in R. Peshin and A. Dhawn (eds) *Integrated Pest Management*, Springer, The Netherlands, pp83–111

Pimentel, D., L. E. Hurd, A. C. Bellotti, M. J. Forster, I. N. Oka, O. D. Sholes and R. J. Whitman (1973) 'Food production and the energy crisis', *Science*, vol 182, no 4111, pp443–449

Pimentel, D., J. Krummel, D. Gallahan, J. Hough, A. Merrill, I. Schreiner, P. Vittum, F. Koziol, E. Back, D. Yen and S. Fiance (1978) 'Benefits and costs of pesticide use in United States food production', *BioScience*, vol 28, no 772, pp778–784

Pimentel, D., H. Acquay, M. Biltonen, P. Rice, M. Silva, J. Nelson, V. Lipner, S. Giordano, A. Horowitz and M. D'Amore (1992) 'Environmental and economic costs of pesticide use', *Bioscience*, vol 42, no 10, pp750–760

Polanyi, K. (2001:1944) *The Great Transformation*, Beacon, Boston, MA

Pretty, J. (1998) 'Supportive policies and practice for scaling up sustainable agriculture', in N. G. Roling and M. A. Wagemakers (eds) *Facilitating Sustainable Agriculture: Participatory Learning and Adaptive Management in Time of Environmental Uncertainty*, Cambridge University Press, New York, NY, pp23–45

Pretty, J. and H. Waibel (2005) 'Paying the price: The full cost of pesticides', in J. Pretty (ed) *The Pesticide Detox*, Earthscan, London, pp39–54

Pretty, J., C. Brett, D. Gee, R. Hine, C. Mason, J. Morison, H. Raven, M. Rayment and G. van der Bijl (2000) 'An assessment of the total external costs of UK agriculture', *Agricultural Systems*, vol 65, pp113–136.

Pretty, J., C. Brett, D. Gee, R. E. Hine, C. F. Mason, J. I. L. Morison, M. Rayment, G. van der Bijl and T. Dobbs (2001) 'Policy challenges and priorities for internalising the externalities of agriculture', *Journal of Environmental Planning and Management*, vol 44, no 2, pp263–283

Pretty, J., A. Ball, T. Lang and J. Morison (2005) 'Farm costs and food miles: An assessment of the full cost of the UK weekly food basket', *Food Policy*, vol 30, pp1–19

Purdue University (1975) *High-Quality Protein Maize*, Dowden, Hutchinson and Ross, PA

Richter, E. D. (2002) 'Acute human pesticide poisonings', in D. Pimentel (ed) *Encyclopedia of Pest Management*, vol 1, Dekker, New York, NY, pp3–6

Rolfe, J. (1999) 'Ethical rules and the demand for free range eggs', *Economic Analysis and Policy*, vol 29, no 2, pp187–206

Robertson, G. P. and S. Swinton (2005) 'Reconciling agricultural productivity and environmental integrity: A grand challenge for agriculture', *Frontiers in Ecology and the Environment*, vol 3, no 1, pp38–46

Röling, N. and E. van de Fliert (1998) 'Introducing integrated pest management in rice in Indonesia: A pioneering attempt to facilitate large-scale change', in N. G. Roling and M. A. Wagemakers (eds) *Facilitating Sustainable Agriculture: Participatory Learning and Adaptive Management in Time of Environmental Uncertainty*, Cambridge University Press, New York, NY, pp153–171

Sen, A. (1989) 'Economic methodology: Heterogeneity and relevance', *Social Research*, vol 56, pp299–330

Settle, R. (1983) 'Evaluating the economic benefits of pesticide usage', *Agriculture, Ecosystems and the Environment*, vol 9, no 2, pp173–185

Shah, S. A. (2010) 'Slackness in cotton market may extend into New York', *Business Recorder*, 29 November, www.brecorder.com/news/cotton-and-textiles/pakistan/1128754:news.html, last accessed 1 December 2010

Shennan, C., T. Gareau and R. Sirrine (2005) 'Agroecological approaches to pest management in the US', in J. Pretty (ed) *The Pesticide Detox*, Earthscan, London, pp193–211

Soares, W. and M. Porto (2009) 'Estimating the social cost of pesticide use: An assessment from acute poisoning in Brazil', *Ecological Economics*, vol 68, pp2721–2728

Spangenberg, J. and J. Settele (2010) 'Precisely incorrect? Monetising the value of ecosystem services', *Ecological Complexity*, vol 7, no 3, pp327–337

Stocking, M. (2003) 'Tropical soils and food security: The next 50 years', *Science*, vol 302, no 5649, pp1356–1359

Tegtmeier, E. M. and M. D. Duffy (2004) 'External costs of agricultural productivity in the United States', *International Journal of Agricultural Sustainability*, vol 2, pp1–20.

Thornes, J., W. Bloss, S. Bouzarovski, X. Cai, L. Chapman, J. Clark, S. Dessai, S. Du,

D. van der Horst, M. Kendall, C. Kidda and S. Randalls (2010) 'Communicating the value of atmospheric services', *Meteorological Applications*, vol 17, pp243–250

Tonsor, G. and N. Olynk (2011) 'Impacts of animal well being and welfare media on meat demand', *Journal of Agricultural Economics*, vol 62, no 1, pp59–72

Van den Bosch, R. (1978) *The Pesticide Conspiracy*, Doubleday, Garden City, NY

Van Der Hoek, W., F. Konradsen, K. Athukorala and T. Wanigadewa (1998) 'Pesticide poisoning: A major health problem in Sri Lanka', *Social Science and Medicine*, vol 46, no 4, pp495–504

van Drunen, M., P. van Beukering and H. Aiking (2010) *The True Price of Meat*, Report W10/02aEN, 4 May, Institute for Environmental Studies, VU University, Amsterdam, The Netherlands, www.ivm.vu.nl/en/Images/haW10-02aEN%20The%20true%20price%20of%20meat_tcm53-161385.pdf, last accessed 23 November 2010

Vedwan, N. (2007) 'Pesticides in Coca-Cola and Pepsi: Consumerism, brand image, and public interest in a globalizing India', *Cultural Anthropology*, vol 22, no 4, pp659–684

Wesseling, C., L. Castillo and C. Elinder (1993) 'Pesticide poisonings in Costa Rica', *Scandinavian Journal of Work, Environment and Health*, vol 19, no 4, pp227–235

Williamson, S. (2005) 'Breaking the barriers to IPM in Africa: Evidence from Benin, Ethiopia, Ghana, and Senegal', in J. Pretty (ed) *The Pesticide Detox*, Earthscan, London, pp165–180

Wolter, B., M. Ellis, S. Curtis, E. Parr and D. Webel (2000) 'Group size and floor space allowance can affect weanling pig performance', *Journal of Animal Science*, vol 78, pp2062–2067

Zhang, W., T. Ricketts, C. Kremen, K. Carney, and S. Swinton (2007) 'Ecosystem services and dis-services to agriculture', *Ecological Economics*, vol 64, no 2, pp253–260

8

Cheap Food, Community and Culture

The relationship between industrialized agriculture and community well-being has been extensively studied. Social scientists have been examining this relationship since the 1930s (see, for example, Tetreau, 1938; Luloff and Krannich, 2002). This expansive body of research has been meticulously reviewed elsewhere (see Lobao and Stofferahn, 2008; see also Swanson, 1988; Lobao, 1990). I will therefore be brief as I summarize this body of literature that spans multiple generations.

The most famous study to look at the costs of industrial agriculture on community well-being was conducted almost three-quarters of a century ago. Overseen by Walter Goldschmidt, a US Department of Agriculture (USDA) anthropologist, and funded by the USDA, the research centres on two California communities during the early 1940s: Arvin, where large absentee-owned, non-family-operated farms were more numerous, and Dinuba, where locally owned, family-operated farms were the norm (the names of both towns are pseudonyms). The final report documents the relationship between industrial agriculture and its negative impact upon a variety of community quality-of-life indicators (Goldschmidt, 1978). Specifically, it notes that, relative to Dinuba, Arvin's population had a smaller middle class, a higher proportion of hired workers, lower mean family incomes, higher rates of poverty, poorer quality schools, and fewer churches, civic organizations and retail establishments. Residents of Arvin also had less local control over public decisions due to disproportional political influence by outside agribusiness interests.

Research has since largely verified the Goldschmidt thesis, in addition to expanding upon the socio-economic costs associated with industrial agriculture. Table 8.1 lists some of those costs, as well as selected studies that speak specifically of them. Not all of the studies find clear links between industrial farming and negative consequences to community well-being. Those that do not, however, generally fail to find any statistical relationship. In other words, they neither support nor refute the Goldschmidt thesis (Lobao and Stofferahn, 2008, p226).

Table 8.1 *Costs to rural communities as farms industrialized*

Greater Inequality
• e.g. Goldschmidt (1978) and Heady and Sonka (1974)

Higher rates of poverty
• e.g. Lobao (1990) and Peters (2002)

Higher unemployment rates
• e.g. Skees and Swanson (1988) and Peters (2002)

Decline in population base
• e.g. Heady and Sonka (1974) and Rodefeld (1974)

Increase in crime rates and civil suits
• e.g. NCRCRD (1999)

Increase in social psychological problems (e.g. depression, stress)
• e.g. Martinson et al (1976) and Schiffman et al. (1998)

Decline in social interaction and trust
• e.g. Constance and Tuinstra (2005) and Smithers et al (2004)

Decline in local control over public decisions
• e.g. Goldschmidt (1978) and McMillan and Schulman (2003)

Decline in civic participation
• e.g. Heffernan and Lasley (1978) and Smithers et al (2004)

Decline in quality of public services (e.g. education)
• e.g. Fujimoto (1977) and Goldschmidt (1978)

Decrease in retail trade and fewer/less diverse retail shops
• e.g. Skees and Swanson (1988) and Smithers et al (2004)

Complied with help from Stofferahn (2006, pp18–19) and Lobao and Stofferahn (2008, pp224–225)

It is important to clarify that 'industrial agriculture' in these studies is usually based on indicators of farm organization and not simply scale. Scale is a measure of an operation's sales or acreage. But scale alone does not capture the organizational characteristics of industrial agriculture (Lobao and Stofferahn, 2008, p224). Features such as absenteeism (where land is leased and the owners live outside the community), contract farming, dependency on hired labour and operation by farm managers (as opposed to owner-operator situations) are far more likely to place communities at risk than if surrounding farms lack these characteristics but happen to be large.

The large-scale family farmers whom I know purchase inputs, farm equipment and services (e.g. custom baling) locally, volunteer their time with local organizations (church groups, school fundraisers, etc.) and regularly interact with, and thus place significant trust in, their neighbours. Granted, more large-scale farms, by definition, means fewer farmers overall. But disappearing farms are only part of the problem. A farm's organizational structure also matters. Is the operator also the farm's owner? Are inputs and equipment brought in from outside the community? How these questions are answered have effects that can be felt beyond the farm gate. The small rural town of roughly 400 people that I grew up in was (and still is) incredibly dynamic, even though the farms surrounding it continue to get larger and fewer in number with each passing year. I think the fact that all the farms, to the best of my knowledge, are owner-operated family farms has a lot to do with my hometown's 'success'. Many rural communities have not been as fortunate.

Biodiversity, Culture and Resilience

While housed in sociology, the classes I teach at the undergraduate and graduate levels focus, to various degrees, on environmental and agricultural issues. Biodiversity is covered extensively in each. The discussions are always lively, as well as substantive. Students tell me that this above-average level of engagement is due to having already talked about biodiversity in other classes. Many 'get' biodiversity – they know what it is and why it needs saving – and are therefore comfortable talking about it.

Eventually the subject of cultural diversity enters the discussion. 'What's its relationship to biodiversity?' I ask. Silence. Students generally express sentiments that cultural diversity is worth saving. But this moral duty is usually rooted in some undefined principle about showing respect for other cultures – a type of 'treat other cultures as you would like your own traditions and practices to be treated' maxim. Biodiversity rarely figures in students' justifications for protecting cultural diversity. The relationship between cultural and biological diversity is not at all obvious. The research, however, is becoming increasingly clear: you cannot have one without the other – or at least not without creating a highly impoverished imposture (see, for example, Maffi and Woodley, 2010; Pilgrim and Pretty, 2010).

It is no coincidence that most of the world's biodiversity hotspots are also cultural hotspots, represented through the density of divergent ethnic groups, linguistic diversity and a multitude of cultural practices and folk knowledge (Sutherland, 2003). Looking at indigenous groups in North America, Eric Smith (2001, p95) finds a correlation between measures of biodiversity within a given space and measures of cultural and linguistic diversity. Studying Native American populations in the south-western portion of the US, Nabhan (1985, 1989) finds a high diversity of folk crop varieties and traditional farming techniques, all of which are maintained through cultural knowledge (Soleri and Cleveland, 1993, describe something similar in their study of the Hopi of northeast Arizona). A study by Veteto (2008) suggests that agro-biodiversity levels in southern Appalachia may even exceed that of many comparable regions in the Global South – a diversity that is inextricably linked to an equally diverse pool of cultural knowledge that provides the people in this region with the proper 'know-how' to raise the heirloom varieties that populate this rural section of the US.[1]

Like biodiversity, cultural diversity is under continual threat. Cultural monocultures: the result of, among other things, cultural assimilation, language and knowledge loss, generational forgetting, and the adoption of non-traditional practices and ways of life (Shiva, 1993; Maffi, 2005). Rural communities, and rural ways of life, are being lost; the transmission of traditional cultural knowledge is declining as younger generations become more likely to dismiss 'the old ways' of doing and knowing things; and agri-*culture* itself is flattening out as management practices from around the world are converging (Rigg, 2005). These trends can all be expressed as a cost to society, recognizing that any decline in knowledge and cultural practices is equally a reduction in the possible solutions that humanity holds to future global challenges (Pretty et al, 2008, p9) (see Box 8.1). Pretty and colleagues (2008, p9) describe this loss as follows:

> The combined loss of biodiversity and ecological knowledge has implications for human health in the future, as we stand to lose the opportunistic uses and future potential of species, for instance in curing human diseases or feeding growing populations. With this loss of knowledge, we are subsequently losing the adaptive management systems embedded in traditional cultures that have sustained natural resource pools through to the present day and may be a key tool in the future of global biodiversity protection.

Take the impending agricultural challenges associated with climate change. Crop scientists are looking to traditional crop varieties and planting techniques as they try to find ways of helping farmers keep up with climate variability. The tremendous genetic and cultural diversity associated with folk crop varieties and traditional agricultural systems provides far more potential responses to climate change than the monocultures – biological and cultural – that define today's commercial farming systems (Veteto and Skarbo, 2009, p77). In the context of food production, while cultural and biological diversity and climate change are rarely linked, their fates are intertwined. Bio-cultural diversity is reduced over the long term through climate change and yet is essential in creating resilience in our food system to a changing climate (Kotschi, 2007). Having said this, I admit that we must not dismiss the possibility that, in the short term, the effects of climate change *could* increase bio-cultural diversity in certain spaces as new ecological niches are created and colonized, especially by invasive species (Hellmann et al, 2008).

There's a saying that the Global North is technology rich but gene poor, while the Global South is technology poor but gene rich. While partially accurate (there

Box 8.1 Local knowledge and food independence in Cuba

The collapse of the Soviet Union in 1989 left Cuba in a precarious position. Not only did Cuba lose a major trading partner, but it lost its main supplier of cheap fertilizers, pesticides, petroleum and grain (in 1989, immediately following the collapse, imports dropped overall by 75 per cent and oil imports by 53 per cent) (Warwick, 2001, p54). This proved disastrous for not only Cuban farmers, but also for the population as a whole as it left the country hungry, unable for the first time since the Cuban Revolution (30 years earlier) to feed itself (Perfecto et al, 2009, p65). While capital-intensive agriculture helped to feed the nation, it also made it heavily dependent upon (and thus vulnerable to) outside forces that were beyond anyone's control.

Searching for alternatives proved difficult as most farmers and agricultural 'specialists' knew only capital-intensive methods of farming – after all, it had been almost 30 years since conventional agriculture took the country by storm. Fortunately, a few traditional farmers, and their knowledge, remained. And with their help the process began of integrating scientific and traditional knowledge to make Cuba more food independent (Perfecto et al, 2009, p65). The transition has been so successful that in 1999 the *Grupo de Agricultura Organica* (GAO), a Cuban organic farming association, which has been at the forefront of the country's transition from industrial to organic agriculture, was awarded the prestigious Right Livelihood Award – or what is more commonly known as the Alternative Nobel Prize.

are, after all, remarkable concentrations of biological diversity in the developing world), this assessment subtly justifies a global division of labour between those with genes (the South) and those with the means to doing something valuable with this biological material (Nazarea, 2005, p57). The genetic material held within seeds, however, is not worth much without corresponding knowledge of the characteristics of the plant (such as its water, soil, nutrient and climate requirements) and a working knowledge of planting techniques. In many cases this knowledge is held in but one repository: the community from whence the genetic material came. Realizing this forces us to re-evaluate the view that the South is but a warehouse for the world's genetic resources.

The most resilient farming systems tend to be found on the margins, where the 'get big or get out' pressures are least felt (Nabhan, 1989, p73; Nazarea, 2005, pp9–11). In her research in Luzon, the Philippines, Nazarea (1998) shows the importance of cultural memory to agro-biodiversity conservation. She talks specifically of the value of 'memory banking' in conservation. Gene banks, which are often viewed as representing the quintessential solution to biodiversity conservation, are by definition only interested in certain things – namely, biological material. As it stands, the systematic collection of cultural data –

which has made possible the existence of biological data housed in today's gene banks – remains largely an afterthought in the conservation community. Memory banks would help to balance out this asymmetry in how we practise conservation. Thus, for instance, to combat diminished freshwater stores due to rising sea levels because of climate change, memory banks would provide a needed resource when seeking knowledge about plants with salinity tolerance. We would be unable to obtain this information looking at genes alone.

My own research has led me to similar conclusions. I have spent time studying a heritage seed bank in northeast Iowa – the Seed Savers Exchange (SSE) (Carolan, 2007, 2011, specifically Chapter 4). Founded in 1975, SSE is a non-profit organization that both saves and sells heirloom fruit, vegetable and flower seeds.[2] On this 360ha (890 acre) farm, which goes by the name of the Heritage Farm, there are 24,000 rare vegetable varieties (including about 4000 traditional varieties from Eastern Europe and Russia), approximately 700 pre-1900 varieties of apples (which represents nearly every remaining pre-1900 variety left in existence out of the 8000 that once existed), and a herd of the rare Ancient White Park cattle (its estimated global population is below 2000).

You do not have to spend much time in this space to realize that SSE is much more than a 'gene bank'. Working at Colorado State University, I walk by almost daily the US Fort Knox of gene banks, the National Center for Genetic Resources Preservation (NCGRP), which is located on campus. SSE practises a noticeably different form of 'conservation' than that practised at NCGRP. SSE is also a memory bank. Its visitor centre (the very fact that it has a visitor centre separates it from the NCGRP), for example, is full of knowledge about the seeds being saved. Here you will find information about, among other things, their seeds' history, phenotypic characteristics, best planting practices, recipes and the taste of their fruit. SSE also practises memory-*making*. It does not just hold knowledge. Recognizing that some knowledge is inextricably wrapped in cultural practice – in actually doing the knowledge in question – SSE provides visitors with the opportunity to put this knowledge to work. Take the Heirloom Tomato Tasting Workshop. Through this event participants get to taste more than 40 different kinds of tomatoes, as well as learn how to save the seeds of their favourite varieties for future planting.

Some cultural knowledge resembles the knowledge of, say, riding a bike. It is knowledge you acquire through *doing* the act of riding a bike. Try teaching someone to ride a bike with only words. You would not get very far. The cultural knowledge tied to traditional crops is very much like this. Many traditional cultures are oral, so you often won't find this folk knowledge written down anywhere. That is why I earlier argued that when rural communities and rural ways of life disappear, this is often followed by a reduction in biodiversity. And once that knowledge is lost, there is a good chance it's gone forever.

Back during the early 1990s, Virginia Nazarea (1998) was in the Philippines studying sweet potato farmers. At one site, production was beginning to become commercialized, while at another it remained predominately at the subsistence level. She had hypothesized that commercialization causes a narrowing of genetic and cultural diversity in relation to sweet potato production. Her hypothesis was confirmed. But she also noticed something unexpected. There was a large disparity between the two sites in terms of the number of varieties known or remembered, compared to the biodiversity that actually existed. At the commercial site, farmers had knowledge about a far lower percentage of sweet potato varieties than at the other site. This suggests a faster erosion of cultural knowledge than genetic diversity itself. Reflecting upon this research, Nazarea (2005, p62) writes how this finding 'signified to me that in the context of agricultural development and market integration, knowledge may actually be the first to go'.

In an article published in the *American Journal of Sociology*, Donald MacKenzie and Graham Spinardi (1995) argue that nuclear weapons are becoming *un-invented* due to global nuclear disarmament trends and nuclear test ban treaties. It is their contention that 'if design ceases, and if there is no new generation of designers to whom that tacit knowledge can be passed, then in an important (though qualified) sense nuclear weapons will have been uninvented' (MacKenzie and Spinardi, 1995, p44). This tacit knowledge – what I've been calling knowledge by *doing* – has been shown to be instrumental for a range of activities, from riding a bike (Polanyi, 1966) to building lasers (Collins, 1974), repairing Xerox copiers (Orr, 1996) and farming (Carolan, 2011, Chapter 6).

By washing away cultural diversity, cheap food policies are causing us to essentially 'uninvent' the practices and folk knowledge currently in use on the margins that makes those biologically diverse spaces possible. Today's food system is about uniformity. Grocery stores are full of bins, bags and boxes of identical fruit, vegetables, eggs and meat. Fields are little different. Growing up in Iowa, I rarely saw a corn field planted with anything other than number two yellow dent. Cheap food is also increasingly about convenience. Half of the money that Americans (the trend among Europeans is no different) spend on food is spent on food outside the home, though the fastest growing foods are ready-made 'home prepared' meals, which typically require nothing more than a little boiling or warming in the microwave (Hayes and Laudan, 2009, p835). The loss of agro-biodiversity means a loss of knowledge of preparing and cooking foods. Note the combined effect of biological and cultural mono-cultures: after losing traditional food staples and the knowledge that goes along with preparing and eating them, we will reach a point, after a generation or two, where we essentially no longer know what we are missing. At this point, what is lost is lost forever.

The 'get big or get out' logic would be impossible were it not for the systematic washing away of diversity. It was no accident that those fields of yellow dent corn that I grew up around in Iowa were full of plants of identical height and maturity. Combines – indeed, most harvesting implements – do not like biological diversity. For mechanization to work – to save labour and thus introduce economies of scale – certain factors must be present. Plants must mature evenly otherwise you risk harvesting some prematurely and others too late. They must also be accessible to the picking mechanism after months of exposure to the elements – a combine doesn't do you much good, for instance, if your corn has been knocked down by the wind.

The creation of biological and cultural monocultures is also a consequence of a highly simplified understanding of 'agro-ecosystem', which, as discussed earlier, gives birth to the view that external controls can non-problematically replace internal interdependencies. Eliminating bio-environmental pest controls, which cannot be avoided when similar or identical crops are grown across hundreds of miles, makes pesticides a necessity, just like breeding plants to tolerate the compact urban life of a modern cornfield necessitates the use of copious amounts of nitrogen. To think about this in cultural terms, this uncoupling from internal agro-ecological processes to create crops 'for all seasons and all reasons' (Nazarea, 2005, p11) places a premium on expert (generalizable) knowledge while marginalizing local knowledge (Carolan, 2006).

As our knowledge base narrows, so shrinks our understanding of 'alternatives'. As Nazarea (2005, p12) notes: 'Cultural memory is relevant to the conservation of biodiversity because it counters the sometimes overwhelming pull to surrender all options to external, prepackaged solutions.' We moderns like to think there is nothing that can be learned from indigenous cultures. It is they, rather, who have much to learn from us. So we're okay with losing certain cultural memories. This arrogance comes at a price. There are a number of instances of indigenous agricultural systems that were highly productive while utilizing internal agro-ecological interactions. One classic example is the raised bed system, which had been utilized for millennia throughout the southern border of North America, Central America, Asia and Africa. The raised bed has been proven highly efficient for irrigation, drainage, soil nutrient management, frost control and plant disease management. There are indications that Pre-Conquest Aztecs at Lake Texoco had 10,000ha of land under cultivation utilizing this raised bed system, feeding a population of 100,000 people (Posey, 2001, pp382–383). Can we really afford to let knowledge of systems like this die off?

Less than 3 per cent of the 250,000 plant varieties available to agriculture are currently in use (Vernooy and Song, 2004, p55). This statistic, however, only tells part of the story. There may be 0.25 million plant varieties housed in gene banks around the world. Yet, how much do we really know about these seeds? Do we still have access to the cultural knowledge that allowed people for generations (perhaps centuries) to put these seeds to work? Gene banks might house amazing levels of biodiversity, but in many cases that biodiversity is saved at the expense of cultural diversity. It's an expense we can ill afford.

Monocultures of the Mouth

So: we need to protect biological diversity and cultural diversity ... but taste diversity? I recently lectured to one of my classes about the Slow Food Movement – an international movement that focuses on the preservation of traditional and regional cuisine at the points of production (e.g. farming), preparation (e.g. cooking) and consumption (e.g. taste) (the subject of slow food is discussed further in Chapter 10). I explained to them that while society places importance on saving things such as biodiversity, language (though the extinction of languages has only recently entered public consciousness) and material cultural artefacts such as paintings, pottery and historical buildings, we have yet to think about food – or, more specifically, taste – in a similar way. I could tell they still weren't buying the argument, so I gave them a personal example. I explained how I feel more connected with my Czechoslovakian heritage than either my Irish or German roots. This connection, I told them, has nothing to do with knowledge of authentic Czech music, clothing or language. Rather, the connection I feel to this heritage is almost entirely due to food. Even my limited grasp of the Czechoslovakian language is mostly confined to either foods (such as the *kolache*) or food-related artefacts (such as *kuchenka*: cooker/stove) – words, coincidently, that were most likely acquired while sitting around the table eating.

We can quite literally taste culture. Thus, as we lose taste experiences we narrow the opportunities through which we can know cultures, others as well as our own. If I never ate a *kolache* again, I'd survive. But my identity, my sense of self, would change. It is a cost I will never be able to attach a dollar sign to. Yet it is a cost I sincerely hope to avoid.

Taste is also deeply political. The power of taste was publicly played out during the Flavr Savr tomato debacle. The Flavr Savr tomato was the first genetically engineered food product to reach the market. Approved for sale in May 1994 by the US Food and Drug Administration (FDA), the tomato was engineered to ripen for a longer period of time on the vine and to retain firmness longer to reduce losses during shipping. A host of variables account for the Flavr Savr tomato's failure in the market (see Pringle, 2005, pp68–77). But one was taste. Many consumers simply did not care for its flavour and overly firm skin (Avise, 2004, p69). Calgene, the company that developed the tomato, promised it would taste better than its competitors. It did not. Had its flavour lived up to the hype, consumers and food companies (Campbell's Soups, for instance, had initially contemplated using the tomato in its soups) might have been more willing to pay the higher market price that this fruit was going for. In the end, few wanted to pay more for a tomato of inferior taste and texture.

As discussed earlier, industrial agriculture has been attempting for decades to breed out variability. This has created not only biological and cultural monocultures, but increasingly taste monocultures as well. Forgetting some tastes while becoming exposed to new ones makes things like the Flavr Savr tomato more palatable, literally, to society (see Box 8.2). Evidence also suggests that cultures new to American fast food take a while before they accept the taste of this food. When studying the introduction of McDonald's food in Moscow, Melissa Caldwell (2004, p15) reports that many respondents initially did not like its taste. One individual went so far as to explain 'that he had tried it and could not understand why a person would eat such food more than to try it once' (Caldwell, 2004, p15).

Box 8.2 The politics of taste

The idea that taste is political can be difficult to grasp precisely because its effects are with us every second of our lives, a fact that can make 'stepping back' difficult. The following exchange is unpublished data – the research was entirely qualitative – conducted a few years back for my recently published book *Embodied Food Politics* (Carolan, 2011). I am interviewing Joe (not his real name), who is explaining to me why he initially preferred industrial foods to fresh fruits and vegetables, until, that is, he 're-socialized' his palate. Since doing this his preference of foods has shifted, away from those cheap foods he earlier 'adored':

Joe: We never got much fresh foods growing up. Most everything that I remember eating as a kid was either in a can or box. I remember one occasion my mom serving us fresh peas one night, right out of the pod. I couldn't eat them. I wanted to know why they tasted so bad. Did she undercook them? Were they old? I didn't know. 'They're fresh', ma said. Turns out the only peas that I had eaten until then were canned ... I adored canned peas.

Interviewer: So what changed? What foods do you prefer now and why do you prefer them?

Joe: It took a while. For one thing, I married someone that liked to garden and liked to eat fresh fruits and vegetables. Really, if she can avoid it, she won't eat foods that have been overly processed. I really didn't care for that stuff at first. I was a canned food man my whole life; that's not something you just drop overnight ... I've managed to re-socialize my taste buds. It's weird but I really can't stomach the stuff I grew up eating.

Interviewer: What do you mean you can't stomach it?

Joe: I just don't like how it tastes or feels in my mouth and, I don't know if it's in my head or not, but it often doesn't seems to agree with my stomach ... Those highly processed foods just don't have the hold on me like they used to.

Interviewer: That's an interesting metaphor. Care to elaborate on what you mean by it, to say industrial foods have a *hold* on people?

Joe: I guess my point is that food companies have a vested interest to shape our tastes early on. Because for most of the people I know, they haven't made the switch like I have. They still prefer processed foods. But I would argue that is not because it inherently tastes any better than whole foods. It's just what people know. Never having been socialized to the tastes and textures of anything else, those other foods just don't stand a chance – that is, unless more people manage to break the spell that food companies seem to have over our taste buds.

France is well known as a site of resistance to industrial foods and the taste matrix it rests upon. The Slow Food Movement has a major presence there (though the movement originated out of Italy). France also has the *Appellation d'Origine Contrôlée* (AOC) ('controlled designation of origin'). Originally applicable to wines, the AOC is now awarded to a variety of agricultural products (cheese, olive oil, etc.). Based upon a rigorous application process, AOC products must be produced in a manner consistent with artisan – versus industrial – practices in designated geographical areas. The French have an international reputation of being more attuned to the tastes of non-industrial food items than, say, those residing in the US (see, for example, Petrini and McCuaig, 2004; Kummer et al, 2008). This position was recently described by Amy Trubek (2008) in her book *A Taste of Place*. Writing on the importance of local food systems in the US, Trubek (2008, p9) looks to the French for inspiration, drawing attention to how they've developed a taste of place (or what the French call *le goût du terroir*):

> It is difficult to translate *terroir* from the French in a way that encapsulates all its meanings ... It is part of people's everyday assumptions about food; it is as fundamental as our assumption that the first meal of the day should include coffee and orange juice but not miso soup. The French are unusual in the attention they place on the role of the natural world in the *taste* of food and drink. When the French take a bite of cheese or a sip of wine, they taste the earth: rock, grass, hillside, valley, plateau. They ingest nature, and this taste signifies pleasure, a desirable good. Gustatory pleasure and the evocative possibilities of taste are intertwined in the French fidelity to the taste of place.

My problem with such arguments is that they gloss over certain socio-historical facts that reveal a French palette that is anything but immutable and fixed. As in other food cultures (see, for example, Diner, 2001; McWilliams, 2005), tastes among the French have been in continual flux over the centuries (see Ferguson, 2004; Freidberg, 2004). An excellent example of this can be found in how French consumers were systematically conditioned – or perhaps a more accurate word is 'tuned' (Carolan, 2011) – to the flavours, textures and preparation practices associated with canned foods. What's interesting about this case is that it shows how even the French taste bud, long exalted for its ability to discriminate, is an artefact of food politics.

As described by historian Martin Bruegel (2002, p113), there was a 'lack of taste' for canned foods in France prior to World War I. At the time, canning industries in other countries – most notably the US, Germany and England – were experiencing steady growth. The French, however, as noted by the editor of the *Journal d'Agriculture Pratique* in 1905, needed 'to overcome the irrational as well as instinctive repugnance [for canned food] among a large part of the population. It would be an arduous task' (as cited in Bruegel, 2002, p113). Some of the initial repugnance was due to health concerns. Defective cans came with the risk of lead poisoning and botulism. One army survey documents approximately 100 cases a year between 1886 and 1904 of canned food-related illnesses (Bruegel, 2002, p118). Initially, improvements to the canning processes did little to make people desire or trust canned foods, as trust first requires experience to build upon. Detailing how the French were slowly made to trust and eventually desire canned foods, Bruegel (2002, p18) points to social institutions and practices. In his own words:

> The acquisition of a new taste requires the intervention of social groups physically to acquaint the newcomer with the merchandise and the technique required to manipulate it, to instigate and organize the first trial, to keep on pressure after the initial rejection, with explanations on the appropriate use of the good, and eventually to sustain routine consumption.

And that's exactly what the French army did. It regularly exposed young men to canned meals, as mandated by military rules, which they had to eat, as required by a strict disciplinary code. This socialized soldiers to the taste of canned food. Daily exposure in the field also accustomed them to the practices of preparation. Something as seemingly simple as opening the can was a major early hurdle to adoption. But once they became accustomed to the practice, soldiers quickly found the foods more palatable.

The socialization process was further enhanced with the Great War. The public was told to 'add a few cans of food to every parcel' sent to soldiers on the front line (Bruegel, 2002, p124). Cans of food purchased to feed the front line were often sold at discounted prices. Getting citizens to go to the store to buy canned food 'proved the last threshold on the way to the quotidian purchase of sterilized comestibles' (Bruegel, 2002, p124). The war also served to create new memories for this otherwise non-memorable food. Canned food began to have its own cultural memories among soldiers, such as of their time around the campfire when the food was eaten and shared with others.

Food halls at fairs, in-store tastings, in-home demonstrations and travelling railcars giving away free products gave civilian consumers the opportunity to experience the food and (perhaps more importantly) begin to learn the practices associated with its preparation and consumption. Meanwhile, producers continually sought ways to improve the palatability of the food. The goal was not to make canned foods taste better than homemade, but to make them taste 'good enough'. Industrial food processors were interested, at least initially, in merely getting their foot in the door – or, more accurately, the mouth – of consumers. Once taste was no longer a limiting issue, it was believed that consumers would choose canned foods for their convenience (it is far easier to buy a can of corned beef than to raise, feed and slaughter a cow), durability and/or price (especially when looking to purchase something out of season) (Fitzgerald and Petrick, 2008, p399).

By the 1920s the canning industry's efforts started paying off. The French were actually showing signs that they wanted canned foods. Although yearly per capita consumption of canned food in France lagged behind comparable industrialized nations at the dawn of the 20th century, by the 1920s the average French mouth appeared to be as socialized to canned foods as its European neighbours.

And the cost of taste monocultures? The costs associated with losing tastes parallel that associated with the loss of cultural diversity. Both represent a shrinking of alternatives and understandings. When we begin to equate, say, the taste of an apple with that of a Red Delicious (the most popular apple variety in the US), we narrow not only our understanding of what an apple should taste like, but our very understanding of 'apple' itself. I remember an apple orchard owner once told me about the difficulty she has in selling her Roxbury Russet variety (Roxbury Russet is the oldest apple variety bred in the US, dating back to the mid 17th century). The apple, you see, has a greenish-yellow skin that is dull and highly blotchy. I'll never forget what she said: 'Apples are supposed to be shiny and red, like a fire truck. One of my daughter's little friends told me that once, and they were right. We've become socialized to hear "apple" and think "shiny red fruit".'

This has clear implications for biodiversity. If 'apple' is reduced to 'Red Delicious' and 'banana' reduced to 'Cavendish' (*the* banana cultivar in the international banana trade), what chance do all the other apple and banana varieties have of remaining in existence? Taste diversity makes protecting biodiversity and cultural diversity more practical. Without it, foods such as the Roxbury Russet, which offer an experience that does not fit with what we have been socialized to expect thanks to cheap food, risk being lost forever.

Whither Local Democracy?[3]

This brings me to a fourth monoculture brought about by today's food system. In addition to biological, cultural and taste monocultures, I conclude this chapter by discussing what I will call a monoculture of geopolitical space. This space was hinted at in the aforementioned Goldschmidt thesis – the finding that as communities become surrounded by industrial farms, there is a decline in local control over public decisions. I will explore this phenomenon in greater detail, using anti-GM (genetically modified) laws as my empirical entry point.

Public concern in the US over biotechnology is nothing new. The so-called anti-GM movement has been around since the 1970s (Schurman and Munro, 2006, p1). Yet it was not until the early 2000s – after some very public cases of GM contamination – that agricultural applications of biotechnology started to face sustained resistance. Perhaps the most widely publicized incident of contamination occurred in 2000 involving Starlink corn. Approved two years earlier for livestock consumption only, Starlink corn containes the Cry9C gene. This gene produces an insect control protein that is toxic to European corn borers (and other insects), thereby minimizing the need for pesticides. The protein represents a potential food allergen for humans, which is why it has not been approved for human consumption. Starlink corn was grown on less than 1 per cent of the total US acreage in 2000 (approximately 362,000 acres, or 146,500ha).

Contamination was first noticed in September 2000, when a testing lab found the presence of Cry9C in a sample of Taco Bell taco shells. After confirming the protein's presence, Kraft Foods Inc., the maker of the contaminated product, recalled all of its shells. Approximately 300 food products, which included over 70 types of corn chips and 80 different kinds of taco shells, were recalled, while 500 million bushels of corn were found to be contaminated due to commingling in grain elevators. To limit further contamination, the USDA launched (in late September) a buy-back programme, giving producers an additional US$0.25 per bushel to keep Starlink from further entering the food chain. The programme did not, however, address the Starlink crop produced the year earlier, which had already been delivered to local grain elevators. In October, an agreement was reached by Arventis – the producers of Starlink corn – and the attorneys general from 13 states to extend compensation to local grain elevators holding the previous year's Starlink crop (Lin et al, 2002, pp31–32). In the end, US$3 billion were spent containing the contamination, though company officials for Arventis admitted it was impossible to remove all traces of Starlink from the corn supply (Hishaw, 2007, p215).

Another widely publicized case of GM contamination occurred in 2002, when tests plots of corn engineered to contain an insulin precursor, trypsin, contaminated neighbouring fields. Adjacent soybean fields to both tests plots – one in Nebraska and another in Iowa – were found, after harvest, to be growing 'volunteer' corn plants from the aforementioned pharma crop. Approximately 500,000 bushels of soybeans had to be destroyed (Fox, 2003, p4; Hishaw, 2007, p216). While not admitting to any wrongdoing or negligence, ProdiGene – the biotech company who designed the corn in question – posted a US$1 million bond and agreed to reimburse the USDA for any containment costs, which ultimately totalled US$3.7 million (Fox, 2003, p3; Hishaw, 2007, p216).

This provides some context for what then occurred in Mendocino County, California. In light of very real concerns over GM contamination, this rural county located approximately 100 miles north of San Francisco, in March 2004, voted to create a GM-free zone within its borders. Measure H, as it was called, passed with 57 per cent of the vote. Measure H can be seen as an attempt to reclaim local control over local agriculture. Multinational agricultural firms (and their surrogates), in contrast, wield greater influence at the regional, national or global levels, where money matters more than boots on the ground. Mulvaney (2008, p153) explains this advantage:

> Activists' formal access to national regulatory discussions is limited to the legal efforts of public interests groups. Access to international dialogues is even more restrictive. Activists are overmatched in lobbying efforts, while government officials work with industry to promote national competiveness and regional economic growth.

Activists, conversely, have greater chance of success when the decision is cast as a local issue. At this level, social relationships are worth many times more, from a mobilization standpoint, than a full war chest. By successfully drawing upon these relationships, activists can often recruit individuals to pound the pavement, make phone calls and knock on doors for free. Local involvement on the Yes-on-H campaign on any given day before the vote was placed at between 150 and 200 volunteers (Walsh-Dilley, 2009, p102). Local activists are also usually better positioned to know what narratives resonate with a majority of voters.

Mendocino County is internationally known for its organic agriculture and biodynamic and organic wines, with over 150 organic farms and wineries (Mulvaney, 2008, p154). Among other things, the measure was an attempt to secure the county's 'brand' as a major organic producer. It was argued that keeping the county free of GM crops would guarantee that their organic products would continue to have access to markets in Europe and Japan (Walsh-Dilley, 2009, p98). The economic considerations of losing organic markets resonated even among farmers who might not have been against biotechnology *per se*. To be shut out of a market for even one year due to contamination was too big of a risk for many, knowing that this could result in losing the market for good if consumer trust for organic products coming out of Mendocino was undermined. Better to err on the side of caution, many supporters of Measure H believed, than tempt fate.

Opponents of Measure H had money on their side, lots of it. On the day of the election, the No-on-H campaign had collected over US$620,000, almost all of which came from CropLife America (a Washington, DC-based consortium of biotech companies, including Dow, DuPont, Monsanto and Syngenta). The No-on-H campaign only raised US$5000 locally. The Yes-on-H campaign, conversely, raised approximately US$135,000. Roughly US$35,000 of this came from sources outside the county (the Center for Food Safety donated US$23,900 and the Organic Consumers Association contributed US$11,500); the rest came from local donors (Walsh-Dilley, 2009, p101).

With few local residents as spokespersons, the No-on-H campaign spent its money primarily on ads and mass mailings and expertise (such as market researchers, legal advisers and out-of-state telemarketing firms to conduct 'push poll' calls) (Walsh-Dilley, 2009, p102). Lacking significant local leadership, the No-on-H campaign was unable to come up with narratives that resonated with voters. Measure H supporters were able to use their opponent's externally funded campaign to their advantage by framing the vote as being an issue of 'Mendocino County versus multinational corporations' (Mulvaney, 2008, p156). Other alliances were formed by anti-GM activists as a result of being able to frame Measure H as helping local fishing communities. Activists showed studies of how herbicide-tolerant crops increased herbicide use and, by extension, degraded water quality as those chemicals made their way into local rivers and streams. This tactic garnered anti-GM activists the support of the Pacific Coast Federation of Fishermen's Associations (PCFFA), the largest commercial fishing association in California (Mulvaney, 2008, p156).

Measure H proponents also drew upon existing institutions and organizations. Local public radio stations, which possessed a level of public trust that commercial radio stations did not, provided the anti-GM movement access to programmes that were not available to No-on-H campaigners. The largest co-operative food market in the area also encouraged customers to donate their member discount to the Yes-on-H campaign. As Walsh-Dilley (2009, p103) explains: 'No amount of money can replace the love and commitment of people who care passionately about the place where they live.' This was certainly the case in Mendocino. While most of the No-on-H actors lived in Washington, DC, the organizational feet of Yes-on-H resided almost entirely in Mendocino County.

The moratorium on GM crops was not limited to Mendocino County. As of January 2009, 12 counties passed ordinances making themselves explicitly GM crop-free zones. This regulatory patchwork in California towards GM crops is unacceptable to the biotechnology regime since it threatens its access to the state's enormous agricultural market (approximately US$32 billion a year). Even though some counties have passed pro-GM crop resolutions, the concept is that the counties with bans could draw international attention to public concerns over GM food and frustrate biotech companies' attempts to enter new markets.

Let's unpack this, in terms of how today's food system works to wash away local decision-making structures. The goal: to produce a smooth, almost universal geopolitical plane to operate on. To produce, in other words, a mono-culture of geopolitical space.

Capital flows most smoothly along a 'flat' regulatory plane. Part of the process of producing a thoroughly capitalist space is the creation of regulations that fit with standards already in place in spaces where capital efficiently operates (Harvey, 1989, p246). A tried and true legal tactic was employed to 'flatten' out the regulatory space of California and to give biotech firms greater mobility: pre-emption.

The legal force of pre-emption in the US is rooted in its constitution, though the practice is not limited to the US (and, arguably, institutions such as the World Trade Organization (WTO) take pre-emption to a whole new level). Pre-emption in the US gives supremacy to the federal level in cases where state and federal policy contradict one another: 'the Laws of the United States, which shall be made in Pursuance thereof ... [the] supreme Law of the Land' (US Constitution, Article VI, clause 2). Increasingly, however, states are employing this tactic to pre-empt policies emanating from the grassroots level (and which, more often than not, disrupt the logics of capital). Given their greater efficacy at state, national and international jurisdictions, firms can thus pre-empt those grassroots battles that they have lost by jumping to another scale. As David Harvey (2007, p66) once astutely argued: 'Neoliberal theorists are ... profoundly suspicious of democracy ... [favouring] governance by experts and elites, ... executive order, and by judicial decision rather than democratic and parliamentary decision-making.' The problem with voter-determined ordinances at the grassroots level is that the 'roots' are sometimes deep enough – as was the case in Mendocino County – to overcome the influence wielded by firms who at other scales are much more influential.

To better understand this neoliberal response, take Senate Bill 1056. SB1056 would have granted the state of California jurisdiction over the regulation of seed and nursery stock. In simple terms, it would have banned county-wide prohibitions on GM crops. Given Monsanto's public backing of the bill, it was widely dubbed 'Monsanto Law'. SB1056 was believed to be a direct response to a successful vote against GM crops in Mendocino County. As Roff (2008, p1427) explains:

> It is not surprising then that within weeks of Mendocino's victory, the Biotechnology Industry Organization (BIO) launched a concerted effort to 'construct vehicles to preempt county bans on the planting of biotech crops'. Financial outlays to prevent subsequent GE-Free campaigns in Butte, Humboldt, San Luis Obispo and Sonoma increased, culminating with record-breaking spending in the Sonoma campaign.

Proponents of the bill attempted to frame the debate in strict neoliberal terms: the need for regulatory harmonization; the importance of private property and the inalienable rights of the property owner to work their land as they see fit; and the belief that the market ought to decide what is grown in the state. Opponents of the bill, conversely, worked to make the debate about democracy, local control and the autonomy of local regulatory bodies (Roff, 2008, p1430).

The bill was eventually tabled and has not been reintroduced to the Senate floor. Over a dozen states have, however, passed legislation that either removes or significantly restricts the ability of local jurisdictions to regulate seed and local agricultural practices. In 2005, for example, a bill was passed in Iowa that pre-empts 'a local governmental entity ... from adopting or enforcing legislation which relates to the production, use, advertising, sale, distribution, storage, transportation, formulation, packaging, labeling, certification, or registration of agricultural seed'.

In 13 US states, it is illegal to talk too disparagingly about industrial agriculture – this in a country with perhaps the most liberal free speech laws in the world. A characteristic example of this is South Dakota's law, which defines disparagement as:

> ... dissemination in any manner to the public of any information that the disseminator knows to be false and that states or implies that an agricultural food product is not safe for consumption by the public or that generally accepted agricultural and management practices make agricultural food products unsafe for consumption by the public. (cited in Donham et al, 2007, p319)

Note the vagueness of this language. It is not like today's food system produces cakes of enriched uranium. This makes *disproving* the safety of a food notoriously difficult in a court of law. A healthy person can eat a meal consisting of Pop-Tarts, Red Bull and peanut M&Ms without worrying about dropping dead afterwards (I give this example because I recently witnessed a student eat just such a 'meal'). Yet what are the risks of eating like this over a lifetime? The evidence, as covered in Chapter 4, is clear. Diets high in cheap processed food negatively affect health in the long term. But is that the same as saying they are not safe for consumption?

Furthermore, all 50 US states have some form of right-to-farm legislation in place. While different from state to state, most legislation includes provisions pre-empting the democratic efforts of local governments. Thus, while this legislation has the veneer of wanting to protect family farms from 'outside' forces, like urban interests that would rather see townhouses rise where corn used to stand, it actually seeks to do just the reverse – namely, stifle the expression of local interests. For example, in a ruling upholding the state's right-to-farm legislation, the Iowa Supreme Court argued that county governments cannot use home rule powers or public health justifications to erect zoning laws that are more restrictive than existing state laws (Donham et al, 2007, p319).

The goal of this legislation: to produce a smooth, monoculture-like geopolitical space that allows for the uninterrupted flow of practices and artefacts associated with large-scale commodity agriculture. Unfortunately, this comes at the expense of democracy itself. Good old-fashioned grassroots democracy: a fatality of neoliberalism. And with this lost, another cost of cheap food.

Yet, sometimes legislation isn't even necessary; or, to put it another way, sometimes agribusiness doesn't need to jump to another scale with pre-emption. Occasionally they find a way to transcend the democratic political arena entirely. This is precisely what terminator technology – or what is less pejoratively known as genetic use restriction technology (GURT) – allows.

For those unfamiliar with so-called terminator seeds, they refer to organisms genetically engineered to germinate into plants that produce sterile seed.[4] A number of criticisms have been levelled at this technology – for example, it raises costs and locks farmers into tightly controlled marketing and licensing agreements (Jumba, 2010, p20); it forces farmers to return annually to the seed companies therefore eradicating the 12,000-year tradition of seed saving, which will minimize biodiversity and increase food security (ETC, 2002, p2; see also Shiva et al, 1999, pp601–602); it will displace locally adapted genetic material, which could affect the resilience and long-term productivity of low-input farm systems (Van Wijk, 2004, pp133–134); and while plants produce sterile seed, they can still generate fertile pollen, which could carry the terminator traits to sexually compatible plants and thereby render them sterile (Pendleton, 2004, p16). My critique of terminator seeds will be a bit unconventional, noting how they allow agribusiness to seize control of geopolitical space that had previously been presided over by a little thing we know as democracy.

The belief that seed saving is a fundamental human right remains widely held within the international community. The International Convention for the Protection of New Varieties of Plants of 1961 (revised in 1972, 1978 and 1991), for example, allows all member states to allow on-farm seed saving of otherwise protected varieties. Australia's Plant Breeder's Rights Act allows farmers to save seed from plant variety protection (PVP) species without having to pay royalties (Muscati, 2005). Following the EU Biotechnology Inventions Directive on Plant Variety Rights, small-scale European farmers can save seed without fear of being taken to court (Muscati, 2005). Around the world, democratically elected bodies have exempted on-farm seed saving, recognizing it in some cases as a fundamental human right (see, for example, Straub, 2006). A variety that terminates, however, makes this impossible. And there is the rub: terminator technology gives multinational corporations the ability to override the will of democratic governing bodies.[5] Those who employ this lock-out technology are acting like an all-powerful sovereign, imposing rules of usage without regard to the broader public interests and outside mechanisms of democratic rule-making (Burke, 2004, p15).

Notes

1   'Appalachia' typically refers to a cultural region in the eastern US spanning as far north as the southern edge of New York and as far south as northern Alabama, Mississippi and Georgia. The Appalachian Mountains extend into Canada and therefore do not exclusively define this cultural region.
2   While there is no universally agreed upon definition, 'heirloom' generally refers to varieties that are capable of being pollen-fertilized and whose existence predates industrial agriculture.
3   This section builds upon an argument made earlier in Carolan (2010, see specifically Chapter 6).
4   Terminator technology comes in two forms: Trait GURT (T-GURT) and Varietal GURT (V-GURT). T-GURT produces viable seed but restricts gene expression of a certain trait that can be switched on with the application on a specific input. V-GURT, which is particularly controversial, restricts the use of the entire plant by rendering seeds sterile.
5   A few points on the argument that terminator technology is, in effect, no different from hybridized corn (which produces inferior seeds) (see, for example, Goeschl and Swanson, 2003, p151). First, while hybrid varieties do not breed true, they still produce fertile progeny. Second, hybridization was explicitly developed to *increase* yields and plant vigour (Sleper and Poehlman, 2006, p175); the non-viability of second-generation seed from hybrid crops is an ancillary effect (though had it not been present, hybridization would probably not have been pursued with the same vigour; see Kloppenburg, 1988, p425). Terminator technology, conversely, is designed for the sole purpose of facilitating monopoly control over the market.

References

Avise, J. (2004) *The Hope, Hype, and Reality of Genetic Engineering*, Oxford University Press, New York, NY

Bruegel, M. (2002) 'How the French learned to eat canned food, 1809–1930', in W. Belasco and P. Scranton (eds) *Food Nations: Selling Taste in Consumer Societies*, Routledge, New York, NY, pp113–130

Burke, D. (2004) 'DNA rules: Legal and conceptual implications of biological 'lock-out' systems', *California Law Review*, vol 92, pp1–35

Caldwell, M. (2004) 'Domesticating the French fry: McDonald's and consumerism in

Moscow', *Journal of Consumer Culture*, vol 4, no 1, pp5–26

Carolan, M. (2006) 'Sustainable agriculture, science, and the co-production of "expert" knowledge: The value of interaction expertise', *Local Environment*, vol 11, no 4, pp421–431

Carolan, M. (2007) 'Saving seeds, saving culture: A case study of a heritage seed bank', *Society and Natural Resources*, vol 20, no 8, pp739–750

Carolan, M. (2010) *Decentering Biotechnology: Assemblages Built and Assemblages Masked*, Ashgate, Burlington, VT

Carolan, M. (2011) *Embodied Food Politics*, Ashgate, Burlington, VT

Collins, H. (1974) 'The TEA set: Tacit knowledge and scientific networks', *Science Studies*, vol 4, pp165–186

Constance, D. and R. Tuinstra (2005) 'Corporate chickens and community conflict in east Texas: Growers' and neighbors' views on the impacts of industrial broiler production', *Culture and Agriculture*, vol 27, pp45–60

Diner, H. (2001) *Hungering for America: Italian, Irish and Jewish Foodways in the Age of Migration*, Harvard University Press, Cambridge, MA

Donham, K., S. Wing, D. Osterberg, J. Flora, C. Hodne, K. Thu and P. Thorne (2007) 'Community health and socioeconomic issues surrounding concentrated feeding operations', *Environmental Health Perspectives*, vol 115, no 2, pp317–320

ETC (Erosion, Technology and Concentration Group) (2002) *Terminate Terminator in 2002*, 19 February, pp1–3, www.etcgroup.org/documents/terminatorbrochure02.pdf, last accessed 2 March 2009

Ferguson, P. (2004) *Accounting for Taste: The Triumph of French Cuisine*, University of Chicago Press, Chicago, IL

Fitzgerald, G. and G. Petrick (2008) 'In good taste: Rethinking American history with our palates', *The Journal of American History*, vol 95, no 2, pp392–404

Fox, J. (2003) 'Puzzling industry response to ProdiGene fiasco', *Nature Biotechnology*, vol 21, pp3–4

Freidberg, S. (2004) *French Beans and Food Scares: Culture and Commerce in an Anxious Age*, Oxford University Press, New York, NY

Fujimoto, I. (1977) 'The communities of the San Joaquin Valley: The relation between scale of farming, water use, and quality of life', in *U.S. Congress, House of Representatives, Obstacles to Strengthening the Family Farm System. Hearings before the Subcommittee on Family Farms, Rural Development, and Special Studies of the Committee on Agriculture, 95th Congress, First Session*, Government Printing Office, Washington, DC, pp480–500.

Goeschl, T. and T. Swanson (2003) 'The development impact of genetic use restriction technologies: A forecast based on the hybrid crop experience', *Environment and Development Economics*, vol 8, pp149–165

Goldschmidt, W. (1978) *As You Sow: Three Studies in the Social Consequences of Agribusiness*, Allanheld, Osmun and Company, Montclair, NJ, originally published 1944

Harvey, D. (1989) *The Urban Experience*, Johns Hopkins University Press, Baltimore, MD

Harvey, D. (2007) *A Brief History of Neoliberalism*, Oxford University Press, New York, NY

Hayes, D. and R. Laudan (2009) *Food and Nutrition*, Marshall Cavendish, Tarrytown, NY

Heady, E. O. and S. T. Sonka (1974) 'Farm size, rural community income, and consumer welfare', *American Journal of Agricultural Economics*, vol 56, pp534–542

Heffernan, W. and P. Lasley (1978) 'Agricultural structure and interaction in the local community: A case study', *Rural Sociology*, vol 43, pp348–361

Hellmann, J., J. Byers, B. Bierwagen and J. Dukes (2008) 'Five potential consequences of climate change for invasive species', *Conservation Biology*, vol 22, no 3, pp534–543

Hishaw, J. (2007) 'Show me no rice pharming: An overview of the introduction of and opposition to genetically engineered pharmaceutical crops in the United States', *Journal of Food Law and Policy*, vol 3, pp209–227

Jumba, M. (2010) *Genetically Modified Organisms: The Mystery Unraveled*, Trafford Publishing, Victoria, British Columbia

Kloppenburg, J. (1988) *First the Seed*, Cambridge University Press, New York, NY

Kotschi, J. (2007) 'Agrobiodiversity is essential for adapting to climate change', *Gaia*, vol 16, no 2, pp98–101

Kummer, C., S. Cushner, C. Petrini and E. Schlosser (2008) *The Pleasures of Slow Food: Celebrating Authentic Traditions, Flavors, and Recipes*, Chronicle Books, San Francisco, CA

Lin, W., G. Price and E. Allen (2002) 'Starlink: Where no Cry9C corn should have gone before', *Choices*, winter, pp31–34

Lobao, L. (1990) *Locality and Inequality: Farm and Industry Structure and Socioeconomic Conditions*, State University of New York Press, New York, NY

Lobao, L. and C. Stofferahn (2008) 'The community effects of industrialized farming: Social science research and challenges to corporate farming law', *Agriculture and Human Values*, vol 25, no 2, pp218–240

Luloff, A. and R. Krannich (eds) (2002) *Persistence and Change in Rural Communities: A 50-Year Follow-Up to Six Classic Studies*, CABI, Wallingford, UK

MacKenzie, D. and G. Spinardi (1995) 'Tacit knowledge, weapons design, and the uninvention of nuclear weapons', *American Journal of Sociology*, vol 101, no 1, pp44–99

Maffi, L. (2005) 'Linguistic, cultural, and biological diversity', *Annual Review of Anthropology*, vol 34, pp599–617

Maffi, L. and E. Woodley (eds) (2010) *Biocultural Diversity Conservation: A Global Sourcebook*, Earthscan, London

Martinson, O., E. Wilkening and R. Rodefeld (1976) 'Feelings of powerlessness and social isolation among 'large-scale' farm personnel', *Rural Sociology*, vol 41, pp452–472

McMillan, M. and M. Schulman (2003) 'Hogs and citizens: A report from the North Carolina front', in W. Falk, M. D. Schulman and A. R. Tickamyer (eds) *Rural Restructuring in Local and Global Contexts*, University Press, Athens, OH, pp219–239

McWilliams, J. (2005) *A Revolution in Eating: How the Quest for Food Shaped America*, Columbia University Press, New York, NY

Mulvaney, D. (2008) 'Making local places GE-free in California's contentious geographies of genetic pollution and coexistence', in M. Goodman, M. Boykoff and K. Evered (eds) *Contentious Geographies: Environmental Knowledge, Meaning, Scale*, Ashgate Press, Burlington, VT, pp147–163

Muscati, S. (2005) 'Terminator technology: Protection of patents or a threat to the patent system?', *IDEA: The Journal of Law and Technology*, vol 45, no 4, pp477–510

Nabhan, G. (1985) 'Native American crop diversity, genetic resource conservation, and the policy of neglect', *Agriculture and Human Values*, vol 11, no 3, pp14–17

Nabhan, G. (1989) *Enduring Seeds: Native American Agriculture and Wild Plant Conservation*, University of Arizona Press, Tucson, AZ

Nazarea, V. (1998) *Cultural Memory and Biodiversity*, University of Arizona Press, Tucson, AZ

Nazarea, V. (2005) *Heirloom Seeds and Their Keepers*, University of Arizona Press, Tucson, AZ

NCRCRD (1999) *The Impact of Recruiting Vertically Integrated Hog Production in Agriculturally-Based Counties of Oklahoma*, Report to the Kerr Center for Sustainable Agriculture, Iowa State University, Ames, IA

Orr, J. (1996) *Talking about Machines: An Ethnography of a Modern Job*, Cornell University Press, Ithaca, NY

Pendleton, C. (2004) 'The peculiar case of 'terminator' technology: Agricultural biotechnology and intellectual property protection at the crossroads of the third Green Revolution', *Biotechnology Law Review*, vol 23, pp1–29

Perfecto, I., J. Vandermeer and A. Wright (2009) *Nature's Matrix: Linking Agriculture, Conservation, and Food Sovereignty*, Earthscan, London

Peters, D. (2002) *Revisiting the Goldschmidt Hypothesis: The Effect of Economic Structure on Socioeconomic Conditions in the Rural Midwest*, Technical Paper P-0702-1, Missouri Department of Economic Development, Missouri Economic Research and Information Center, Jefferson City, MO

Petrini, C. and W. McCuaig (2004) *Slow Food: The Case for Taste*, Columbia University Press, New York, NY

Pilgrim, S. and J. Pretty (eds) (2010) *Nature and Culture: Rebuilding Lost Connections*, Earthscan, London

Polanyi, M. (1966) *The Tacit Dimension*, Doubleday, Garden City, NY

Posey, D. (2001) 'Biological and cultural diversity: The inextricable, linked by language and politics', in L. Maffi (ed) *Biocultural Diversity: Linking Language, Knowledge and the Environment*, Smithsonian Institute Press, Washington, DC, pp379–396

Pretty, J., B. Adams, F. Berkes, S. Ferreira de Athayde, N. Dudley, E. Hunn, L. Maffi, K. Milton, D. Rapport, P. Robbins, C. Samson, E. Sterling, S. Stolton, K. Takeuchi, A. Tsing, E. Vintinner and S. Pilgrim (2008) 'How do biodiversity and culture intersect?', Plenary paper for Conference on Sustaining Cultural and Biological Diversity in a Rapidly Changing World: Lessons for Global Policy, Organized by American Museum of Natural History's Center for Biodiversity and Conservation, IUCN/Theme on Culture and Conservation and Terralingua, 2–5 April 2008

Pringle, P. (2005) *Food Inc.: Mendel to Monsanto – The Promises and Perils of the Biotech Harvest*, Simon and Schuster, New York, NY

Rigg, J. (2005) 'Land, farming, livelihoods, and poverty: Rethinking the links in the rural South', *World Development*, vol 34, no 1, pp180–202

Rodefeld, R. (1974) *The Changing Organization and Occupational Structure of Farming and the Implications for Farm Work Force Individuals, Families, and Communities*, PhD thesis, University of Wisconsin, Madison, WI

Roff, R. (2008) 'Preempting to nothing: Neoliberalism and the fight to de/re-regulate agricultural biotechnology', *Geoforum*, vol 39, pp1423–1438

Schiffman, S., E. Slatterly-Miller, M. Suggs and B. Graham (1998) 'Mood changes experienced by persons living near commercial swine operation', in K. Thu and E. P. Durrenberger (eds) *Pigs, Profits, and Rural Communities*, State University of New York Press, Albany, NY, pp84–102

Schurman, R. and W. Munro (2006) 'Ideas, thinkers, and social networks: The process of grievance construction in the anti-genetic engineering movement', *Theory and Society*, vol 35, pp1–38

Shiva, V. (1993) *Monocultures of the Mind: Perspectives on Biodiversity and Biotechnology*, Natraj Publishers, Dehra Dun

Shiva, V., A. Emani and A. Jafri (1999) 'Globalization and threat to food security: Case of transgenic cotton trails in India', *Economic and Political Weekly*, vol 34, no 10/11, pp601–613

Skees, J. and L. Swanson (1988) 'Farm structure and rural well-being in the South', in L. Swanson (ed) *Agriculture and Community Change in the US: The Congressional Research Reports*, Westview Press, Boulder CO

Sleper, D. and J. Poehlman (2006) *Breeding Field Crops*, Blackwell, New York, NY

Smith, E. (2001) 'On the coevolution of cultural, linguistic and biological diversity', in L. Maffi (ed) *Biocultural Diversity: Linking Language, Knowledge and the Environment*, Smithsonian Institute Press, Washington, DC, pp95–117

Smithers, J., P. Johnson and A. Joseph (2004) 'The dynamics of family farming in north Huron County, Ontario. Part II: Farm–community interactions', *The Canadian Geographer*, vol 48, pp209–224

Soleri, D. and D. Cleveland (1993) 'Seeds of strength for Hopis and Zunis', *Seedling*, vol 10, no 4, pp13–18

Stofferahn, C. (2006) *Industrialized Farming and Its Relationship to Community Well-Being: An Update of a 2000 Report by Linda Lobao*, Report prepared for the State of North Dakota, Office of the Attorney General, for Case State of North Dakota versus Crosslands, North Dakota District Court, September

Straub, P. (2006) 'Farmers in the IP wrench – how patents on gene-modified crops violate the right to food in developing countries', *Hastings International and Comparative Law Review*, vol 29, no 2, pp187–214

Sutherland, W. (2003) 'Parallel extinction risk and global distribution of languages and species', *Nature*, vol 423, pp276–279

Swanson, L. (ed) (1988) *Agriculture and Community Change in the US: The Congressional Research Reports*, Westview Press, Boulder, CO

Tetreau, E. D. (1938) 'The people of Arizona irrigated areas', *Rural Sociology*, vol 3, no 2, pp177–187

Trubek, A. (2008) *The Taste of Place: A Cultural Journey into the Terroir*, University of California Press, Los Angeles, CA

Van Wijk, J. (2004) 'Terminating piracy or legitimate seed saving', *Technology Analysis and Strategic Management*, vol 16, no 1, pp121–141

Vernooy, R. and Y. Song (2004) 'New approaches to supporting the agricultural biodiversity important for sustainable rural livelihoods', *International Journal of Agricultural Sustainability*, vol 2, no 1, pp55–66

Veteto, J. (2008) 'The history and survival of traditional heirloom vegetable varieties in the Southern Appalachian Mountains of western North Carolina', *Agriculture and Human Values*, vol 25, no 1, pp121–134

Veteto, J. and K. Skarbo (2009) 'Sowing the seeds: Anthropological contributions to Agrobiodiversity Studies', *Culture and Agriculture*, vol 32, no 2, pp73–87

Walsh-Dilley, M. (2009) 'Localizing control: Mendocino County and the ban on GMOs', *Agriculture and Human Values*, vol 26, pp95–105

Warwick, H. (2001) 'Cuba's organic revolution', *Forum for Applied Research and Public Policy*, summer, pp54–58

9

Cheap Food: Who Wins?

It is not unusual to find the Colorado State University campus teeming with livestock producers, representatives from the feeder industry or generally anyone involved in the animal protein business. Thus, the fact that there was a public forum on campus on 27 August 2010 concerning the concentration of the meat industry didn't surprise me. Who was hosting the event, however, did. On this sunny, warm Friday, approximately 1500 people crammed into the university's student centre to participate in a forum co-hosted by the US Department of Agriculture (USDA) and the US Department of Justice (DOJ). The sessions held on that day were jointly chaired by none other than US Attorney General Eric Holder and US Secretary of Agriculture Tom Vilsack, the 'top cop' in the US and the face of US cheap food policy, respectively.

In 2010, the DOJ and USDA hosted five public forums to examine the consequences of concentration within the food system. The first workshop (held in Iowa) examined concentration in the seed and hog industries. The second workshop (held in Alabama) looked at concentration within the poultry industry. Workshop number three (held in Wisconsin) focused on consolidation within the dairy industry. The forum held in Fort Collins was the forth. The final public forum occurred in Washington, DC, in December 2010, and focused on producer-to-retail price margins.

Time will tell if anything meaningful comes out of these meetings. Yet the very fact that the USDA is co-hosting such forums with the policing branch of the US government (the DOJ) – a branch of government with a bite that matches its bark – is quite astounding. Its suggests that, for perhaps the first time in decades, some at the very highest levels of food policy are starting to wonder if the costs of cheap food are becoming too great. The very thought of, for example, Earl Butz, the late, great advocate of cheap food, sponsoring forums that question the validity of the 'get big or get out' mentality would have been anathema to 1970s food policy.

I have spent considerable time writing about the various costs of cheap food. Yet, clearly, no system can survive for very long without benefiting someone.

Cheap food produces a lot of losers, contributing handily to the displacement of millions of smallholder farmers, fuelling conflicts, retarding the development of entire swathes of the globe, costing the environment, and expanding waistlines while simultaneously increasing food insecurity. There are some, however, who benefit considerably from these policies. Cheap food has its winners – big winners. This chapter discusses who these people, organizations and sectors of the economy are that have a huge interest in maintaining the status quo.

The food system 'hourglass' refers to the demographic shape of the commodity chain. The farm and fork 'ends' are, relative to the 'middle', well populated. In the US, there are roughly 2.2 million farmers and 300 million consumers (see Figure 9.1). The points of the commodity chain connecting farm and fork, however, are considerably more concentrated. In the US, for example, there are approximately 25,000 food processors and manufacturers and 112,600 food and beverage retailers. This narrow middle, like an hourglass, reflects a decades-long trend of market concentration.

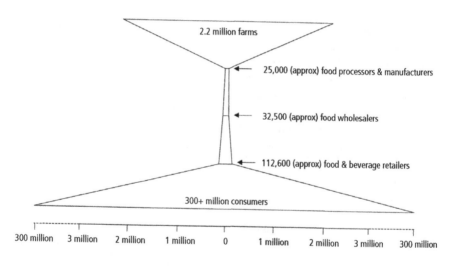

Figure 9.1 *The US food system 'hourglass'*

Source: Author

The following sections investigate this hourglass. The winners mentioned above are scattered throughout, though some links have bigger winners than others. Rather than jumping around from country to country, the primary focus of this chapter is the US hourglass. This allows me to conduct a more systematic critique of the aforementioned concentration. The *look* of the hourglass may differ from country to country; the *reasons* driving this concentration, however, do not. References are therefore made to other countries to show the reader that the US case is generally representative of what's happening throughout much of the world (see Box 9.1).

Box 9.1 Market concentration in the developing world: Brazil

From the mid 1970s, the rapid concentration of markets in the middle of the hourglass has helped to keep profits high for a handful of transnational firms. More recently, attention has turned to the domestic markets of developing countries, where high population growth rates and rapid urbanization are creating ideal conditions for global food corporations. Early on, Latin American countries were the main focus of investments. More recently, attention has shifted towards Asia, where many countries are experiencing sustained population, economic and dietary growth (McCullough et al, 2008, pp5–6; Wilkinson, 2009).

Markets in developing economies began concentrating as early as the 1980s with the consolidation of domestic firms. By the 1990s, however, pressure from international lending agencies, such as the International Monetary Fund (IMF) and the World Bank, and the World Trade Organization (WTO) began to force domestic markets further open, allowing for the free flow of international capital into these countries. Large domestic firms thus began to be displaced by large international companies. In Brazil, for example, the domestic company Ceval grew throughout the 1980s to become the largest soy processor in Latin America, responsible for 20 per cent of the region's total production (Wilkinson, 2009). In 2001, Ceval was bought by Bunge. Today, approximately half of the country's soy crushing – what has been referred to as 'the Brazilian soybean crushing complex' – is handled by four massive global firms – namely, Bunge, Cargill, ADM (Archer Daniels Midland) and Coinbra (a Brazilian subsidiary of Louis Dreyfus) (Goldsmith and Hirsch, 2006).

The story is the same with the seed industry. Early on, domestic seed firms dominated the Brazilian market. Biotechnology and strict patent enforcement changed that, quickly transforming the Brazilian private seed sector. Today, the sector is dominated by three firms: Monsanto, Syngenta and DuPont. Monsanto alone controls over 60 per cent of the Brazilian seed market (McMichael, 2005, p120). Similar trends are also found in the Brazilian coffee sector. According to Brazilian-based political economist John Wilkinson (2009), within a decade, this sector, which until recently was almost exclusively domestic, will be dominated by global firms. Even Brazil's retail sector is coming to be dominated by a handful of large transnational firms. Three companies now account for roughly 40 per cent of the country's retail sales (Wilkinson, 2009). Walmart's operation ranks third, behind France's Carrefour and CBD (the Companhia Brasileira De Distribuição, a domestic firm partly owned by French retailer Casino). In 2009, Walmart generated US$11.3 billion in sales (up 56 per cent from the previous year), while the other two firms each made more than US$13.3 billion (Bussey, 2010).

Farming: Getting Big Doesn't Get You What It Used To

There is a general misperception that small farmers are the primary beneficiaries of agricultural subsidies. These payments must be for those too small to make it otherwise – why help those least in need of assistance? Most are shocked after finding out whose pockets this taxpayer money is lining. And what a significant amount of pocket lining it is: a total support estimate, which includes direct payments, money for research and infrastructure development, and consumer support programmes (e.g. food stamps), for all 30 Organisation for Economic Co-operation and Development (OECD) countries is roughly US$365 billion annually (or US$1 billion a day) (Peterson, 2009, pxv). Yet, recognizing the various forms that agricultural subsidies take (as described in Table 9.1), even this estimate is conservative.

Who gets this money? As illustrated in Figure 9.2, the US, from 1995 to 2009, doled out US$211 billion in farm payments, 88 per cent of which (or US$186.5 billion) went to 20 per cent of farms. That left US$24.5 billion to be distributed (very unevenly) to the remaining 1,760,000+ farms. The Environmental Working Group, based in Washington, DC, maintains a database of major recipients of farm subsidies (see http://farm.ewg.org). The top recipient of agriculture subsidy payments from 1999 to 2009 was Riceland Foods Inc. (Stuttgart, Arkansas), at over US$554 million. The top three payment recipients – Riceland Foods Inc., Producers Rice Mill (Wynne, Arkansas) and Farmers Rice Coop (Sacramento, California) – collectively received during this period over *US$1 billion* dollars. Figure 9.2 gives some context to Key and Roberts (2007) finding that farm subsidies appear to have contributed to increases in the average size of US farms. Between 1999 and 2005, US$1.1 billion were paid to US farmers who had died (Peterson, 2009, p147). A *Washington Post* article estimates that between 2000 and 2006, the government paid out US$1.3 billion in direct payments to individuals who do not even farm. Pointing to thousands of acres previously used to grow rice in Texas, but now used for suburban housing, the newspaper notes that the landowners are paid by the government as if the land is under cultivation (Morgan et al, 2006).

The US case is far from unusual in this regard (see Box 9.2). For example, while over 500,000 farms in France receive European Union (EU) farming subsidies, no independent farmer is listed amongst the top 20 recipients. Over 80 per cent of these funds go to large industrial food-processing businesses and charitable organizations. The largest recipient is the chicken production and processing firm Doux, having received 62.8 million Euros between October 2007 and October 2008.[1] But when looking at the EU as a whole, Doux is a small fish. So who gets the largest pieces from the 55 billion Euros pie that the EU provides each year to European producers? Farmsubsidy.org, a non-profit group that campaigns for transparency in the reporting of subsidies, maintains a database so that we can now answer this question (a couple of years ago, EU member countries didn't publish their farm subsidy data).

Table 9.1 *Nine Types of US Farm Subsidies*

Direct Payments: Cash subsidies for producers of 10 crops: wheat, corn, sorghum, barley, oats, cotton, rice, soybeans, minor oilseeds, and peanuts (the last three were added in the 2002 Farm Bill).

Marketing Loans: A price-support programme initially created during the New Deal. While originally a short-term loan programme, today they provide large payments by guaranteeing minimum prices for crops (covering wheat, corn, sorghum, barley, oats, cotton, rice, soybeans, minor oilseeds and peanuts, and recently expanded to include wool, mohair, honey, dry peas, lentils and chickpeas).

Countercyclical Payments: Provides a 'safety net' in the event of low crop prices. Payments are issued if the price for a commodity is below the target price for the commodity. The programme covers the same 10 commodities as the direct payments programme plus dry peas, lentils and chickpeas.

Conservation Subsidies: About US$3 billion annually are dispensed in this form, most through the Conservation Reserve Program (CRP). Created in 1985, the CRP pays farmers to idle millions of acres of farmland.

Insurance: 'Yield' and 'revenue' insurance are available to farmers to manage risk of adverse weather, pests, and low market prices. This subsidized insurance allows farmers to pay roughly one-third the full cost of the policy.

Disaster Aid: This is a form of emergency crop relief due to 'acts of God' (possibly for crops already covered by subsidized crop insurance).

Export Subsidies: A major export subsidy programme is the Export Enhancement Program (EEP). The EEP's objectives are to challenge unfair trade practices and to expand US agricultural exports. Commodities eligible under EEP initiatives are wheat, wheat flour, semolina, rice, frozen poultry, frozen pork, barley, barley malt, table eggs and vegetable oil. The government spends over $1 billion in its EEP and its Dairy Export Incentive Program (though the EU spends close to US$10 billion annually in export subsidies).

Agricultural Research and Statistics: The USDA spends approximately US$3 billion annually on agricultural research, statistical information services and economic studies. In some instances, this taxpayer-funded research results in patented products that consumers (and farmers) must pay for again in the form of higher commodity (e.g. seed) prices.

Indirect Subsidies: Subsidies that indirectly reduce the cost of production (e.g. subsidized electricity or undervalued water) or increase demand for a commodity (e.g. subsidies for ethanol processing plants).

From the USDA (2006, p2), Edwards (2009) and Cairns Group (2010)

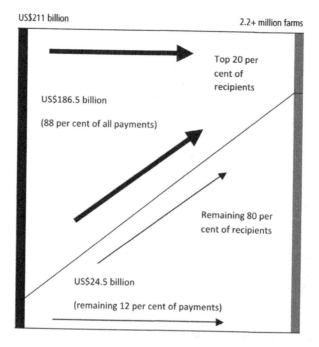

US$211 billion

2.2+ million farms

Top 20 per cent of recipients

US$186.5 billion

(88 per cent of all payments)

Remaining 80 per cent of recipients

US$24.5 billion

(remaining 12 per cent of payments)

Figure 9.2 *Asymmetries in US Department of Agriculture (USDA) farm payments, 1995–2009*

Source: compiled from the databank maintained by the Environmental Working Group (http://farm.ewg.org/progdetail.php?fips=00000&progcode=total&page=conc)

Box 9.2 Subsidizing drug traffickers and top government officials

A 2010 article in the *Los Angeles Times* exposes the major recipients of farm subsidies in Mexico (Wilkinson, 2010). What began as a programme (called Procampo) to protect small farmers in the age of NAFTA has since become a source of government welfare for families of notorious drug traffickers (most notably three children of billionaire drug lord Joaquin 'El Chapo' Guzman, head of the powerful Sinaloa cartel) and several senior government officials, including the country's agriculture minister. The *Times* also reports that one of Guzman's top associates, Victor Emilio Cazares, received more than US$100,000 from Procampo to subsidize his cattle operation (most large drug traffickers in Mexico maintain legitimate farming operations). Other benefactors include Mexico's agriculture minister, Francisco Javier Mayorga Castaneda. 80 per cent of the US$1.3 billion that was paid out in 2009 went to the largest 20 per cent of Mexican farmers.

Topping the list is the Dutch firm Campina (a Dutch dairy co-operative that merged with Royal Frieslands Foods in 2008). Between 1997 and 2009, Campina received 1.6 billion Euros in taxpayer money. In second place is the Denmark-based firm Arla Foods Amba (formed by the 2000 merger between Sweden's Arla and Denmark's MD Foods). Between 1999 and 2009, Arla Foods Amba received cheques totalling 952 million Euros. In third place: the London-based company Tate and Lyle Europe, who received 826 million Euros between 1999 and 2009. Three companies; ten years; 3.378 *billion* Euros: just a small fraction of the real cost of cheap food in Europe.

Understanding the asymmetrical distribution of farm subsidies – favouring those firms in least need of help – contextualizes why agriculture has taken its current shape. The high commodity prices of 2008 bring the impacts of this asymmetry into sharp focus. Driven by significant price increases from the previous year (the price of corn, soybeans and wheat were up 50, 60 and 50 per cent, respectively), net farm income in 2008 was 50 per cent higher than the previous ten-year average. A record average net farm income was recorded in 2008: US$86,864. Not a bad year, it seems, for US farmers. Disaggregate the data, however, and a different picture emerges. The year 2008 was a winner for some farmers, but definitely not all. In fact, thanks to the work of Timothy Wise and Alicia Harvie (2009, p1), we know 'that mid-sized family farmers actually saw lower incomes from farming operations in 2007 than they did in 2003, with high costs and reduced government support outpacing the rise in income from farm sales'.

How can this be, that farms can actually lose money during a time of record high farm gate prices? During times of record prices, incomes from farm sales were uniformly up for all farms, regardless of size. From 2003 and 2007, 'higher sales' farms – sales from US$100,000 to US$249,999 – saw a sizeable bump in income from US$11,795 to $17,303 (the vast majority of household income even for 'higher sales' operations comes from off the farm, which tells you something about how 'big' you need to be in order to live off the farm's income alone). But input costs increased too. Between 2003 and 2007, the costs of fuel and fertilizer increased 67 and 100 per cent, respectively. Moreover, as market prices increased, government payments decreased. According to Wise and Harvie (2009), the average farm payment dropped during this period from US$17,453 to US$8712. The drop in farm payments and the increase in input costs were collectively more than enough to offset increases in commodity prices for mid-sized family farms in 2007 and 2008.

Meanwhile, commercial farmers – sales greater than US$500,000 – enjoyed a net increase in income from farm sales, from US$130,263 in 2003 to US$189,547 in 2007. This roughly US$60,000 raise was more than enough to offset the US$12,196 reduction in farm payments, which dropped from US$46,675 in 2003 to US$34,479 in 2007 (Wise and Harvie, 2009, pp2–3).

There is a good chance you are familiar with the statistic concerning how much (or, more accurately, how little) of each dollar spent on food actually goes to the farmer. According to the USDA, in 2006, 19 cents of every dollar spent on US-grown food went to the farmer.[2] This figure was close to 50 cents at around the time of World War II. This is, however, a problematic statistic. For one thing, we spend considerably more dollars on food today than we did back in the 1940s. While the percentage of our annual income spent on food has dropped, we nevertheless spend more money on food (and buy more food) today than people did 70 years ago.

Rather than the cents on the dollar statistic, I prefer to look at price indices. A price index, quite simply, is a normalized average of prices for a commodity. It can be quite useful when seeking to compare prices over time and across geographical locations. When we compare the price indices for both agricultural inputs and outputs (farm gate prices), we find a rather startling divergence, as noted in Figure 9.3. The cost of inputs and farm gate prices held fairly steady until the mid 1960s. After that, the costs of farming greatly outpaced farm gate prices. Figure 9.3 shows that not only did farmers a half century ago see more of every dollar spent on food, they also kept more of what came back to them because the cost of doing business was considerably less.

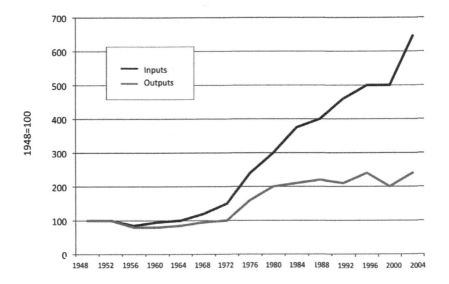

Figure 9.3 *Price index for agricultural outputs and inputs, 1948–2004 (US)*

Source: Fuglie et al (2007, p3)

To further understand the dynamics of Figure 9.3 requires that we move 'into' the hourglass, to the next link in the agri-food commodity chain: food processors and manufacturers. While subsidies represent a poster child of sorts for what's wrong with our food system, they are but a piece of the puzzle. As argued in earlier chapters, subsidies are quite effective at deafening producers to market signals. When market prices are high there is an incentive to produce fencerow to fencerow in order to maximize units produced. When market prices are low, there is also an incentive to produce fencerow to fencerow in order to maximize farm payments. Few farmers, however, really benefit from this constant overproduction (though, as detailed in Chapter 2, subsidies certainly give US growers a comparative advantage over their international competition). So who does?

Food Processors: Monopsony Has Its Privileges

Considerable market concentration lies in the 'link' just beyond the producer (though, as indicated in Table 9.2, the seed market is the most concentrated of the lot). This concentration is illustrated in Table 9.2. A popular measure of market concentration is the CR4, the four-firm concentration ratio. The CR4 is defined as the sum of the market shares of the top four firms for that industry. When reviewing potential mergers, the Federal Trade Commission (FTC) looks at the industry's current concentration ratio and estimates how it will change if the merger is given the green light. A standard rule of thumb is that when the CR4 reaches 20 per cent, a market is considered concentrated, 40 per cent is highly concentrated, and anything past 60 per cent indicates a significantly distorted market (Wise and Trist, 2010, p5). If this rule of thumb is to be believed, the market is anything but free at this point in the food system.

Table 9.2 *Concentration of US agricultural markets*

Sector	CR4*
Beef packers	83.5%
Soybean crushing	80.0%
Steer and heifer slaughter	79.0%
Pork packers	67.0%
Broilers	58.5%
Turkeys	55.0%
Pork producers	37.3%
Ethanol production	31.5%
Corn seed	CR1 80.0%
Soybean seed	CR1 93.0%
GE cotton seed	CR1 96.0%

* unless otherwise stated

Developed from Hendrickson and Heffernan (2007), Domina and Taylor (2010), Paarlberg (2010, p130), and Wise (2010)

CR4 statistics, however, reference macro-market conditions and ignore regional market realities, transportation restrictions and other issues that might restrict access to markets. For example, while the national CR4 for the broiler industry is a high 58.8 per cent, in some parts of the US the actual measure of buyer concentration is CR1 = 100 per cent (Domina and Taylor, 2010). In some regional markets, Tyson is the *only* game in town for farmers looking to sell their chickens.

The CR4 statistic is a measure of 'horizontal' concentration. It is therefore silent on the subject of 'vertical' concentration. For example, Smithfield Farms have captured (CR1) a 31 per cent market share of the pork packing industry. Additionally, Smithfield Farms raise 19.7 per cent of all hogs in the US. This duel concentration – horizontal and vertical – gives firms unique 'efficiencies' by allowing them to operate, at least as a buyer, under market conditions that are somewhat of its own choosing. This arrangement can eliminate what economists call 'price discovery'. As the company's cost for their livestock remains internal to the firm, market forces are not allowed to dictate ('discover') price. Packers who produce a significant quantity of hogs can also manipulate the market by flooding it with their own supply when lower prices would benefit them.

The recent 'food crisis' crippled urban poor around the world. The record high prices, as just discussed, failed to place any more money in the pockets of most mid-sized US growers. In light of this, it is rather surprising how well agribusinesses did when so many others were doing so poorly. In 2007, Cargill's, ADM's and Bunge's profits rose 36, 67 and 49 per cent, respectively. During the first quarter of 2008, at the very heart of the food crisis, Cargill, ADM and Bunge saw their net earnings rise 86, 55 and 189 per cent, respectively. And as fertilizer costs skyrocketed, the fertilizer industry watched its net quarter earnings multiply – for example, in the first quarter of 2008, Potash Corporation's net income rose 186 per cent while Mosaic's net income rose more than 1200 per cent (McMichael, 2009, p241). What's going on here?

'Monopoly' is a term most are familiar with. It refers to seller dominance in a market to the point where the monopolist is largely able to set the price of what it's selling (like Standard Oil did before its break-up in 1911). It's a statement of seller power. What we have in the food processing and manufacturing sector is buyer power, also known as monopsony. Under monopsony conditions, market concentration has reduced the number of potential buyers to the point that the seller has little option but to accept the price dictated by the buyer. This buyer power is expressed not only in the price that producers receive for their products, but also in buyers being able to set the conditions of almost every aspect of the farming operation while leaving the seller liable for most of the risk (Wise and Trist, 2010, p8).

Farmers are particularly susceptible to buyer power given the nature of what they produce. Many agricultural commodities are perishable. This is why animal protein (especially meat and milk) producers are particularly vulnerable to buyer power. Livestock producers rely on selling their animals at optimum weight (cattle are at their ideal slaughter weight for just a few weeks) and, in the case of hogs, in a timely manner to make room for the next litter. For dairy producers, the bulk tank only holds so much milk and must be emptied daily (occasionally multiple times a day), making it impossible to hold out for a better price. Livestock producers also have an incentive to avoid, when at all possible, distant markets. Shipping live animals long distances can be prohibitively expensive, increase animal mortality and carcass shrinkage, and result in a drop in the quality of their meat. When faced with these realities, farmers report feeling as though they frequently have no choice but to accept the 'take it or leave it' price offered by buyers (*Congressional Record*, 2000, p9320).

A review of the 2007 DOJ-approved Smithfield Farms–Premium Standard Farms merger concludes that it did not leave North Carolina pork producers with just one buyer, as many critics had argued (Heyer and Hill, 2008). This conclusion was based on the fact that independent hog sellers still had access to a national market. The nearest potential buyer for some sellers, however, was 400 miles (644km) away. So in addition to the risk of compromising the quality of their product, independent hog producers wishing to sell to someone other than Smithfield Farms must also incur the costs of shipping their animals great distances. Recognizing that a load of 200 hogs averages US$1.50 per head per 100 miles (160km) shipped, North Carolina hog producers must ask themselves if a slightly higher price, in light of these additional costs, is really worth sending their animals across state lines (Wise and Trist, 2010, p18).

When my grandpa took a load of hogs to market during the 1950s, he had a lot of options. There were at least ten buyers, each supplying a different slaughterhouse or butcher, within 10 miles (16km) of his farm.[3] As recently as 1976, the CR12 for pork packers was 27 per cent (Wise and Trist, 2010, p4). If my grandpa were still alive today and farming, he'd have far fewer buyers to pick from and those buyers would be owned by one or two processors. The earlier mentioned public forums hosted by the USDA and DOJ are full of stories like this – of farmers looking to sell their commodities in a regional market with CR1 and CR2 statistics approaching (or at) 100 per cent. Critics of any sort of regulation directed at opening these markets up will point to the prevalence of, say, grain elevators throughout the mid Midwest as evidence that monopsony conditions do not exist and that farmers do have a choice in who they sell their commodities to (US Senate, 2002, p62). But when these elevators are all owned by the same one or two firms, what sort of 'choice' is this?

Rising concentration in the retail sector also shapes the degree to which buyer power is aggressively pursued and utilized once attained. As discussed in the following section, large retail firms – such as Walmart and Kroger – are dealing increasingly with just a handful of very large packers, bypassing the wholesale sector entirely. Retail firms do this, in part, to exploit the buyer power held by large processing firms. Large retailers, in pursuit of the best price possible, are usually able to get a better deal from large firms such as Smithfield Farms (and large processors, in turn, benefit from massive orders and nation-wide visibility of their brand) than from smaller firms who do not wield the same influence over producers. The processors then pass the tighter margins on to producers, rather than absorbing them. This partially explains the growing gap between what producers are paid and retail prices, and why farmers receive less of every dollar spent on food with each passing decade. A study from 2004 found that the difference between the price paid to farmers and the prices faced by consumers increased by 149 per cent between 1970 and 1998 (Marsh and Brester, 2004).

Take the beef industry. The prices that producers receive for their beef cattle have fallen steadily over the last 20 years. Between 1981 and 1994, the net returns for fed cattle averaged US$36 a head. Between 1995 and 2008, this figure dropped to US$14 a head (Hauter, 2009, p20). One model calculates that the rapid escalation of the CR4 in the meat-packer market over the last couple decades has reduced profits to cattle producers by 31 per cent, while allowing packers to capture 55 per cent more profit than if the markets were not concentrated (Sexton, 2000, p1091). Given the tight margins that producers work with today, these are significant price impacts that can mean the difference between profitability and insolvency. Between 1999 and 2008, real consumer prices for ground beef increased by 24 per cent, going from a monthly average price (in 2009 dollars) of US$1.89 per pound to $2.34 per pound, whereas farm gate prices for beef cattle rose by 8.5 per cent (Hauter, 2009, p20).

Horizontal concentration is just one piece of the puzzle. Another is vertical integration. Smithfield Farms (the hog empire) has a presence at every 'link' of the commodity chain, from farm to fork, extending from production to packing, processing and, finally, the branding of the final retail product. An important linking mechanism, allowing processors to become involved in the production sector, is the contract. Contracts are most prevalent in livestock markets, where 94.2 per cent of poultry and egg production, 76.2 per cent of hog production and 17.6 per cent of beef cattle production were under contract in 2005 (Hauter, 2009, p10). Yet, contracts are not limited to the livestock market. In 2005, contracts covered 63.6 per cent of fruit production, 54.3 per cent of vegetable production, 19.6 per cent of corn production and 18.4 per cent of soybean production (Hauter, 2009, p10).

Under livestock production contracts, the contractor (the processing firm) owns the animals. Producers are responsible for building and managing the facilities – both usually to the contractors' specifications – and, in turn, receive all inputs from their buyer: animals, feed, veterinary services and transportation when it is time for the animals to be slaughtered. While an argument could be made that contracts benefit both parties, they unquestionably benefit contractors more.

First, the structure of the food system (as illustrated in Figure 9.1) leaves farmers with little negotiating power. Whereas processors have their pick of growers, farmers, conversely, may have only one or two processors from whom to obtain a contract. In such an asymmetrical relationship, growers lack what is known as 'exit power' – the power to walk away from the negotiating table knowing that other bidders are out there who will buy your product. When only one or two firms are handing out contracts, growers can't just hold out for someone better to come along offering a higher price.

Contracts also limit transparency – the all-important means towards price discovery – in the market. Most contracts come with confidentiality clauses that keep the terms and conditions from seeing the light of day (as well as from the eyes of other growers, who might be getting less). The evidence pointing to the fact that contracts distort market prices is clear. Looking at the hog market, for example, a study prepared for the USDA notes that for every 1 per cent increase in the number of hogs under contract, the open market price (what is known as the 'spot price') paid to producers dropped by 0.88 per cent (RTI International, 2007, pES10).

The almost universal use of contracts in some markets – most notably hogs – also creates near insurmountable barriers to entry for new farmers. Processors prefer dealing with large-scale, capital-intensive operations. Operations of this size, however, require tremendous amounts of capital. Yet, at the DOJ/USDA co-sponsored public forum that took place in Iowa, attendees complained that loans are becoming difficult to secure without already being in possession of a contract (DOJ/USDA, 2010a, p113). Chicken producers can invest as much as US$500,000 to US$1 million in facilities that have a 20- to 30-year economic life with no practical alternative use (Domina and Taylor, 2010). And then there are the required upgrades, which can cost a poultry grower an additional US$50,000 to $100,000 (*Meat Trade News Daily*, 2010). A medium-sized finishing hog operation with six 1100-head hog houses costs, on average, between US$600,000 and $900,000 to build (Hauter, 2009, p25).

Robert Taylor, a professor of agricultural economics at Auburn University, notes that the more indebted a grower is, the less likely they are to complain about their terms and risk bankruptcy. In a recent interview he notes: 'Growers are not getting a competitive return for their labor, management and risk. The relationship can work for a long time, but when push comes to shove, the integrator [contractor] holds all the cards' (*Meat Trade News Daily*, 2010). These points were brought to life during the DOJ/USDA co-sponsored public forum held in Alabama, when a poultry producer explained that 'when you have that kind of debt load over you, of course, you're going to choose to sign the contract. You feel that there's no other option when you owe, you know, a half a million dollars or a million dollars' (DOJ/USDA, 2010b, p85).

In a study prepared for the DOJ and USDA, Professor Robert Taylor and antitrust attorney David Domina (Taylor and Domina, 2010, p9) note that between 1995 and 2009 poultry growers in Alabama witnessed a negative net return in 10 of those 15 years. This translates to average cumulative losses for each grower in excess of US$180,000. Almost ten years ago Taylor (2002, p4) argued that 'contract producers who once had acceptable income from their poultry operations now put up a few hundred thousand dollars of equity, and borrow several hundred thousand more to hire themselves at minimum wage with no benefits and no real rate of return on their equity. Yet integrators continue to earn 10–25% rates of return on equity.' Since then the situation appears to have only gotten worse for growers.

The asymmetrical relationships just described are not unique to US growers. In India, for example, it has been reported that contracts drafted by Frito Lay for potato farmers allowed buyers to reject the farmers' product for any reason, even though sellers were responsible for transportation costs to the drop-off facility where the decision to purchase is made (McCullough et al, 2008, p26). Pea farmers in contract with Birds Eye in the UK are prohibited from selling their peas to anyone else, even those rejected by the company (Stuart, 2009, p117). Vertical integration is a worldwide reality, repeated around the world, from Brazil to Bangladesh to Thailand (see Box 9.3) (see, for example, Begum, 2005; Jabbar et al, 2007; Jepson et al, 2010).

Box 9.3 Contract poultry farming in Thailand

Thailand was an early leader in contract farming in Asia. This was due in large part to Thailand's largest business conglomerate, agribusiness firm Charoen Pokphand (CP), who established contracts with poultry farmers as early as the 1970s. The active promotion of contract farming by the Thailand government beginning in the 1980s has helped to rapidly expand the practice (Delforge, 2007, p4).

While contracts provide farmers with some short-term security, their one-sided nature when negotiated often makes them exploitive. One study looking at contract farming in Thailand found that many of the contracts were, indeed, one sided, favouring the buyer. For example, contracts with Frito Lay Thailand had farmers agreeing to sell their produce only to the company, while the company retained the right to reject any and all that the producers raised and forbid the farmers from selling to another party any unsold product (see, for example, Singh, 2006, p251). An investigation into contract farming by the Thai Senate Committee on Agriculture and Cooperatives reached similar conclusions. While the report acknowledges the potential of contract farming to expand and develop Thailand's agricultural sector while helping its small farmers, it admits that 'most of the contracts exploit farmers and producers. Farmers have to follow the conditions set by the processing factory which are not equitable' (cited in Delforge, 2007, p5). As contract farming has become more prevalent in Thailand since the 1980s, average farm income has been steadily on the decline (Kurian, 2004, p12).

Thai farmers under contract are also at risk of acquiring tremendous debt. The average debt among chicken producers interviewed for a 2004 study was approximately 241,034 baht, or US$6025 (Delforge, 2007, pp5–6). That same year, the average farm income was US$700 (Kurian, 2004, p12). And as long as they remain in debt, they become even more vulnerable. For, ultimately, an in-debt farmer is a farmer who can't afford to risk rocking the boat and upsetting their buyer, especially in Thailand, where Charoen Pokphand so thoroughly dominates the poultry market.

The last point I'll make about contracts is the additional barriers to entry that they can create for competing processors – something that is particularly a problem in livestock markets. The ease by which firms can enter and exit a market is an often-used indicator of its 'openness', recognizing that the more difficult the entry, the more concentrated the power is among firms already in the market. Packers looking to enter and successfully compete in a new market require a large plant running at full capacity. Obtaining that capacity, however, can be difficult if most of the surrounding producers are already locked into a multi-year contract with a competitor (Wise and Trist, 2010, p13). These barriers to entry, however, are more easily overcome by large market-diverse firms, who have the ability to cross-subsidize operations by operating in multiple markets of the food system. Highly profitable markets where firms are able to exert buyer power allow for the temporary absorption of losses in other markets that are being entered into (Wise, 2007, p78). In short, market diversity brings with it a type of buyer power that single market firms will never have. This explains why, for example, Tyson is a giant in pork, poultry and beef production. Similarly, JBS, the Brazilian multinational firm (who purchased Swift in 2007 and bought a majority stake in Pilgrim's Pride in 2009), not only dominates beef packing in South America, the US and Europe, but also casts a large shadow on the US poultry industry. Or take Cargill: a long-time world leader in the processing of grains who has, during the last 20 years, also become a major player in the pork packing industry (Wise and Trist, 2010, p6).

What do we mean by 'market efficiencies'? Too often, market efficiencies are equated to producing rock bottom prices. I heard this rationale played up a lot on the part of industry at the DOJ/USDA co-sponsored public forum on livestock concentration. The food system, industry spokespeople would say, is just doing what we want it to do, which is produce the cheapest food possible. If small independent producers can't compete, that is their problem. Why burden the system with their inefficiencies? Yet, this position naively ignores the inefficiencies that buyer and/or seller power introduces into a market.

When an industry begins consolidating, the goal is usually economic efficiencies, most notably economies of scale. The gains, however, are diminishing. After a certain level of consolidation, the rationale for any additional consolidation becomes market power gains (which are 'efficiencies' attained at the expense of market forces). The efficiency = cheap food logic limits redundancies, endangers public health and creates economic conditions inhospitable to mid-sized family farms (see Box 9.4). An example of how today's lean, mean and highly concentrated food system equals a vulnerable food system: in August, 2010, one of the largest egg producers in the US recalled some 380 million eggs after being linked to an outbreak of salmonella poisoning in Colorado, California and Minnesota. When a merger lowers prices simply because monopsony power is being exercised, this cannot be seen as a true efficiency. Indeed, as buyer power depresses farm gate prices below competitive levels, farmers have less of an incentive to produce. This risks leading to a situation where too few resources are available compared to what would be available in a more 'open' market (Domina and Taylor, 2010).

Box 9.4 Honey laundering

Approximately 1.2 million tonnes of honey are produced annually, 400,000 tonnes of which are traded on the international market. The largest exporting country is China, followed by Argentina and Mexico, whereas the largest importers of honey are the US, the EU and Japan (Michaud, 2005, p52). In what has been called 'the largest food fraud in US history' (Leeder, 2011), honey, it turns out, is at the centre of what has the makings of a major international conspiracy.

A five-month investigation by the newspaper *Seattle P-I* reveals that international honey traders are profiting widely from bee colony die-off in countries such as the US by 'resorting to elaborate schemes to dodge tariffs and health safeguards in order to dump cheap honey on the market' (Schneider, 2008). Among the *P-I*'s findings:

- Large shipments of tainted honey from China are laundered in other countries – an illegal practice known as 'trans-shipping' – to avoid US import fees, protective tariffs and/or taxes imposed on foreign products that intentionally undercut domestic prices (China has gotten into trouble in the past for 'dumping' honey in the US market). Tonnes of honey are produced in China but marked as tariff-free products from Russia.
- Mysteriously, many thousands of tonnes of honey enter into the US each year from Asian countries with no record of producing honey for export.
- The US government has yet to adopt a legal definition of 'honey'.
- For many years, China has used the animal antibiotic chloramphenicol to manage the health of their colonies. The US Food and Drug Administration (FDA) has banned the drug in any food product.
- More recently, two other antibiotics – iprofloxacin and enrofloxacin – have been found in imported Chinese honey (and blends of honey syrup).
- Some Chinese honey has had sugar or corn syrup mixed into it to improve its taste.

The difficulty of tracking the international movement of honey lies in the growing multinational reach of honey processors. With subsidiaries in countries such as Indonesia, Malaysia and Taiwan, giant Chinese honey processors can easily avoid the 'Made in China' label (Leeder, 2011).

In many countries (and especially in the US), for mergers to be halted (or monopsony power broken up) the burden of proof is on society to prove that concentration harms competition. The alternative would be to place the onus on the firms to prove their actions do not harm competition. Let me give an example. Chicken farmer Tom Terry started to ruffle feathers at Tyson Foods back in 2005 after asking to witness his chickens getting weighed. Plant managers repeatedly refused this request. Tom Terry further claimed he never received feed that Tyson says was delivered to his farm. And when birds in his houses got sick and died, he believed the mortalities were due to the quality of birds Tyson supplied him. Tyson, in response to the deaths, demanded that Tom Terry upgrade his chicken houses. Tyson eventually dropped its contract with Mr Terry. The battle then moved to the courts. With two rulings against him, five empty broiler houses that are now owned by the bank, and a US$57,000 bill (he must pay Tyson's legal fees), Tom Terry is running out of options. It's an uphill legal struggle for all growers who take firms such as Tyson on in a US federal court. Growers must prove that their injury resulted in an injury to competition, rather than showing that the individual growers themselves were harmed by the actions of the buyer. Given the current structure of the livestock sector, this is an impossible standard for growers. As Mr Terry explains: 'An individual cannot show harm to competition because their production is so infinitesimal to the whole of the industry' (*Meat Trade News Daily*, 2010).

Retail: From Push to Pull

A lot of ink has been spilled over the last decade in describing the power of agro-processors, of giants such as ADM and Cargill (see, for example, Troughton, 2005; Wilkinson, 2009; Domina and Taylor, 2010). The influence wielded in this link of the commodity is undeniable. But this influence still does not compare to the buyer *and* seller power held among the largest food retailers. In their book *Food Policy*, Tim Lang and colleagues (2009, p166) succinctly describe the shifting centre of power within food systems for all Western nations over the last 100 years. Prior to the 1900s, farmers were the key players. From 1900 to 1950, processors assumed the dominant position, which they continued to hold through the 1970s along with wholesalers. And from 1980 to the present, the centre of power moved to the retail sector, where it looks to stay for quite some time.

Figure 9.4 depicts sales in three sectors of the food system – the input industry, food processors and traders, and retail – during 2004 and 2006. As we can see, the 'pie' increased considerably in just two years, from US$1.177 trillion to US$1.540 trillion. Each sector's 'slice' also increased in size. Yet the rate of increase differed considerably across the three sectors. The retail sector is capturing more and more of every dollar spent on food. Looking at US sales (in 2005) among the top companies for the sectors of food manufacturing, grocery wholesalers, food-service wholesalers, grocery retailers and food-service provides another indicator of the size of 'slice' taken by the retail sector. Tyson Foods led food manufacturing, with US$23.9 billion in sales. Among grocery wholesalers, the leading firm was C & S Wholesale, with US$15.2 billion in sales. Sysco led the food-service wholesale industry with US$31.4 billion in sales. The food-service industry was led by McDonald's at US$26.9 billion in sales.

And with sales greater than all sector leaders *combined*, Walmart led grocery retailers with US$98.7 billion in sales, followed by Kroger (at US$58.5 billion), Albertsons (at US$36.3 billion) and Safeway (at US$32.7 billion) (Martinez, 2007, p46). Just in case you're wondering: Walmart's global sales are expected to once again top, for the third year in a row, US$400 billion (up from approximately US$250 billion in 2003) (Walmart, 2010).

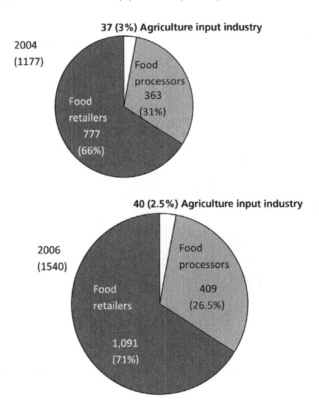

Figure 9.4 *Sales of top ten companies for three sectors of the global food system (in US$ billion)*

Source: developed from von Braun (2007, p4)

The power of processors lies largely in their ability to manipulate prices as a buyer. As sellers, however, their power is diminished considerably. Smithfield Farms, for example, would not dare to raise the price they charge Walmart for their pork. Walmart holds buyer power over Smithfield Farms, just as it does over nearly everyone it buys from. Unlike some aspects of the food system, however, which are highly concentrated at the global level (e.g. four privately owned companies – Cargill, ADM, Louis Dreyfus and Bunge – are responsible for most of the global grain trade), retail concentration remains highly variable across countries.

For example, independent grocers still represent 85 per cent of retail sales in Vietnam and 77 per cent in India (von Braun, 2007, p5). A CR2 statistic approaching 80 for the retail sector in Australia indicates remarkable concentration (Wardle and Baranovic, 2009, p477), while Indonesia, with a CR5 below 4, seems to have a thriving independent grocer market (von Braun, 2007, p5). The US lies somewhere between these two extremes, with a CR4 of 48.7 (and rising). Looking 'within' that figure, as indicated in Figure 9.5, we find Walmart with the largest market share, followed by Kroger, Safeway and Costco. CR3 ratios for the retail sectors in Sweden, The Netherlands, France, Spain, Greece and Italy are 95, 83, 64, 44, 32 and 32 per cent, respectively (Lang et al, 2009, p164).

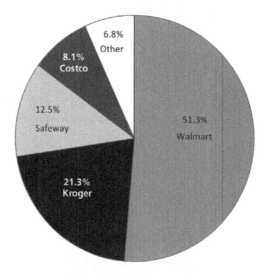

Figure 9.5 *US market share of top four supermarkets, 2008 (CR4 of 48.7)*

Source: redrawn from Richards and Pofahl (2010)

Large retail firms have tremendous buyer power because of the sheer volume in which they deal. They also prefer to deal directly with manufacturers and processors, reducing the power and influence of traditional wholesalers. General Mills and Kraft Foods, for instance, generate about 19 and 15 per cent, respectively, of their revenue through Walmart retail sales (*Bloomberg*, 2009). When so much revenue is dependent upon one contract, this gives the buyer tremendous negotiating power. As retailer insiders have explained in interviews, Walmart tells sellers what they're willing to pay and it's the sellers' problem to make that price profitable for them (see, for example, *Frontline Transcript*, 2004). Producers of commercial goods make the price work by shipping their production facilities overseas to where labour is cheaper, environmental regulations more lax and every day is a tax holiday. Food processors don't have to look overseas. They can preserve their margins by passing these price cuts on to farmers.

Between July 2007 and June 2009, the real price that farmers in the US received for milk fell by 49.3 per cent, but the retail price for fresh whole milk decreased only half as fast (declining by 22.6 per cent), while the price of cheddar cheese actually increased by 5.8 per cent (Hauter, 2009, p34). While consumer prices for dairy products, citrus fruits and peanuts tend to rise when farm gate prices increase, they rarely fall as quickly or as far when farm gate prices drop (Sexton et al, 2003). Over the past 20 years, the average farm gate price for tomatoes, potatoes and lettuce – the three vegetables that represent a significant share of US consumer vegetable purchases – has fallen, by 24.3, 15.7 and 3.5 per cent, respectively. During this period, however, real consumer prices for tomatoes and potatoes rose. For tomatoes, the real margin between farm gate and retail rose by 25.5 per cent, whereas the real margin for potatoes increased 8.2 per cent. And while the real price for lettuce fell by a few cents, this drop was about half the decline in its real farm gate price (Hauter, 2009, p36).

Time to challenge the one thing cheap food policy has going for it: producers' cheap – price-wise – food. Take corn. Nominally speaking, its price has rarely dropped below US$2 a bushel or gone above US$4 a bushel since 1970 (the US$7+ a bushel in 2008 was an anomaly). Looking at the real (inflation-adjusted) price of corn, however, shows it getting progressively cheaper with each passing decade. In 1974, the farm gate price, in 2008 dollars, was a whopping US$15.67 a bushel. By 2005, it had dropped to roughly US$2.50 a bushel. Even during the 'food crisis' in 2008, the real farm gate price of corn was less than half of what it was in 1974.[4] So, in terms of farm gate prices, 'food' today is incredibly cheap. But we don't eat, directly at least, field corn (or soybeans, wheat and cotton). It has to be processed first.

What about retail prices? Surely they, too, must have dropped over the decades. I know proponents of today's food policies like to point to the statistic that indicates how citizens in the US spend less of their annual income today on food than in past decades (indeed, throughout much of the world this figure is dropping). That may be. Yet, one can spend a lower percentage of one's annual income on food and still spend *more overall* on food than past generations. And that is precisely what is happening. Not only does the cheapness of this food rest on masking all the various 'costs' discussed throughout this book – as it turns out, this food isn't even as inexpensive we've been led to believe. Figure 9.6 shows the consumer price index for food in US cities from 1913 to 2006.

Two things stand out in Figure 9.6. First, the real retail price of food today is much higher than it was a century ago. Second, note specifically when the index price for retail food began to really take off. Food was becoming more expensive right around the time when cheap food policy was gathering steam during the 1970s, on the heels of our embracing the 'get big or get out' philosophy in the production sector and when market concentration in the processing sector began picking up speed. In recent decades, a significant price

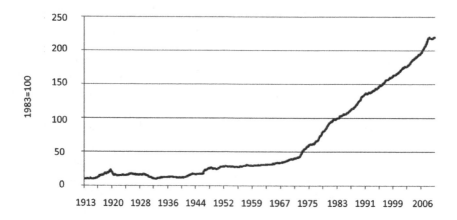

Figure 9.6 *Consumer price index for US urban consumers, food*

Source: data obtained from the Federal Reserve Bank of St Louis,
http://research.stlouisfed.org/fred2/series/CPIUFDNS/downloaddata?cid=9

accelerant has been the extra-value added products lining supermarket shelves. Today, more than half of total annual food purchases in the US are for food that is fully prepared (Tillotson, 2004, p618). Similarly, because the average American now eats one out of every five meals in his or her car (Trubek, 2008, pxiii), another layer of 'value' for many retail foods is portability.

Further concentrating power in the hands of the retail sector is the establishment of 'private labels', which refers to in-store brands. In light of the massive growth in private labels over the last decade, some scholars go so far as to argue that, with the exception of certain iconic brands such as Coca-Cola, 'the primacy within the food system of long-standing, mass-produced branded food lines is coming to an end' (Burch and Lawrence, 2007, p102). The top 30 private label grocery retailers in North America generate more than US$150 billion in private label sales, which constitutes 21.6 per cent of total grocery sales. Walmart dominates the private label grocery sector, generating US$32.4 billion in sales in 2009 (Planet Retail, 2010). For comparison, private labels constitute an even greater brand market share in Europe. In Switzerland, Germany, the UK, Spain, Belgium and France, for example, private labels make up 45, 30, 28, 25 and 24 per cent, respectively, of food retail sales (Vorley, 2007, p253). Some grocers rely heavily upon private label sales. Approximately 95 per cent of all sales for Aldi are of their private label, while for Tesco the figure is 45 per cent (Vorley, 2007, p254). As Figure 9.7 illustrates, the number of new foods and beverages introduced yearly in the US under a private label has increased considerably since 2005.

Private labels are sourced from a variety of suppliers. This again places retail chains in the envious position of being among 'the few' who can choose among 'the many' when looking for manufacturers for their store brands. Most large

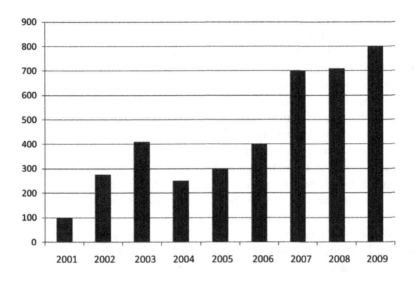

Figure 9.7 *Yearly food and beverage private label product introductions, 2001–2009 (US)*

Source: US Economic Research Service, USDA, www.ers.usda.gov/briefing/foodmarketingsystem

food manufacturing companies, such as Heinz, Pepsi Co and Unilever, now also produce private labels for large grocers. Food manufacturers have learned that saying 'No' only harms their own bottom line. Better to compete against a private label that you manufactured than compete against a private label that your competitor made. Take what happened to the then-Australian canned pineapple firm Golden Circle (acquired in 2009 by Heinz) when the nation's leading supermarket, Coles, requested that they produce a store brand line for the grocery chain. Out of concerns that such a contract would cut into their bottom line, Golden Circle turned down Coles's request. In response, Coles (and other Australian retailers) began looking overseas (specifically, Thailand and Indonesia) for their canned pineapple. It wasn't long until Golden Circle re-evaluated the offer and agreed to supply the supermarket chain with their own line of this canned fruit (Vorley, 2007, p107).

A survey taken in 2006 found that 41 per cent of shoppers already described themselves as frequent buyers of private labels, while only 12 per cent did in 1992 (PLMA Consumer Research Report, 2006). Marketing products that are nearly identical to national brands allows retailers to build loyalty for their own store. Yet, most importantly, it moves the pricing power to retailers (Richards and Hamilton, 2006; Richards and Pofahl, 2010).

Manufacturers are dependent upon grocery stores for their shelf space.[5] Yet, this dependency is not reciprocal, as retailers can shop around when looking for suppliers for their private label. Manufacturers have little negotiating leverage in this relationship. Another expression of this buyer power is what is known as slotting fees. These are essentially a pay-to-play (and then pay-to-stay) arrangement where food manufacturers pay grocers to stock their product (and/or stock their product in a prominent place). Slotting fees can cost as much as US$25,000 to US$40,000 per *item* per *store* (Shimp, 2009, p463). In some cases, slotting fees for new products have been known to exceed the product's sales revenue for the first year (Hauter, 2009, p41). Estimates place total US slotting fees paid by manufacturers at around US$10 billion annually (Burch and Lawrence, 2007, p108). Or, to put it another way, these fees reduce consumer and/or producer welfare by about US$10 billion annually because that's money that could have otherwise been used to increase farm gate prices 'upstream' or reduce retail prices 'downstream'.

While the buyer power of the retail sector is not in doubt, the question of seller (aka monopoly) power remains unaddressed. This, I admit, is harder to empirically document, especially when trying to establish monopoly power industry-wide. As I have already shown, the increasing buyer power held within the retail sector might have passed some cost savings onto consumers – though, remember, in absolute terms we are still spending more on food today than past generations. Those really helped by the buyer power, however, are large retail firms.

There is evidence that market retail concentration in the US is associated with higher grocery store prices (see, for example, Cotterill, 2006; Anders, 2008). One study sponsored by the USDA found that retailer buyer power enabled supermarkets to pay shippers of grapefruit, apples and lettuce less than what they would have received in a more competitive market. The study further concludes that consumer retail prices for these foods were inflated due to high levels of retail market concentration (Dimitri et al, 2003). In the UK, retailer mergers are estimated to have increased grocery prices by as much as 7 per cent (Cotterill, 2006). Another study, looking at concentration in the German food retail sector, calculates that retail market power accounts for 0.5 to 11 per cent of the retail unit margins of beef and pork and higher consumer prices (Anders, 2008).

As in the processing sector, industry-wide CR statistics tell us little about specific local and regional market conditions. While the CR4 ratio for the US retail sector as a whole is around 50, some markets have experienced much higher levels of concentration. Figure 9.8 shows the CR4 ratios for five major US retail markets: Atlanta, New York, Dallas, Chicago and Los Angeles. While all are above the national CR4 figure, Atlanta is shown to be particularly concentrated. In markets like this, where four firms essentially represent the extent of the 'choice' that consumers have for their retail food, consumers risk paying more for their food and thus becoming 'losers'.

There is evidence even at the neighbourhood level, in the absence of competition, of seller power at work. For example, in a Nebraska town a shopper bought identical items at two Walmart stores in the same community.

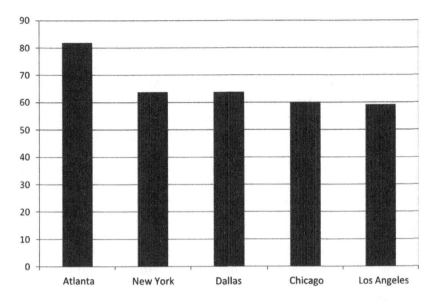

Figure 9.8 *Supermarket CR4 ratios in five major US markets, 2007*

Source: redrawn from Richards and Pofahl (2010)

The items in one shopping cart cost 17 per cent more than the other. The more expensive Walmart also happened to be in a neighbourhood where it had already put surrounding competitors out of businesses, which might explain the higher retail prices (Patel, 2009, p235). I have mentioned this example to others who are quick to dismiss it as a meaningless coincidence. After all, the shoppers living near the more expensive Walmart can just drive to the cheaper store. The expensive store, they tell me, is just pricing itself out of the market and will probably not remain in business long. Yet, this ignores a well-documented retail behavioural fact: shoppers tend to shop at their nearest store, irrespective of its ownership and, to some degree, even price (Guy, 2007, p178). Another factor at play in the Nebraska example is the simple fact that most people living in the expensive Walmart neighbourhood would never think to compare prices with the other Walmart, assuming that the two are identical in almost every way.

Large retail firms are also heavily subsidized by the government. Farmers are arguably paying twice for concentration in the retail sector, once through lower farm gate prices and then again through taxes. It has been estimated that public coffers lose more than US$1 billion a year through what is known as 'vendor discounts', of which over US$70 million goes to Walmart. A vendor discount is basically a service fee designed to compensate store owners for recording sales tax collections and remitting them to revenue agencies. When this practice was first adopted, shopkeepers kept records by hand, whereas today much of this 'work' is electric and automatic. Of the 26 states that provide compensation, 13 place ceilings on the amount that any individual store or chain can receive, ranging from US$1000 to $240,000 a year. The states without a ceiling give substantial amounts away to retail firms. Leading all states in this category is Illinois, losing roughly US$126 million annually (Mattera, 2008, pi).

An article by Jesse Drucker (2007) in *The Wall Street Journal* describes how a Walmart subsidiary pays rent to a real-estate investment trust owned by another Walmart subsidiary. The trust hands the rent to the second subsidiary in the form of a dividend that cannot be taxed. Moreover, Walmart counts the initial rental payment as a business expense, thereby allowing it to be deducted from taxes in the state where the store is located. In one four-year period, Drucker (2007) reports that Walmart avoided US$350 million in taxes using this strategy.

A study published by UC Berkeley's Labor Center concludes that California taxpayers are spending US$86 million annually providing healthcare and other public assistance (such as food stamps and subsidized housing) to the state's 44,000 Walmart employees (Dube and Jacobs, 2004). Two factors account for much of this expense: the company's low wages and the fact that 23 per cent fewer Walmart workers are covered by employer-sponsored health insurance than those of other large retail firms. The study estimates that if competing retailers were to adopt Walmart's wage and benefit levels, California's taxpayers would have to pay an additional US$410 million a year in public assistance.

US Representative George Miller, in a report published in 2004, arrived at remarkably similar conclusions. In addition to describing how the company blocks union-organizing efforts, pays employees an average US$8.23 an hour (compared to more than US$10 for an average supermarket worker), extracts off-the-clock work, and provides unaffordable healthcare packages for employees, the report contained the following remarkable figure: US$420,750 a year. According to Congressman Miller's report, a Walmart store with 200 workers costs federal taxpayers US$420,750 a year. The following are some of the expenses listed in the report for qualifying Walmart families, paid for by taxpayers:

- US$36,000 a year for free and reduced lunches;
- US$42,000 a year for Section 8 housing assistance;
- US$125,000 a year for federal tax credits and deductions;
- US$100,000 a year for the additional Title I (educational) expenses;
- US$108,000 a year for the additional federal healthcare costs of moving into State Children's Health Insurance Programs (S-CHIPs).

As the saying goes, there is no such thing as a free – or cheap – lunch. Walmart, and other firms like them, may be famous for their 'Always low prices'. But those low prices come at an expense – like at the expense of public coffers.

The Globalization of Retail

Compared to the processing sector, the internationalization of retail is a relatively recent phenomenon. Compared to markets in affluent countries, which have become saturated and highly price competitive, the retail sector in transitional economies is seen as holding great promise for global retail giants. While domestic firms continue to dominate the retail sector in many developing countries, transnational firms are putting their deep pockets to work buying access to these markets, such as through acquiring national retail chains, although, as Reardon and colleagues (undated) point out, domestic and state investment, especially in Asia, has helped to keep national firms competitive. A few recent examples:

- In May 2010, Walmart announced a US$1.3 billion deal to acquire stores in the UK from Denmark's Netto.
- In November 2010, Walmart announced a US$2.32 billion deal to buy 51 per cent of South Africa's Massmart Holdings Ltd.
- And in December 2010, Walmart announced that it had bought one of China's largest online retailers (called 360buy) for more than US$500 million.

In 2000, the top four global retailers (in terms of sales) – Walmart (US), Carrefour (France), Metro (Germany) and Tesco (UK) – operated slightly more than 13,500 outlets worldwide (Biles, 2008). At the end of 2010, that number had grown to 29,593.[6] And Walmart's recent aforementioned acquisitions indicate considerably more concentration for the sector globally in the years to come.

Take the case of Mexico. Prior to NAFTA, supermarkets accounted for roughly 20 per cent of all food sales in Mexico (Reardon and Bergedue, 2002). That figure now exceeds 60 per cent (Biles, 2008). Although a handful of foreign retail chains formed partnerships with Mexican firms during the 1990s, Walmart was the first to have its presence felt. Walmart's relationship with the Mexican consumer stretches back a generation, to 1981, when it acquired 49 per cent ownership in Futurama, a Mexican food and general merchandise combo chain. Yet it wasn't until 1991 with its joint venture agreement with Cifra that Walmart entered into the Mexican food retail sector in a big way (*Discount Store News*, 1999). By the turn of the century, after acquiring a majority stake in Cifra, all stores had changed their name to Walmart de Mexico. At least 3 out of every 10 pesos spent on food in Mexico now go to Walmart (Biles et al, 2007, p58).

The growing internationalization of retail firms is changing the structure of the commodity chain in ways that harm the livelihood of small-scale producers. Dolan and Humphrey (2004), for example, document how the production and distribution of fresh produce in Kenya initially displayed characteristics that truly enhanced the security of the country's smallholders. The rising influence of UK food retailers during the 1990s within this African country, however, has changed things. Their size and market dominance (buyer power) allows these firms to bypass traditional wholesale suppliers that purchase from whoever is looking to sell. In their stead, international retail firms, for reasons discussed earlier, are working increasingly with large-scale suppliers who have vertically integrated within the production sector through contracts.

To be clear, I am not criticizing the processes of globalization *per se*. Globalization is not an unalterable, exogenous process that operates uniformly at all scales in all places for all times (Kellner, 2002). Nor is it inherently destructive. For example, as discussed in Chapter 6, we need globalization if we wish to continue eating fresh fruits and vegetables (which we ought to eat more of, not less), while minimizing our ecological footprint. Globalization benefits some, marginalizes others (see, for example, Zimmerer, 2006). Let's return briefly to the case of Mexico to tease this point out.

Since 2000, when the conservative National Action Party (*Partido Acción Nacional*) assumed control of the Mexican presidency, agricultural policies primarily sought to improve the competitiveness of the agriculture sector within global markets. Rural economic development was, nevertheless, given considerable lip service by the ruling party. The primary policy vehicle implemented to aid the rural poor was the *Sistema Producto* initiative. Its explicit purpose: to strengthen supply-network linkages and improve access to market opportunities for small-scale producers. As a result of this initiative, workshops and public meetings were convened between small-scale farmers, buyers, food processors and other key players in the regional food system to improve integration and competitiveness of supply networks. The secretary of industrial and commercial development even sponsored trade fairs to link local growers with Walmart and large domestic supermarket chains (Biles et al, 2007, p65).

In the end, state government officials found it easier to focus their energies on a handful of key intermediaries as a means of improving 'market integration', rather than committing to the admittedly arduous task of offering small-scale producers the requisite training and resources to penetrate supply networks and export markets directly (Biles et al, 2007, p70). James Biles and colleagues (2007, p70) give the example of a large intermediary who distributes and processes chilli habaneros and who, in 2004, procured 80 per cent of its chillies from small growers. Shortly thereafter, retailers began demanding better standardized production processes, fixed prices (contracts) and uniform quality standards. Without aid from state agencies, small-scale farms lacked the technology, resources and knowledge to comply with these demands. The intermediary firm was eventually forced to look to a handful of large-scale commercial growers who were easily able to supply significant quantities at the predetermined price, while meeting the strict quality-control demands. By July 2006, the firm was sourcing over 75 per cent of its chilli habaneros from large-scale producers. In this case, it was not so much globalization *per se* that failed small-scale farmers as it was, for a variety of reasons, state officials.

And the cheap food lesson of this story? It tells us that affordable food is possible. It's just going to take a lot of work – on the part of people, private industry and, yes, public officials – to switch to a food system that is focused on 'affordability' rather than 'cheapness'. I think it's time we talk about what such a system might look like ...

Notes

1 See www.french-property.com/news/french_life/eu_farming_subsidies.
2 See www.ers.usda.gov/Briefing/FoodMarketingSystem/pricespreads.htm.
3 My dad tells the story, to further highlight how times have changed, of grandpa taking a load – 20 animals – of hogs to market in 1949 and returning with a brand new 1949 Plymouth purchased entirely with the sale's proceeds, plus having about US$500 left over.
4 See http://inflationdata.com/inflation/inflation_Articles/Corn_Inflation.asp.
5 Manufacturers, consequently, are looking to break free from their historical dependency on grocery stores. Kraft Foods, for example, which has been supplying vending machines for decades with its cookies and crackers, has expanded into direct-to-consumer distribution by establishing its own network of automated dispensers. This move not only allows Kraft Foods to sell directly to consumers, but to sell its products at a premium price.
6 See http://supermarketnews.com/profiles/top25-2010/top-25.

References

Anders, S. (2008) 'Imperfect competition in German food retailing: Evidence from state level data', *Atlantic Economic Journal*, vol 36, no 4, pp441–454

Begum, I. (2005) 'An assessment of vertically integrated contract poultry farming: A case study in Bangladesh', *International Journal of Poultry Science*, vol 4, no 3, pp167–176

Biles, J. (2008) 'Wal-Mart and the "supermarket revolution" in Mexico', *Geographische Rundschau International Edition*, vol 4, no 2, pp44–49

Biles, J., K. Brehm, A. Enrico, C. Kiendl, E. Morgan, A. Teachout and K. Vasquez (2007) 'Globalization of food retailing and transformation of supply networks: Consequences for small-scale agricultural producers in southeastern Mexico', *Journal of Latin American Geography*, vol 6, no 2, pp55–75

Bloomberg (2009) 'Wal-Mart's store-brand groceries to get new emphasis', *Bloomberg*, 19 February, www.bloomberg.com/apps/news?sid=afVJJxZ4oCtY&pid=newsarchive, last accessed 26 September 2010

Burch, D. and G. Lawrence (2007) 'Supermarket own brands, new foods and the reconfiguration of agri-food supply chains', in D. Burch and G. Lawrence (eds) *Supermarkets and Agri-Food Supply Chains: Transformations in the Production and Consumption of Foods*, Edward Elgar, Northampton, MA, pp100–128

Bussey, J. (2010) 'Latin 500: Walmart revs up regional presence', *Latin Trade*, 12 August, http://latintrade.com/2010/08/latin-500-walmart-revs-up-regional-presence, last accessed 30 December 2010

Cairns Group (2010) *Export Subsidies, Fact Sheet*, Cairns Group, Cairns, Australia, www.cairnsgroup.org/factsheets/export_subsidies.pdf, last accessed 22 September 2010

Congressional Record (2000) 'Agricultural Risk Protection Act of 2000 – conference report', *Congressional Record*, vol 146, part 7, pp8991–10402

Cotterill, R. (2006) 'Antitrust analysis of supermarkets: Global concerns playing out in local markets', *The Australian Journal of Agricultural and Resource Economics*, vol 50, no 1, pp17–32

Delforge, I. (2007) *Occasional Papers 2: Contract Farming in Thailand – A View from the Farm*, Report for Focus on the Global South, May, Bangkok, Thailand, www.esocialsciences.com/data/articles/Document11512200998.102053E-02.pdf, last accessed 31 December 2010

Dimitri, C., A. Tegene and P. Kaufman (2003) *US Fresh Produce Markets: Marketing Channels, Trade Practices, and Retail Pricing Behavior*, Agricultural Economic Report No (AER-825), September, www.ers.usda.gov/Publications/AER825, last accessed 27 September 2010

Discount Store News (1999) 'A partnership for the long haul – Walmart's involvement in Mexico', *Discount Store News*, October, http://findarticles.com/p/articles/mi_m3092/is_1999_Oct/ai_57578926, last accessed 1 January 2011

DOJ/USDA (Department of Justice/US Department of Agriculture) (2010a) *Public Workshops Exploring Competition Issues in Agriculture: A Dialogue on Competition Issues Facing Farmers in Today's Agricultural Marketplace*, Des Moines Area Community College, Ankeny, Iowa, www.justice.gov/atr/public/workshops/ag2010/iowa-agworkshop-transcript.pdf, last accessed 24 September 2010

DOJ/USDA (2010b) *Public Workshops Exploring Competition in Agriculture: Poultry Workshop*, Normal, Alabama, www.justice.gov/atr/public/workshops/ag2010/alabama-agworkshop-transcript.pdf, last accessed 24 September 2010

Dolan, C. and J. Humphrey (2004) 'Changing governance patterns in the trade in fresh vegetables between Africa and the United Kingdom', *Environment and Planning A*, vol 36, no 3, pp491–506

Domina, D. and C. R. Taylor (2010) 'The debilitating effects of concentration markets affecting agriculture', *Drake Journal of Agricultural Law*, vol 15, no 1, pp61–108

Drucker, J. (2007) 'Wal-Mart cuts taxes by paying rent to itself', *The Wall Street Journal*, 1 February, http://online.wsj.com/article/SB117027500505994065.html?mod=hps_us_pageone, last accessed 27 September 2010

Dube, A. and K. Jacobs (2004) *Hidden Cost of Wal-Mart Jobs: Use of Safety Net Programs by Wal-Mart Workers in California*, UC Berkeley Labor Center, Briefing Paper Series, 2 August, UC Berkeley, Berkeley, CA, http://laborcenter.berkeley.edu/retail/walmart.pdf, last accessed 27 September 2010

Edwards, C. (2009) *Agricultural Subsidies*, CATO Institute, Washington, DC, www.downsizinggovernment.org/agriculture/subsidies, last accessed 22 September 2010

Frontline Transcript (2004) 'Is Walmart good for America?', Public Broadcasting Service, www.pbs.org/wgbh/pages/frontline/shows/walmart/etc/script.html, last accessed 26 September 2010

Fuglie, K., J. McDonald and E. Ball (2007) *Productivity Growth in US Agriculture*, Economic Brief Number 9, United States Department of Agriculture, Economic Research Service, Washington, DC, www.ers.usda.gov/publications/EB9/eb9.pdf, last accessed 21 September 2010

Goldsmith, P. and R. Hirsch (2006) 'The Brazilian soybean complex', *Choices*, vol 21, no 2, www.choicesmagazine.org/2006-2/tilling/2006-2-11.htm, last accessed 30 December 2010

Guy, C. (2007) *Planning for Retail Development: A Critical View of the British*

Experience, Routledge, New York, NY

Hauter, W. (2009) *Agriculture and Antitrust Enforcement Issues in Our 21st Century Economy*, Comments submitted to the US Department of Justice and US Department of Agriculture on Agriculture and Antitrust Enforcement Issues in Our 21st Century Economy (74 Fed. Reg. 165 43725-43726), Food and Water Watch Institute, Washington, DC, 31 December

Hendrickson, M. and W. Heffernan (2007) *Concentration of Agricultural Markets, 2007*, Department of Rural Sociology, University of Missouri, Columbia, MO, www.foodcircles.missouri.edu/07contable.pdf, last accessed 22 September 2010

Heyer, K. and N. Hill (2008) 'The year in review: Economics at the antitrust division, 2007–2008', *Review of Industrial Organization*, vol 33, no 3, pp247–262

Jabbar, M., H. Rahman, R. Talkder and S. Raha (2007) *Formal and Informal Contract Farming in Poultry in Bangladesh*, Food and Agriculture Organization of the United Nations, Rome, Italy, www.fao.org/Ag/againfo/home/events/bangkok2007/docs/part3/3_6.pdf, last accessed 25 September 2010

Jepson, W., C. Brannstrom and A. Filippi (2010) 'Access regimes and regional land change in the Brazilian Cerrdo, 1972–2002', *Annals of the Association of American Geographers*, vol 100, no 1, pp87–111

Kellner, D. (2002) 'Theorizing globalization', *Sociological Theory*, vol 20, no 3, pp285–305

Key, N. and M. Roberts (2007) *Commodity Payments, Farm Business Survival and Farm Size Growth*, Economic Research Report, No ERR-51, Economic Research Service, USDA, Washington, DC, www.ers.usda.gov/Publications/ERR51/ERR51ref.pdf, last accessed 22 September 2010

Kurian, M. (2004) *Institutions for Integrated Water Resources Management in Upland Watersheds of Southeast Asia: A Comparative Analysis of Thailand and Lao PDR*, Working Paper 81, International Water Management Institute, Bangkok, Thailand, www.iwmi.cgiar.org/Publications/Working_Papers/working/WOR81.pdf, last accessed 31 December 2010

Lang, T., D. Barling and M. Caraher (2009) *Food Policy: Integrating Health, Environment, and Society*, Oxford University Press, New York, NY

Leeder, J. (2011) 'Honey laundering: The sour side of nature's golden sweetener', *Global Food Reporter* www.theglobeandmail.com/news/world/honey-laundering-the-sour-side-of-natures-golden-sweetener/article1859410, last accessed 11 January 2011

Marsh, J. and G. Brester (2004) 'Wholesale-retail marketing margins behavior in the beef and pork industries', *Journal of Agricultural and Resource Economics*, vol 29, no 1, pp45–64

Martinez, S. (2007) *The US Food Marketing System: Recent Developments, 1997–2006*, Report No 42, May, USDA, Economic Research Service, Washington, DC

Mattera, P. (2008) *Skimming the Sales Tax: How Wal-Mart and other Big Retailers (Legally) Keep a Cut of the Taxes We Pay on Everyday Purchases*, White Paper, November, Good Jobs First, Washington, DC, www.goodjobsfirst.org/pdf/skimming.pdf, last accessed 27 September 2010

McCullough, E., P. Pingali and K. Stamoulis (2008) 'Small farms and the transformation of food systems: An overview', in E. McCullough, P. Pingali and K. Stamoulis (eds) *The Transformation of Agri-Food Systems: Globalization, Supply Chains, and Smallholder Farmers*, Earthscan, London, pp3–46

McMichael, P. (2005) 'Biotechnology and food security: Profiting on insecurity', in L. Beneria and S. Bisnath (eds) *Global Tensions: Challenges and Opportunities in the World Economy*, Taylor and Francis e-Library, London, pp113–127

McMichael, P. (2009) 'Banking on agriculture: A review of the World Development Report 2008', *Journal of Agrarian Change*, vol 9, no 2, pp235–246

Meat Trade News Daily (2010) 'USA poultry processors under pressure', *Meat Trade News Daily*, 1 June, www.meattradenewsdaily.co.uk/news/010610/usa___poultry_processors_under_pressure_.aspx, last accessed 24 September 2010

Michaud, V. (2005) 'Antibiotic residues in honey', *APIACTA*, vol 40, pp52–54

Miller, G. (2004) *A Report by the Democratic Staff of the Committee on Education and the Workforce*, US House of Representatives, 16 February, http://wakeupwalmart.com/facts/miller-report.pdf, last accessed 30 September 2010

Morgan, D., G. Gaul and S. Cohen (2006) 'Farm program pays $1.3 billion to people who don't farm', *The Washington Post*, 2 July, www.washingtonpost.com/wp-dyn/content/article/2006/07/01/AR2006070100962.html, last accessed 21 September 2010

Paarlberg, R. (2010) *Food Politics: What Everyone Needs to Know*, Oxford University Press, New York, NY

Patel, R. (2009) *Stuffed and Starved: The Hidden Battle for the World Food System*, Melville House Publishing, Brooklyn, NY

Peterson, E. W. (2009) *A Billion Dollars a Day: The Economics and Politics of Agricultural Subsidies*, Wiley-Blackwell, Malden, MA

Planet Retail (2010) *Private Labeling in North America: Fertile Ground for Growth Summary*, Planet Retail, www.planetretail.net/OnlineShop/RetailShopHome.aspx?Section=2&PageID=6&ProfileTypeID=1&tocProductType=3&tocCatalogueID=61062, last accessed 27 September 2010

PLMA Consumer Research Report (2006) *Star Power: The Growing Influence of Store Brands in the US*, Ipsos MORI Survey for Private Label Manufacturers Association, Chicago, IL

Reardon, T. and J. A. Bergedue (2002) 'The rapid rise of supermarkets in Latin America: Challenges and opportunities for development', *Development Policy Review*, vol 20, no 4, pp371–388

Reardon, T., C. P. Timmer and B. Minten (undated) 'Supermarket revolution in Asia and emerging development strategies to include small farmers', *Proceedings of the National Academy of Sciences of the United States of America*, 6 December, www.pnas.org/content/early/2010/12/01/1003160108.full.pdf, last accessed 14 March 2011

Richards, T. and S. Hamilton (2006) 'Rivalry in price and variety among supermarket retailers', *American Journal of Agricultural Economics*, vol 88, no 3, pp710–726

Richards, T. and G. Pofahl (2010) 'Pricing power by supermarket retailers: A ghost in the machine?', *Choices*, vol 25, no 2, www.choicesmagazine.org/magazine/issue.php, last accessed 18 September 2010

RTI International (2007) *GIPSA Livestock and Meat Marketing Study*, vol 1, contract no 53–32KW-4-028, RTI International, Research Triangle Park, NC, http://archive.gipsa.usda.gov/psp/issues/livemarketstudy/LMMS_Vol_1.pdf, last accessed 24 September 2010

Schneider, A. (2008) 'Honey laundering: A sticky trail of intrigue and crime', *Seattle P-I*, 30 December, www.seattlepi.com/local/394053_honey30.asp, last accessed 11 January 2011

Sexton, R. (2000) 'Industrialization and consolidation in the US food sector: Implications for competition and welfare', *American Journal of Agricultural Economics*, vol 82, no 5, pp1087–1104

Sexton, R., M. Zhang and J. Chalfant (2003) *Grocery Retailer Behavior in the Procurement and Sale of Perishable Fresh Produce*, Report No 2, September,

USDA, Economic Research Service, http://ddr.nal.usda.gov/bitstream/10113/32806/1/CAT30930093.pdf, last accessed 27 September 2010

Shimp, T. (2009) *Advertising Promotion and Other Aspects of Integrated Marketing*, South-Western Cengage Learning, Mason, OH

Singh, S. (2006) *Contract Farming and the State: Experiences of Thailand and India*, Kalpaz Publications. Delhi, India

Stuart, T. (2009) *Waste: Uncovering the Global Food Scandal*, Norton, New York, NY

Taylor, R. (2002) *Restoring Economic Health to Contract Poultry Production*, Agriculture and Resource Policy Forum, College of Agriculture, Auburn University, Auburn, AL, www.competitivemarkets.com/library/academic_reports/2002/5-poultry.doc, last accessed 24 September 2010

Taylor, R. and D. Domina (2010) *Restoring Economic Health to Contract Poultry Production*, Report prepared for the Joint US Department of Justice and US Department of Agriculture/GIPSA Public Workshop on Competition Issues in the Poultry Industry, 21 May 2010, Normal, AL, www.dominalaw.com/ew_library_file/Restoring%20Economic%20Health%20to%20Contract%20Poultry%20Production.pdf, last accessed 24 September 2010

Tillotson, J. (2004) 'America's obesity: Conflicting public policies, industrial economic development, and unintended human consequences, *Annual Review of Nutrition*, vol 24, pp617–643.

Troughton, M. (2005) 'Fordism rampant: The model and reality, as applied to production, processing, and distribution in the North American agro-food system', in S. Essex, A. W. Gilg and R. B Yarwood (eds) *Rural Change and Sustainability*, CABI, Cambridge, MA, pp13–27

Trubek, A. (2008) *The Taste of Place: A Cultural Journey into the Terroir*, University of California Press, Los Angeles, CA

US Senate (2002) *Hearing Before a Subcommittee of the Committee on Appropriations United States Senate*, US Senate, 17 May

von Braun, J. (2007) *The World Food Situation: New Driving Forces and Required Actions*, Food Policy Report, December, International Food Policy Research Institute, Washington, DC, www.ifad.org/events/lectures/ifpri/pr18.pdf, last accessed 26 September 2010

Vorley, B. (2007) 'Supermarkets and agri-food supply chains in Europe: Partnership and protest', in D. Burch and G. Lawrence (eds) *Supermarkets and Agri-Food Supply Chains: Transformations in the Production and Consumption of Foods*, Edward Elgar, Northhampton, MA, pp243–265

Walmart (2010) *Walmart Appoints Vice Chairman Eduardo Castro-Wright President and CEO of Global.com and Global Sourcing; Bill Simon is Promoted to President and CEO of Walmart U.S*, Press Release, http://investors.walmartstores.com/phoenix.zhtml?c=112761&p=irol-newsArticle&ID=1442604&highlight=, last accessed 27 September 2010

Wardle, J. and M. Baranovic (2009) 'Is lacking of retail competition in the grocery sector a public health issue?', *Australian and New Zealand Journal of Public Health*, vol 33, no 5, pp477–481

Wilkinson, J. (2009) 'The globalization of agribusiness and developing world food systems', *Monthly Review*, September, www.monthlyreview.org/090907wilkinson.php, last accessed 24 September 2010

Wilkinson, T. (2010) 'Mexico farm subsidies are going astray', *Los Angeles Times*, 7 March, http://articles.latimes.com/print/2010/mar/07/world/la-fg-mexico-farm-subsidies7-2010mar07, last accessed 31 December 2010

Wise, T. (2007) 'Policy space for Mexican maize: Protecting agro-biodiversity by promoting rural livelihoods', *GDAE Working Paper No. 07-01*, February, http://ase.tufts.edu/gdae/Pubs/wp/07-01MexicanMaize.pdf, last accessed 14 March 2011

Wise, T. (2010) 'Monopolies are killing our farms: Editorial', *East Texas Review*, 13 April, www.ase.tufts.edu/gdae/Pubs/rp/WiseMonopoliesAndFarms13Apr10.pdf, last accessed 22 September 2010

Wise, T. and A. Harvie (2009) *Boom for Whom? Family Farmers Saw Lower On-Farm Income Despite High Prices*, Global Development and Environment Institute, Tufts University, Medford, MA, www.ase.tufts.edu/gdae/Pubs/rp/PB09-02BoomForWhomFeb09.pdf, last accessed 22 September 2010

Wise, T. and S. Trist (2010) *Buyer Power in the US Hog Markets: A Critical Review of the Literature*, Global Development and Environment Institute, Working Paper No 10-04, Tufts University, Medford, MA

Zimmerer, K. (ed) (2006) *Globalization and New Geographies of Conservation*, University of Chicago Press, Chicago, IL

10

Making Food Affordable

Cheap food, by nature of its design – a design that socializes many of its costs – is not affordable. We've got lots of cheap food – far too much, in fact. Conventional wisdom says the surest way to improve global food security is with cheap (and, if possible, cheap*er*) food. I hope I have turned this 'wisdom' on its head. Cheaper food is not the solution. It's the problem.

What is the ultimate goal of any food system? Everyone – *everyone* – I have posed this question to ultimately makes some reference to the ability to feed the world well into the foreseeable future. Cheap food policy will not – cannot – fulfil this goal. Too much of the world is undernourished in large part because of our infatuation with cheap food, not in spite of it. How can a system with so many 'losers' – from small farms to developing economies, the environment, animals, the poor, future generations, rural communities and taxpayers – be viewed as a success?

There is no easy to way make food affordable. Yet it has to be done. The market is not going to help us. Not, at least, without some coaxing from policymakers, politicians and concerned citizens, who need to roll up their sleeves and instruct the market in how to more accurately value the world. Politicians, if left unpressured, will not be fountainheads of social change either. Take the case of agricultural subsidies, which over the last century have been guided by a cheap food mentality. Individual gains among those benefiting from subsidies outweigh individual consumer and taxpayer losses that go along with maintaining the status quo. Consequently, the former have far greater incentive to act on behalf of maintaining subsidies than the latter do for seeking their elimination (Harvey, 2004, p266). Without sufficient mass mobilization from 'below', I do not see politicians taking on the immensely powerful interests that benefit from keeping food cheap. Critiquing cheap food in terms of its *real* costs offers, I believe, a narrative that many can get behind, regardless of their political and ideological affiliations.

I would like to offer in this chapter some suggestions about what an affordable food system might look like. I use the word 'system' in the previous sentence with some trepidation because one thing I can say unequivocally is that there is no one-size-fits-all glove when it comes to affordable food policy. A truly affordable food system will in the end be a composite of multiple, well-coordinated food systems, which vary in scale (big and small), length (global and local) and production technique.

Affordable Meat

Affordable meat is possible. Indeed, I see livestock playing a critical role in any affordable food system. It is the *amount* of animal products that those in the livestock industry think ought to be produced and consumed that is problematic, as are the methods that give us cheap meat.

A recently published study in the journal *Nature* addresses nitrous oxide (N_2O) emissions in one specific method of beef production (Wolf et al, 2010). It is commonly thought that pasture-fed ruminants raise atmospheric nitrous oxide levels because grazing disrupts grass's ability to draw nitrogen into the soil. A recently published *Nature* article, however, challenges this assumption, showing that grazing can actually *reduce* N_2O emissions. The research site was Inner Mongolia, where harsh winters alternate with temperate summers. Grazing keeps grasses short. Ungrazed tall grass traps the snow, forming an insulting layer that warms the soil. Grazed land, conversely, does not hold snow, thus exposing the ground to winter's full onslaught. This exposure kills many of the soil's microbes, including those that emit nitrous oxide. Come spring, the soil beneath ungrazed grass has a high population of N_2O emitting soil microbes, as well as significant moisture from snowmelt, which further amplifies microbial activity and, thus, nitrous oxide emissions. The London *Telegraph* published a story irresponsibly entitled 'Cows absolved of causing global warming with nitrous oxide' (Gray, 2010) when this article in *Nature* first appeared. Clearly, this is an overreach – if not an outright distortion – of the article's conclusions. The *Nature* study is a reminder, however, of how ecological 'hoof prints' vary considerably across the livestock industry (see also Liebig et al, 2010).

Management-intensive rotational grazing (MIRG) has received considerable attention in recent years for its ability to produce meat sustainably, humanely, and, yes, even profitably. A key component of MIRG systems is the utilization of short grazing episodes on relatively small parcels of pasture. Cattle are rotated between small plots, allowing plants sufficient time to recover and grow before the next grazing. This is especially important for the survival of high-quality and high-yielding foliage, which, under continuous pasture conditions, are eaten first and usually given insufficient time to recover before lower-quality (and lower-yielding), invasive plants (weeds) take over.

A growing body of research is touting the value of MIRG (e.g. Dartt et al, 1999; Mariola et al, 2005; National Research Council, 2010). Continuous grazing conditions often lead to soil compaction, diminished soil quality, reduced ground cover and the elimination of high-quality (and high-yielding) forage (Brummer and Moore, 2000). MIRG systems have a considerably lighter ecological hoof print, even though livestock numbers per hectare of land tend to be higher than in traditional pastoral grazing systems, as grazing is carefully monitored for any sign of diminished ecological value (Taylor and Neary, 2008; National Research Council, 2010). Rotational grazing schemes also rely upon much lower levels of artificial fertilizers and fossil fuels by allowing for the direct recycling of nutrients between livestock and pasture (Taylor and Neary 2008; National Research Council 2010).

MIRG systems compare favourably to conventional systems, offering equal or greater profitability per cow or unit milk output and higher quality of life levels for the animals (Mariola et al, 2005, pp28–36; National Research Council, 2010, p235). A study by the Center for Dairy Profitability at the University of Wisconsin-Madison compares MIRG dairies (where half of the herds' forage needs were obtained from grazing) with traditional confinement (TC) systems (50 to 75 cows) and large modern confinement (LMC) systems (more than 250 cows) (Kriegl and Frank, 2005). Spanning ten years, the study provides compelling economic data showing the profitability of an alternative to the total confinement model.

Figure 10.1 plots out the basic cost (which refers to paid labour, paid management, debt, investment level and depreciation) per hundredweight (cwt) for all three systems across the ten-year period. Note MIRG's lower costs, especially in relation to LMC systems. Lower capital requirements mean fewer barriers to entry for new dairy farmers (e.g. lower credit demands). Lower operating costs also typically means wider profit margins per unit produced, thus taking pressure off the producer to follow the 'get big or get out' mentality.

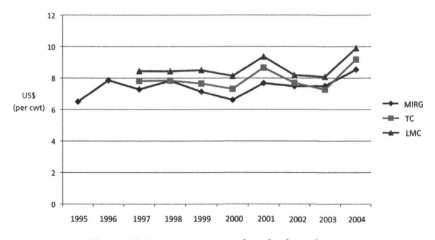

Figure 10.1 *Basic costs per hundredweight (cwt)*

Notes: Data not available for confinement farms in 1995 and 1996.

cwt = hundredweight; MIRG = management-intensive rotational grazing; TC = traditional confinement system; LMC = large modern confinement system.

Source: based on Kriegl and Frank (2005, p2)

Figure 10.2 shows the pounds of milk produced per cow annually in each of the three systems. Not surprisingly, calorically dense feed rations (formulated specifically for maximum milk yield), coupled with reduced mobility, puts 'traditional' and 'modern' confinement systems ahead of MIRG. It deserves mentioning, however, that these impressive yields are not without cost. Annual milk yields exceeding 20,000 pounds have been associated with decreased fertility, increased leg and metabolic problems, reduced longevity, mastitis (and other production diseases), higher disease incidence, and behaviour modifications indicative of a decline in animal welfare (Oltenacu and Broom, 2010).

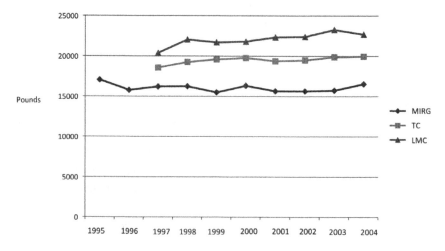

Figure 10.2 *Pounds of milk sold per cow annually*

Notes: Data not available for confinement farms in 1995 and 1996.

MIRG = management-intensive rotational grazing; TC = traditional confinement system; LMC = large modern confinement system.

Source: based on Kriegl and Frank (2005, p3)

As Figure 10.3 illustrates, however, LMC systems *need* such yields to give economic legitimacy to their existence. The figure depicts the net farm incomes for each of the three dairy systems per hundredweight equivalent of milk sold. As shown, MIRG systems greatly outperform TC and LMC systems when compared according to this economic parameter. Note also the remarkable income variability per unit for both TC and LMC systems. In 2002, for example, LMC systems recorded a net profit of *47 cents* per hundredweight of milk sold, whereas MIRG dairies rarely made less than (net) US$3 per hundredweight. While the authors do not explain this income variability, it seems a significant portion of it is due to the volatility of the feed market. This is both a curse and a blessing for MIRG dairies. As a result of satisfying a significant portion of the herds' dietary needs through grazing, MIRG dairies are insulated from high feed prices. At the same time, especially during years when grain prices are depressed, they miss out on being able to purchase high-yielding feed well below the cost of production.

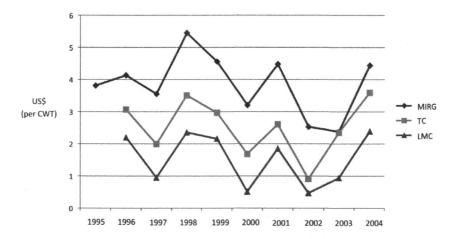

Figure 10.3 *Net farm income per hundredweight (cwt) equivalent of milk sold*

Notes: Data not available for confinement farms in 1995 and 1996.

cwt = hundredweight; MIRG = management-intensive rotational grazing; TC = traditional confinement system; LMC = large modern confinement system.

Source: based on Kriegl and Frank (2005, p4)

Finally, Figure 10.4 shows total net farm incomes for MIRG, TC and LMC systems. As one would expect (given levels of capital investment and total output), LMC systems generate a sizeable income. Yet, as the study's authors point out, while 'most of the farms in the MIRG and TC groups are owned and operated by one owner/operator family ... Many of the LMC farms are owned and operated by more than one owner/operator family.' Thus, depending upon the number of owners/investors, the actual average family income for LMC systems could be lower than that depicted in the figure. Note also the enormously irregular annual incomes associated with LMC operations, no doubt due, in part, to their dependence upon price-volatile inputs. A rather dramatic profit swing – of US$149,734 – occurred between 2001 (when LMCs averaged US$209,361) and 2002 (when they averaged a paltry US$59,616).[1]

In addition to being profitable, well-managed pasture dairy systems can also be remarkably *ecologically* efficient. Professor John Webster (University of Bristol, UK), former president of both the Nutrition Society and the British Society for Animal Science, calculates that dairy cows on a 70 to 80 per cent

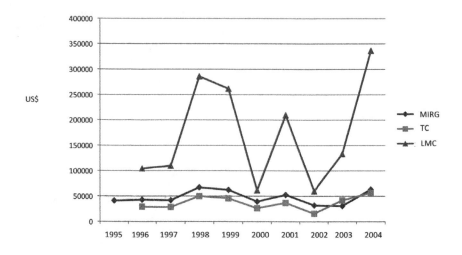

Figure 10.4 *Total net farm income between different dairy systems*

Notes: Data not available for confinement farms in 1995.

MIRG = management-intensive rotational grazing; TC = traditional confinement system; LMC = large modern confinement system.

Source: based on Kriegl and Frank (2005, p6)

pasture-based diet give back *more* calories than they take in – quite a bit more, in fact (Webster, 2010). Webster places the return at 170 per cent. Of course, this is due to a wonderful adaptive trait that allows ruminants to digest plant-based foods that we can't. As cellulose is the most common polymer in nature (Tudge, 2010, p12), finding a way to 'consume' it just makes sense when the goal is global food security.

Conversion efficiency matters less when talking about pasture-fed livestock. Arguably *any* conversion ratio of cellulose to animal protein is better than nothing in food-insecure regions rich in pasture but poor in arable farm land. Table 10.1 details population and land use by region. Note the variability in food-supporting land types. Some regions are well endowed with arable land, such as Europe, Russia and North America. Other countries are arable land poor but rich in permanent pasture. In sub-Saharan Africa, for example, only 17 per cent of land is arable, with most of the remaining in permanent pasture. It would be irresponsible to deny this region the ability to raise livestock given that, in some respects, animals are the only human food capable of 'growing' there.

Notice also the lack of good agricultural land in the tropics. Tropical soils are generally of a poorer quality than temperate zone soils (Achard et al, 2002). In the temperate zone, most of the life is 'down below', in the soil, while in the tropics it is 'up above', in foliage. This agricultural constraint means that tropical soils are easily depleted. Estimates place the rate of agricultural lands lost in the tropics at between 5 and 15 million hectares a year (Schade and Pimentel, 2010, p248). While less than 20 per cent of the land in Latin America and the Caribbean is (poor-quality) arable farmland, this region has an abundance of permanent pasture. Given these ecological conditions, affordable meat (and milk) might be the easiest food to produce. Livestock may even be able to improve the quality of tropical soils. Poultry and pigs could be used to root through the soil (pigs are especially good at this), picking up any loose ends while simultaneously cultivating and fertilizing.

Table 10.1 *Population and land use by region (2006)*

Region	Population	Ag. land (ha)	% arable	% permanent pasture
Sub-Saharan Africa	750,500,000	947,000,000	17.0	80.8
China and India	2,480,300,000	736,000,000	41.2	55.9
Asia (other)	1,503,000,000	635,000,000	24.8	68.7
Australia/New Zealand	24,700,000	457,000,000	10.8	88.7
Europe	588,100,000	267,000,000	60.4	34.0
Russia	143,200,000	216,000,000	56.7	42.5
Latin America/Caribbean	565,000,000	726,000,000	19.7	77.6
North Africa	408,800,000	458,000,000	19.1	78.4
North America	335,500,000	477,000,000	45.9	52.3
World	6,592,800,000	4,973,000,000	28.2	69.0

Data obtain from Peterson (2009, p54)

A beef producer and friend once told me how he hopes we reach a point when everyone in the world eats beef at least three times a day. 'Who's to say we can't?', was his final salvo in attempting to justify this position. 'The laws of nature', I told him. You see, while there is a place for animals in an affordable food system, we cannot expect everyone to eat like the average US citizen, who consumes on average a whopping 330 grams of meat a day (Stokstad, 2010, p810). This is not a matter of could we or should we. We simply can't.

Figure 10.5 illustrates average greenhouse gas emissions per kilocalorie of food for different categories. As depicted, animals vary considerably in terms of their greenhouse gas hoof print, with beef cows emitting the most per kilocalorie. For some contrast, a non-animal protein source – beans – has been included in the figure. The same variability lies in the ability of animals to convert grain to animal protein, with the beef cow, again, proving to be the least sustainable animal. The feed conversion ratio for beef cattle is at least 7 to 1 (though, as mentioned early, some estimate this ratio to be considerably higher). The ratio for pork is 3.5–4 to 1, while for poultry it is 2–3 to 1. Eggs are produced at a 2.01 to 1 ratio. And tilapia have a conversion ratio of 1.6–1.8 to 1 (FAO, 2006; Patnaik, 2009). A recent study in the journal *Climate Change* estimates the amount of land needed to produce 100kg of protein from various sources. For beef the figure was 0.6ha, while pork, pulses and milk require 0.36ha, 0.2 ha and 0.1ha of cropland, respectively (Stehfest et al, 2009, p83). In some parts of the world, such as desert communities, having a milk cow or two around may well be the one thing keeping a household food secure. Similarly, a beef cow living in a region rich in permanent (well-managed) pastureland is a magical creature of evolution, converting the inedible (grass) into the edible (meat).

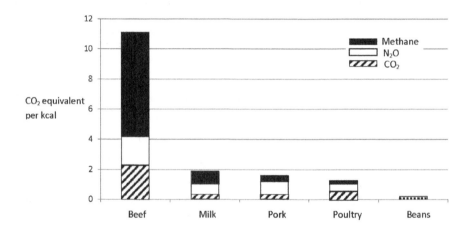

Figure 10.5 *Greenhouse gas emissions per unit (kcal) of food produced at farm gate level*

Note: Soil carbon balances are not included.

Source: based on Wirsenius and Hedenus (2010, p240)

Ruminants are not the only animals that possess the power to transform non-food into food. Monogastrics, too, have this ability. Pigs and poultry can thrive on food waste, which, as described in Chapter 6, there is plenty of in this world. Poultry and pigs fattened on food waste end up giving us *more* than what we put into them. One of the more famous examples of this comes from Egypt. In May 2009, the Egyptian government killed some 300,000 pigs in an attempt to control swine flu, even though no cases at the time had been reported within the country's borders. Soon after the mass-culling, however, Egyptian officials began to regret their decision. A mistake had been made, some said, as the trash that was once fed to pigs began to accumulate in putrid mounds of waste (Rogers, 2009). Prior to this action the Zabaleen – a primarily Christian community of garbage collectors who reside in slums on the outskirts of Cairo – would go door-to-door and collect organic waste. The animals consumed roughly 60 per cent of the collected waste. The Zabaleen ate some of the animals (religion forbids Muslims from eating pork), sending the rest to market where they brought their keepers a good price. Sadly, few families have been adequately compensated for the loss of this important income stream. The government originally promised to set aside US$5.4 million to compensate pig owners. Those who have been paid report receiving US$7 for each pig – a fraction of the US$90 to $180 that a pig sold at market would receive. Many pig owners, however, have received nothing for their destroyed animals (Stack, 2009).

The Nebraska Corn Board recently issued a press release denigrating grass-finished beef on account of its inability to feed the world, noting that there is insufficient pasture land to satisfy global beef demand, while ignoring the fact that there is not enough grain either: 'The fact is, in order to meet the demand of a growing world, which is expected to increase to a population of 9 billion by the year 2050, modern beef production is needed to feed the world this powerful protein' (Pope, 2010). No one would claim *financial* security lies in an investment that costs at least US$7 to earn $1. So how do we become *food* secure by putting in at least 7kg of grain to get 1kg of beef in return, especially in a world where grain, water and land are becoming increasingly precious resources?

A highly cited (and equally controversial) paper published in *The Lancet* argues that to stabilize greenhouse gas emissions from the livestock sector by 2050 relative to its contribution in 2005, the global per capita consumption of meat needs to be reduced from the current average. The global per capita consumption of meat today: 100 grams. The recommended average: (at most) 90 grams, out of which less than 50 grams can be derived from ruminants (McMichael et al, 2007). Both averages are well below per capita meat consumption levels of affluent countries – again, the average US mouth masticates roughly 330 grams of meat a day (Stokstad, 2010, p810). To reach the target level of 90 grams would require significant dietary restrictions in some parts of the world, and increases in meat consumption in others. Restrictions on meat consumption make sense from both ecological as well as public health standpoints. An influential study conducted by geophysical scientists at the University of Chicago calculates that a vegan diet in the US produces 1.5 tonnes fewer greenhouse gas emissions annually than the average US diet (Eshel and Martin, 2006, p13). Elsewhere it has been calculated that 45,000 early deaths could be prevented and UK£1.2 billion saved each year if the average UK diet consisted of less meat (specifically the authors based their calculations on diets where meat was eaten three times a week). This could cut annual deaths from heart disease by 31,000, from cancer by 9000, and from strokes by 5000 (see Box 10.1) (Friends of the Earth, 2010, p3).

Box 10.1 The social cost of individuals choosing to consume animal proteins

Oxford University's British Heart Foundation Health Promotion Research Group analysed the health implications of three diet scenarios: 'current diet trends', 'less meat' and 'fair less meat' (Friends of the Earth, 2010, p8). The diets are defined as follows:

- Current diet trends – the level of meat and dairy that will be eaten in the UK if nothing changes (177.7 grams, or 6oz, of meat and 332.2 grams, or 11oz, of milk daily.
- Less meat – consuming 70 grams (2.5oz) of meat and 142 grams (5oz) of milk daily (and more fruits and vegetables).
- Fair less meat – assuming a *fair* distribution across the UK, consuming 31 grams (1.1oz) of meat and 57 grams (2oz) of milk daily (and more fruits and vegetables).

A 'less meat' diet was found to reduce UK government expenditures by UK£0.85 billion annually. The breakdown of these savings is as follows: UK£0.57 billion from reduced heat disease, UK£0.07 billion from reduced stroke incidents and UK£0.20 billion from reduced cancer rates. A 'fair less meat' diet, on the other hand, would save British taxpayers UK£1.20 billion annually: UK£0.80 billion, £0.10 billion and £0.30 billion from reduced heart disease, strokes and cancer, respectively.

Some countries may need to rely more on the consumption of animal products because of a lack of insufficient arable farmland. In cases like this, the consumption of farmed animal products may, in fact, produce the least greenhouse gas emissions compared to alternatives that would need to be imported from great distances (Deckers, 2010, p499). Discussions about eating less meat also need to address the question of what people would replace meat with. As noted recently in the journal *Science*, eating less meat could backfire and make food insecurity worse (Stokstad, 2010, p810). Researchers modelling different reduced meat consumption scenarios found that when consumers in developed countries replaced meat with pasta and bread, world wheat prices rose considerably, which produced increased food insecurity in countries such as India that rely heavily on wheat.

Then there is the question of how to get people to eat less meat. One way to do this would be to put a price – such as a tax or emission trading scheme – on greenhouse gas emissions linked to animal agriculture. More accurately pricing the cost of meat would undoubtedly lead to the adoption of less greenhouse gas-intensive livestock-rearing practices, while providing an incentive for developing new technologies and implementing existing ones (Popp et al, 2010, p459). Placing an economic value on an animal's greenhouse gas hoof print would also make consumers 'choose' to eat further down the food chain, thus freeing up grain for people and more efficient monogastrics (such as poultry and fish).

An emission trading scheme also has tremendous economic development potential in poor countries. A recently produced study by the International Livestock Research Institute finds that reductions in greenhouse gases could be worth US$1.3 billion to poor farmers if they are allowed to sell saved carbon credits in international markets (Thornton and Herrero, 2010). According to the study, these emissions reductions could be achieved with relative ease in tropical countries by modifying production practices (such as switching to more nutritious pasture grasses), restoring degraded grazing lands, planting trees to both trap carbon and produce leaves that cows can eat, and adopting more productive breeds (see also International Livestock Research Institute, 2010).

More Food, Less Fuel

'You care more about feeding your cars than you do people'. That's what I was told about a year ago by someone who was visiting my university from Nigeria. He was referencing the love affair that Western countries appear to have towards biofuels (though this affection appears to be spreading as countries such as China jump headlong into biofuel investments). Of course, biofuels have their critics, even in countries like the US and China, where hundreds of millions of dollars are spent annually on agro-fuels. Yet, the European Union (EU) decision in 2008 to adopt a 10 per cent target (10 per cent of transport fuels should come from renewable sources by 2020) and the very recent decision in the US to bump the E10 blend (10 per cent ethanol/90 per cent gasoline) standard up to E15 (15 per cent ethanol/85 per cent gasoline) shows those critics to be in the minority in agro-energy policy circles (though there are indications that the EU is contemplating reversing its 2008 decision; see, for example, Harrison and Dunmore, 2010). Is the gentlemen from Nigeria correct: by promoting biofuels, are we tacitly (if not explicitly) choosing cars over the hungry?

The question is meant to be rhetorical, as the situation is not so black and white. But biofuel policy is the offspring of cheap food policy, for if the price of agricultural commodities better reflected their real cost, there would be no economic justification for diverting grain and/or arable land away from food and towards fuel production. Two-thirds of the world's biofuel demand is expected to be met by the production of jatropha, sugar cane and palm oil in developing countries (many of whom are food insecure) (Vidal, 2010). Just to meet the EU 10 per cent target would require 17.5 million hectares, well over half the size of Italy (Rice, 2010, p4). In the US, roughly one quarter of the domestic corn crop goes into the making of ethanol for cars, which is sufficient grain to feed 330 million people for one year at average world consumption levels (Humphries, 2010).

The claim that biofuels are sustainable is also, at best, questionable (see Box 10.2). The majority of biofuels require copious amounts of nitrogen and

Box 10.2 How 'green' are biofuels?: A review of the literature

A search was conducted on four peer-reviewed journals – *Science*, *Natural Resources Research*, *Renewable Energy* and *Journal of Cleaner Production* – from 2006 to 2008 for original research looking into the virtues and drawbacks of biofuels. These journals were selected because they represent well-respected outlets for research on this topic. Two key terms were searched: 'biofuels' and 'ethanol' (searching 'biofuels' alone often missed articles on ethanol). In the journals *Science* and *Natural Resources Research*, a title search was conducted using these terms, while in *Renewable Energy* and *Journal of Cleaner Production*, a title, abstract and keyword search was used (the divergence in parameters is an artefact of the search properties offered by each journal). These search criteria produced ten articles, which were then analysed and coded for specific costs and benefits linked to biofuels. Fourteen such dimensions were recorded. Articles were also categorized according to whether they examined biofuels derived from

Table 10.2 *Summary of review findings*

	Agricultural foodstock							
Article ID #	1	2	3	4	5	6	7	8
Biodiversity	↓	X	X	X	X	X	X	X
Global warming	X	X	↑	↑	X	↓	↓	X
CO_2 (net)	↑	↑	↑	↑	↑	X	↓	↓
CO_2 (auto)	X	X	X	X	X	↓	X	X
SO_x	X	X	X	X	X	↓	↑	X
NO_x	X	X	X	X	↑	NC	↑	↓
CO	X	X	X	X	X	X	X	↓
Ecological toxicity	X	↑	↑	↑	X	X	X	X
Hydrocarbons	X	X	X	X	X	X	X	X
Water insecurity	X	X	↑	↑	X	X	X	X
Net energy	X	X	↑	↑	X	X	X	↑
Food displacement	↑	X	↑	↑	X	X	X	X
Soil erosion	X	X	↑	↑	X	X	X	X
VOCs	X	X	X	X	X	↑	↓	X

↑   Increase impact for biofuel
↓   Decrease impact for biofuel
NC   No significant change reported
X   Not assessed
Source: Carolan (2009), permission to reproduce obtained from publisher

foodstock or waste/non-foodstock (one article did not specify the biomass source of the gasoline–alcohol blend examined). To clarify, 'foodstock' refers to biofuels developed using products that otherwise would have entered into the food system (either as food or feed); 'waste/non-foodstock' refers to fuels derived from products (typically) not part of the food system (e.g. biomass waste, perennials). Table 10.2 offers a summary of the results. Two general comments can be made about the findings:

1   Biofuels cannot be treated monolithically; each has its own virtues and drawbacks (though some clearly have more of one than the other).
2   Second-generation non-foodstock biofuels seem to have fewer costs than the food-based biofuels of today.

				Waste/non-foodstock				Unclear
9	10	11	Average (1–11)	12	13	14	Average (12–14)	15
X	X	X	↓	X	↑	↑	↑	X
↑	X	X	↑	↑	↓	↓	↓	X
↑	X	X	↑	↑	↓	NC	mixed	X
X	NC	↓	↓	X	X	X	X	↓
X	X	X	mixed	X	X	X	X	X
X	X	X	↑	X	X	X	X	X
X	X	X	↓	X	X	X	X	X
X	X	X	↑	X	↓	X	↓	X
X	↑	↑	↑	X	X	X	X	↓
X	X	X	↑	X	↓	X	↓	X
X	X	X	↑	X	↑	X	↑	X
X	X	X	↑	X	NC	NC	NC	X
X	X	X	↑	X	↓	X	↓	X
X	X	X	mixed	X	X	X	X	X

phosphorus fertilizer: the former releases nitrous oxide into the atmosphere (a greenhouse gas with an atmospheric warming potential some 300 times greater than carbon dioxide) (see Chapter 5); the latter is a life-giving element that is quickly running out (see Chapter 6); both are responsible for the eutrophication of the world's rivers, lakes and oceans (see Chapters 6 and 7). Large-scale biofuel plantations also increase carbon dioxide emissions. This can occur either directly, by cutting down and ploughing up forests or cultivating other carbon-rich habitats (thereby releasing previously trapped CO_2 into the atmosphere), or indirectly, by displacing farmers, such as by way of land grabs (see Chapter 3), onto previously uncultivated land (Fargione et al, 2008; Humphries, 2010).

From Green Revolution to Rainbow Evolutions

The gains of the Green Revolution have not been evenly distributed. Over the last two decades, the number of poor in Africa has doubled to 300 million – that's over 40 per cent of the continent's population (World Bank, 2008, p26). More than 2 billion people in the world today are deficient in key vitamins and minerals, most notably vitamin A, iodine, iron and zinc (UNICEF et al, 2007, p1). Half of the world's population is food insecure (Dybas, 2009, p646).

That's today. We will need to produce anywhere between 70 (FAO, 2009) to 100 (Schade and Pimentel, 2010, p247) per cent more food by 2050. Some 2.3 billion more mouths – for a total of 9 billion – will need to be fed by 2050. Unfortunately, most of this growth will occur in developing countries (such as Africa), where food security is already tenuous. Annual average global crop production growth is expected to slow from 2.2 per cent (as recorded between 1997 and 2007) to 1.3 per cent between now and 2030, slowing still further to 0.8 per cent between 2030 and 2050. In developing countries, yield growth looks to slow even faster, from 2.9 to 1.5 to 0.9 per cent during these respective periods (Bruinsma, 2009; see also Balmford et al, 2005). So if all goes as planned, yield increases (intensive means) will take care of about half of the world's projected needs by 2050 (Schade and Pimentel, 2010, p247). And the rest?

In 2002, then UN Secretary General Kofi Annan asked an appointed panel of experts from such countries as Brazil, China, Mexico and South Africa how a Green Revolution could be achieved in Africa. After more than a year of study, the group had their answer. Foremost, they questioned the one-size-fits-all approach to food security taken by the Green Revolution: 'The diverse African situation implies that no single magic "technological bullet" is available for radically improving African agriculture' (InterAcademy Council, 2003, pxviii). The panel's strategic recommendations explain that 'African agriculture is more likely to experience numerous "rainbow evolutions" that differ in nature and extent among the many systems, rather than one Green Revolution as in Asia' (InterAcademy Council, 2003, pxviii).

The Green Revolution is publicly about increasing yields. Yet, increased yields of one or two strains of one or two crops – what is known as 'monoculture within monoculture' – will do little to make Africa more food secure. The diversity of African ecological and farming systems makes single magic bullet remedies all the more unrealistic. Rather than initiatives that focus solely on monocultures within monocultures, Africa needs a wide range of programmes, which encourage intercropping onto permaculture and that value both traditional ecological knowledge and modern plant genetics (Thompson, 2007).

Less publicly, the Green Revolution is also about reducing biodiversity. Covering the world with high-yielding varieties will *inevitably* make the world less biologically diverse. Yet, food security, especially in regions such as Africa, requires *greater* biodiversity, not less. To wipe out the roughly 2000 plants that currently feed the continent and replace them with a handful of high-yielding varieties with certain soil, water and input requirements would make conditions worse rather than better (Thompson, 2007). Consequently, the Food and Agriculture Organization of the United Nations (FAO) now promotes the practice of African farmers breeding a variety of different seed (*in situ* conservation), viewing it as a better method of protecting biodiversity than collecting and storing seed in large gene banks (*ex situ* conservation) (Thompson, 2007).

It has already been established that yield increases alone will come nowhere close to feeding the world in 2050. If we cannot increase yields (intensive means), then the next logical option would be to increase the amount of the land under cultivation (extensive means). To make up for the yield shortfall, we are going to need anywhere between 200 million and 750 million additional hectares of land by 2050 (Tilman et al, 2001; Vance, 2001; Balmford et al, 2005; Schade and Pimentel, 2010).

A number of studies have settled on the figure of 1.5 billion as the amount of additional hectares available to be brought under cultivation (Koning et al, 2008; Bruinsma, 2009, p10). Much needs to be accomplished, however, before land can be brought into production – land rights have to be settled, credit must be available, and an infrastructure and market must be in place. These constraints explain why arable land worldwide has grown by a net average of 5 million hectares per year over the last two decades (Rabobank, 2010, p14). It also means that it will be decades until a sizeable amount of arable land is prepared for agriculture. More problematic still is the projected slowdown in the annual growth of arable land, as potential arable land becomes increasingly marginal (the land easiest to convert has already been brought into production). The annual growth of arable land will slow from 0.30 per cent between 1961 and 2005 to 0.10 per cent between 2005 and 2050. This calculates out to an average annual net increase of arable area of 2.75 million hectares per year between 2005 and 2050 (Rabobank, 2010, p14), or a total of 120 million additional hectares – well below the most optimistic estimates that claim only 200 million hectares will be needed by mid century to satisfy global food demand. At the same time, arable land in developed and transitional countries will continue to decline, losing an estimated 0.23 per cent a year between 2005 and 2050 (Bruinsma, 2009, p13).

The distribution of this available arable land is also very 'lumpy'. 66 per cent of available new arable land is located in developing countries and roughly 80 per cent of this amount is concentrated in Latin America and sub-Saharan Africa. Conversely, agricultural expansion in South Asia, the Middle East and North Africa – regions with some of the highest population growth in the world – is next to impossible as most available arable land has already been brought into cultivation (Rabobank, 2010, p14).

Out-of-the-box thinking will be required in the decades to come if we hope to feed the world in a way that's fair and just for all involved, including animals and the environment (see Box 10.3). The one-size-fits-all approach to food security that the Green Revolution rests upon is unacceptable. To prescribe, then, a *Greener* Revolution as the elixir to our food problems (see, for example, Beddington, 2010) makes me wonder if we have not learned anything the first time around (it reminds me of my four-year-old who, after getting a stomach ache from eating too much, thinks eating more would make the pain go away). The term 'rainbow evolutions' alters the scope of the discussion. First, it turns our attention away from looking for that magic bullet – hence, the term 'rainbow' rather than 'green'. But it also emphasizes – which is why the word 'evolutions' is so apt – how agricultural development cannot be divorced from the social, economic and agro-ecological conditions of place. The remainder of the chapter provides examples of what some of these evolving rainbows might look like.

Box 10.3 Ocean agriculture

While the oceans cover 70 per cent of Earth and hold 97 per cent of its water, they yield just a fraction of our food. Capture fisheries and marine aquaculture produce about 1.2 and 0.5 per cent by weight, respectively, of the world's total food supply (Forster, 2010, p96). The idea of seaweed/algae food security is nothing new. Considerable ink was spilled during the 1970s on the subject as a world oil crisis forced many to critically re-evaluate cheap food policy (see, for example, Chedd, 1975, p574). The most recent productivity data from these cellulose sea-farms is nothing short of remarkable. Chinese seaweed farms producing *Laminaria* (a type of brown algae), for example, averaged 19.4 metric tonnes dry weight per hectare in 2004. At this level, it would require a scant 0.9 per cent of the ocean's surface to produce enough seaweed to equal all the food from plants currently produced by land-based agriculture (Forster, 2010, p100).

Old McDonald Had a Farm ... In the City

Approximately 70 per cent of the world's population is expected to live in urban areas in 2050 (in 2011 that figure is around 50 per cent) (Rabobank, 2010, p6).

While rural poverty is rife in the developing world, the number of urban poor in these countries is expected to rise rapidly in decades to come. In these swiftly urbanizing nations, the poor are moving to the city at a faster rate than the population as a whole (Ravallion et al, 2007, p2). If current trends continue, the world's slum population – those living in irregular settlements, without access to suitable food, water, shelter and sanitation – will increase 50 per cent between now and 2020 (Mougeot, 2005, p1). The urban poor are highly vulnerable to food market volatility, which makes food security – if ever attained – a precarious state at best.

Most urban poor in the developing world spend the majority of their incomes on food. An estimated 60 per cent of all household income generated by the poor in Asia goes to buying food, compared to Canada and the US, where the figure is closer to 10 per cent (Redwood, 2010, p5). Food insecurity among the urban poor is further exacerbated by diets composed heavily of tradable commodities (versus traditional foods) and a lack of space to grow their own food (Redwood, 2010, p5).

Whenever I talk about urban agriculture – whether in the classroom, among agriculture professionals or in the context of public talks – few seem to view it as a viable strategy for promoting food security. Indeed, it is surprising how few know the role that urban agriculture *currently plays* in keeping millions from falling over the edge into abject hunger and poverty. A recently published study, spanning 15 developing countries located in Africa, Asia, Eastern Europe and Latin America, assesses the role of urban agriculture in promoting food and economic security and diet diversity among the urban poor (Zezza and Tasciotti, 2010). I was even surprised to learn just how significant urban agriculture is in the lives of some of the world's less fortunate city dwellers.

Participation in urban agriculture among poor households in the countries studied was as high as 81 per cent (see Figure 10.6). Urban agriculture was also shown to be of major economic significance, making up over 50 per cent of the income for poor households in certain countries (see Figure 10.7). Two-thirds of the countries analysed were shown to have a correlation between active participation in agricultural activities and greater household dietary diversity, even after controlling for economic welfare and other household characteristics. There was also a tentative link between urban agriculture participation and the consumption of fruits and vegetables (Zezza and Tasciotti, 2010).

Urban gardening (and urban livestock) requires minimal capital investment (save for perhaps the land), making this a practical strategy for improving food and financial security amongst the urban poor. But we are going to need more than a couple of raised beds on every city block if our goal is to affordably feed the world. This is where other forms of urban agriculture come into play. One form, which, unlike urban gardens, would require significant capital investment turns (or perhaps I should say stands) agriculture on its head: vertical farming – the practice of growing 'up' rather than 'out'.

The vertical farm has long been a pipedream, something discussed in science fiction novels rather than in books on food security. Until recently, I viewed it

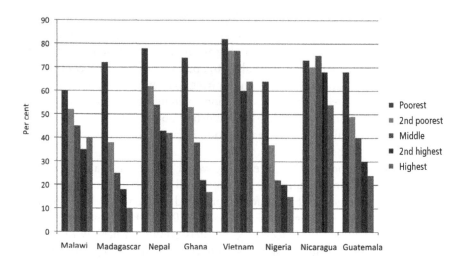

Figure 10.6 *Percentage of households participating in urban farming for selected countries by quintile*

Source: based on Zezza and Tasciotti (2010, p269)

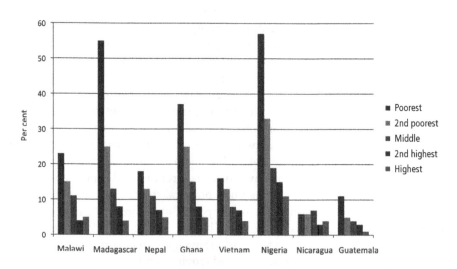

Figure 10.7 *Share of household income from urban farming for selected countries by quintile*

Source: based on Zezza and Tasciotti (2010, p269)

this way too. Technological advances coupled with food crises, impending water and environmental calamities, and projections of future food, feed and biofuel demands have caused me, however, to reassess the practicality of (and need for) vertical agriculture. It is true: we will not feed the world any time soon with vertical farming. And that is precisely why it deserves a closer look, because without serious study vertical agriculture will remain a future possibility rather than a practical option in our Rainbow Evolution.

Roughly US$1 billion a day are spent globally to subsidize food in some way (Peterson, 2009). Saudi Arabia spent US$40 billion during the 1980s to develop its agricultural sector, much of which went to subsidizing water (Myers and Kent, 2001, p135). China has earmarked US$104.8 billion to be spent on rural development in 2010. Much of this money will be used to either directly (agricultural subsidies) or indirectly (infrastructure construction and improvements) prop up the country's food production (*People's Daily*, 2009). Countries already spend mind-boggling amounts of money to achieve some degree of food security. If just a fraction of that were directed towards vertical agriculture, the returns, as others have argued (see, for example, Despommier and Ellingsen, 2008; Ehrenberg, 2008; Despommier, 2010), could be enormous. The question is: do we want to continue to prop up a system that is clearly unsustainable or are we serious in our convictions in wanting to achieve global food security, in which case we will have to alter how we are spending taxpayer money.

Earlier I mentioned how arable land is not evenly distributed around the world. Many countries import food not because they lack water, but because they lack sufficient arable land that can be put to cultivation (Kumar and Singh, 2005, p759). Other than the land 'footprint' of the facility itself, vertical farms require no land. Contrary to what we've been taught in grade (primary) school, plants do not require soil to survive. Soil is a support system, a carrier of nutrients and a holder of water. Hydroponics – from the Greek words *hydro* (water) and *ponos* (labour) – involves growing plants using mineral nutrient solutions in water, without soil. The practice has been around since the 1930s and is used in nurseries around the world for seed germination. More recently, aeroponic agricultural techniques have emerged. Invented in 1982, aeroponics is essentially an ultra-efficient version of hydroponics, where tiny nozzles spray a nutrient-rich mist onto the plant's roots (Roberto, 2003, p24–25). The National Aeronautics and Space Administration (NASA) and the European Space Agency are at the forefront of aeroponic and hydroponic agriculture, working to make these systems entirely self-contained for long-distance space exploration (see, for example, Finetto et al, 2010). If these ultra water-efficient systems – compared to, say, gravity or sprinkler irrigation, which are 40 and 60 to 70 per cent efficient, respectively – could be successfully scaled up, it could mark an armistice of sorts on the water war front.

Plants also do not need all the energy in sunlight to grow to maximum yield. Light-emitting diodes (LEDs) have recently been engineered just for plants (that emit energy stripped of the broader spectrum of light not used by the plants). LEDs are also incredibly efficient. Still in the development phase are organo light-emitting diodes (OLEDs) (Yersin and Finkenzeller, 2007). This technology will offer an even narrower spectrum of light, thus further reducing the energy demands of vertical farms (Cox, 2009; Despommier, 2010, pp186–187). In those regions of the world where sunlight is plentiful, vertical farms could take advantage of this free and abundant resource. To still further minimize the ecological energy footprint of vertical farms, buildings could be powered utilizing renewable energy sources, such as wind turbines and photovoltaic panels. We would also do well to reconsider the ecological value of composting for purposes of methane generation and capture – what is known as anaerobic digestion. The microbes 'take' between 80 to 90 per cent of the energy contained in rotting organic waste, leaving us 10 per cent in return in the form of methane. Incinerating biomass – namely, post-harvest crop residue – might be a more efficient use of resources, especially in light of new technologies (such as plasma arc gasification devices) that produce minimum pollutants (Despommier, 2010, pp196–197).

The amount of money lost annually to crop failures is another of those mind-boggling statistics that ought to be included in any discussion about the benefits of vertical farming. The following are some figures for crop losses in the US for the year 2005: US$843 million in oranges, US$805 million in tomatoes, US$433 million in sugar cane, US$213 million in bell peppers, US$208 million in grapefruit, US$127 million in watermelons, US$73.7 million in cucumbers, and US$68.4 million in tangerines (Despommier, 2010, p149). Dickenson Despommier (2010, pp149–150), in his book *The Vertical Farm*, tells the story of a strawberry farmer who replaced his 30 acre (12ha) farm after it was destroyed by a hurricane with a 1 acre vertical farm utilizing hydro-stackers (a hydroponic vertical gardening system). He now produces as much on that 1 acre as he did prior to the hurricane when he had 30 acres in production. Since farming 'up', the farmer also reports that the remaining 29 acres have returned to a more biologically diverse 'wild' state.

Making Agriculture a Goods and Service Industry

In Chapter 7 I discuss how food and fibre production is not possible without the supporting, regulating and cultural services that are provided, free of change, by the environment. Yet, these services continue to be undervalued, if not outright ignored. It is not that farmers fail to care about the environment. But, like the rest of us, they've got bills to pay. And if the incentive structure is such that it rewards certain unsustainable practices (such as the over-application of nitrogen, especially when fertilizer prices are low), while penalizing those who are just trying to do the ecologically right thing, how can we blame farmers who are just doing what most of us would do if in their shoes? Without a market to reward producers who take the time (and incur the expense) to restore and maintain ecosystem services, the incentive structure will continue to undermine this supportive 'base'.

Most programmes designed to increase provision of environmental services from agricultural land rarely pay directly for the services maintained and/or produced. Instead, the incentive takes the form of either a subsidy or some form of cost-sharing for implementing best management practices (BMPs) (or the farmer is paid to not farm the land) (Bohlen et al, 2009, p46). These initiatives, however, still undervalue ecosystem services; ultimately, any incentive that they provide is offset by a market perpetually blind to the productive base of the environment.

There is no market silver bullet to the problem of cheap food. Yet, as long as there are markets, we must get them to value more than just the feed and fibre provisioning services of the environment. This means finding creative ways to value and trade ecosystem services that farmers can provide and therefore avoid many of the disservices that have been produced in lock step with cheap food. In Chapter 7 I discuss some of the metrics used by economists to better value the ecological base upon which all life is predicated. I will now offer some specific examples of what these valuations schemes might look like.

Nitrogen farming

Watersheds in many parts of the world carry excessive amounts of nitrogen to the sea, producing massive 'dead zones' (such as that found in the Gulf of Mexico just south of the Mississippi River Delta). A significant portion of this nitrogen has been linked to agriculture (Beman et al, 2005). One proposed solution to this is 'nitrogen farming' (Hey, 2002). Nitrogen farms could 'grow' wetlands for the explicit purposes of removing excess nitrogen while providing the operator the option of raising an additional crop. For maximum impact, the practice could be set up in such a way as to target farmers with land in floodplains and bottomlands. A nitrogen market could be similar to those of more conventional agricultural commodities, allowing farms to enter or leave nitrogen farming depending on the price of nitrogen credits (Hey, 2002, p1).

Carbon farming

Another option is 'carbon farming' – a practice that involves carbon capture and emission reduction (Dempster, 2007; Khan and Hanjra, 2009). Carbon farming is a practice most farmers could participate in, as carbon can be stored many different ways, such as through grazing management, pasture cropping, low and no till, organic and biodynamic agriculture, and composting (Khan and Hanjra, 2009, p136). Like nitrogen farming, carbon farming requires market instruments to support its existence, which could eventually be integrated within international carbon trading schemes.

Australia has real aspirations for carbon farming. The AU$46 million Carbon Farming Initiative in Australia hopes to reward landholders for sequestering carbon while reducing environmental degradation. Australian Prime Minister Julia Gillard believes the market could be worth AU$500 million over ten years. The scheme includes reforestation, fire management, excess manure control and so-called 'legacy waste' emissions from landfill sites. The plan is to slowly grow the carbon market, issuing credits to farmers over time. The government has promised that there will be no limit on the number of credits that a farmer would be able to make. The initiative also provides institutional support for landowners by connecting them with brokers who better understand

Box 10.4 Florida ranchers 'raising' ecosystem services[2]

Total phosphorus (P) concentrations in Lake Okeechobee – a massive (1890 square kilometre/730 square mile) freshwater lake located in the south-central tip of Florida – has more than doubled during the last 30 years, leading to more frequent and longer algal blooms and the eutrophication of the lake and surrounding watershed. In an attempt to reduce annual P loads to the lake, in 2000, the Florida state legislature passed (and, in 2007, later strengthened) the Lake Okeechobee Protection Act. The passing of this act has created a sizeable public demand for improved water retention and reduction of nutrient runoff within the lake. Part of this demand will be met with public works projects (e.g. reservoirs and underground storage), cost-sharing for agricultural best management practices, and land acquisitions for the purposes of conservation. The remainder of the demand will be satisfied by ranchers. A payments for environmental services (PES) programme will pay ranchers to maintain certain ecosystem services targeted by the Lake Okeechobee Protection Act.

Cow–calf operations are the dominant form of land use in the Lake Okeechobee watershed (some pasturelands are in excess of 50,000ha). These permanent pastures provide important wildlife corridors that are critical to wildlife movement, support water recharge and storage, and offer habitat to a variety of threatened and endangered species. Yet the future of these pastures is in jeopardy, as narrow economic margins are making intensive forms of agriculture more attractive to these cow–calf farmers. Also, at least before the financial meltdown during the autumn of 2008, Florida's real-estate market was booming. The pressure to sell one's land to developers has been increasing with each passing decade, representing yet another incentive for ranchers to give up on this ecologically important grassland. It has been estimated, for example, that Florida will lose an additional 2,813,886ha of land to residential or urban development by 2060. The additional revenue stream offered by this PES programme, it is hoped, will not only reduce total P concentration in Lake Okeechobee, but also offer enough financial incentive to keep ranchers (and their pastures) in business. Ranchers are paid to utilize the extensive canals, ditches and berms – originally built for drainage and irrigation – to retain more water, replenish drained wetlands and reduce P loads in the Lake Okeechobee watershed. Taxpayers, too, will benefit from this arrangement, as it appears these ecosystem services can be 'bought' from ranchers at a cost either competitive with or lower than the cost of securing the services through large, expensive public works projects.

the domestic and international carbon market. Some Australian farmers and landowners have already earned AU$20 per tonne of stored carbon on the open market (Howlett, 2010).

Payments for environmental services (PES) programmes

One persistent critique of government programmes designed to incentivize the adoption of certain management practices is that little (if any) attention is paid to results (Knowler and Bradshaw, 2007; Wunder et al, 2008). They usually fail to ask, in other words, if the desired environmental benefits are being achieved. An alternative is to pay directly for ecosystem services produced, what are known as payments for environmental services (PES) programmes (see Box 10.4) (Ferraro and Kiss, 2002; Pagiola et al, 2007; Wunder et al, 2008). For this to happen, environmental services must first be identified and ascribed a value by buyers willing and able to pay for them. There is always the risk, of course, that these services will continue to be undervalued (though undervalued is arguably better than unvalued). PES programmes could be a potential game changer in terms of broadening our definition of 'agriculture' and the 'commodities' that farmers produce. The underlying complexity of the ecological processes involved makes it very difficult to quantify these services. A hurdle for any PES programme is therefore a cost-effective way to document what is being 'produced', which can assure buyers that they are getting what they have paid for and sellers that they are being fairly compensated for what they have produced (Ruhl, 2008, p447; Bohlen et al, 2009, pp46–47). A baseline and upper limit of services must also be established so that realistic production goals can be set.

Growing clean water

Billions of dollars are spent annually in countries such as the US to deal with their never-ending streams of liquid and municipal waste. Following the saying 'In nature there is no such thing as "waste"', some have suggested having farmers 'grow' clean water (Despommier, 2010, pp7–8, pp174–175). In this scenario, farms would literally raise wastewater treatment *plants*. Through the process of transpiration, plants take up water through their roots and then leaves, where it is eventually released through small pores into the atmosphere as pure H_2O. Elements – from nitrogen to the more pernicious (e.g. lead) – drawn up the root system during this process become incorporated within the plant's growing tissue. In a large vertical farm, hydroponic technology could be utilized to feed plants a steady stream of grey water. The pure water from transpiration could then be easily captured (through dehumidification) before being released into the atmosphere (versus traditional greenhouses where humid air is drawn out by fans). The city of New York dumps 1.5 billion gallons of treated grey water daily into the Hudson River.[3] The system used to treat this ocean of wastewater cost New York City taxpayers billions to build and maintain (New York's Bureau of Wastewater Treatment has an annual operating budget of US$262 million) (New York City, undated). It is conceivable to envision a scenario where farms profit from selling purified water. And while from a public health standpoint you wouldn't want to eat the plants on this farm, the harvested biological material could be incinerated and the energy generated sold.

The Epistemic Value of Food

Cheap food makes us forget. This is not to suggest that the typical Western diet leads to, say, Alzheimer's (though there is some evidence linking the two; see Knopman, 2009). The comment builds upon issues discussed in Chapter 8, where links between cultural monocultures and cheap food policies were highlighted. My point is this: once we've developed a *taste* for industrial monocultures – the food base for the average Western diet, after all, is derived primarily from 12 plants (Thompson, 2007) – will we still want to eat from an alternative food system/s?

A leader of the Slow Food Movement describes a lesson that occurred in an Australian primary school (Hayes-Conroy and Hayes-Conroy, 2008, p467). The students were given a blind tasting of homemade and commercial strawberry jams. Before the experiment they were asked which jam they thought they would like most. Most answered the homemade jams. Their taste buds, however, betrayed their expectations. The vast majority of students reported actually preferring the jam they grew up eating – commercial jam. After describing this story, the Slow Food proponent remarked: 'It illustrates the need to educate their tastes ... or they won't be able to appreciate other tastes and they will always go for the industrial products' (cited in Hayes-Conroy and Hayes-Conroy, 2008, p467).

The concept of 'slowness' has been talked about during recent years by social and political theorists as being an important strategy for resisting the quickening pace of everyday life (see, for instance, Wolin, 1997; Mackenzie, 2002; Stengers, 2002; Connolly, 2003). Slow food, slow living and slow knowledge (to name but a few 'slow' examples; see Honore, 2004, for a more detailed list): such 'movements', at their heart, are about the search for more concrete sensual experiences. In the case of the slow food movement (see, for example, Petrini and Watson, 2001; Petrini and McCuaig, 2004), the goal, or at least one of them, is to reclaim the 'gastronomic aesthetic' (Miele and Murdoch, 2002, p313) – a recognition that our understandings of 'quality', 'goodness', 'freshness' and 'tastiness', as they pertain to food, could someday (if not already) be shaped in the image of cheap food and cheap food policy. Indeed, that aforementioned blind taste test involving Australian schoolchildren and jam suggests that this is already occurring.

We have all heard the arguments before, about how 'We are hardwired to love the taste of fat, salt, and sugar' (Wansink, 2007, p180). The love so many display towards sugary, salty, highly processed food is in our genes, or so we're told (see also Allman, 1995, p50; Cartwright, 2000, p47). While I have no reason to doubt that there is some hardwiring involved – after all, it makes sound evolutionary sense to desire sweet, fatty things – I have no desire to essentialize taste either.

I have conducted extensive research on the subject of how we come to know, understand and, ultimately, taste food. Specifically, I have studied the process of becoming 'tuned' to industrial food and how phenomena such as community-supported agriculture (CSA), farmers' markets, heritage seed banks and backyard chicken coops carry in them the potential to turn these tastes around (Carolan, 2011). My research indicates that by instilling within people alternative understandings of 'quality', 'goodness', 'freshness' and 'tastiness', these alternative modes of food provisioning have the ability to make cheap food less appealing to our senses (see Box 10.5).

Box 10.5 Making cheap food feel out of tune

The following exchange is unpublished data for research I conducted for my recently published book *Embodied Food Politics* (Carolan, 2011). I am talking with Betty (not her real name) about how her experiences with local community-supported agriculture (CSA) have shaped her feelings – literally – towards industrial foods. Betty used to eat a lot more canned and frozen foods. She claims her experiences with a local CSA programme have changed her 'tastes' – broadly defined – towards food:

> *Betty:* It's not just about taste. I mean, I don't care much for the taste of most of the produce at big-box stores. But it's more than taste. You know what I mean?

> *Interviewer:* Could you elaborate on this, please?

> *Betty:* Well, having gotten used to fresh, whole fruits and vegetables and those fresh eggs, the canned and frozen stuff really doesn't compare. On taste alone I'd have a hard time going back. But it's more than just taste. I know what a fresh melon or tomato or asparagus ought to look like now. Before I didn't, so I wasn't entirely comfortable spending time in my store's 'fresh' [she makes air quotes] produce section. It's also about food's texture in my mouth and its smell. It's all just different when you get it right from a farm. It's just different.

> *Interviewer:* And why is that difference significant?

> *Betty:* Because it makes you feel differently about the stuff you get from a store. I used to not really care. I do now. I think because of my positive experiences with local fresh foods, like what I get from my CSA.

A popular argument made in favour of more local food systems – of practices such as CSA, farmers' markets, farm-to-school programmes and u-pick – is that they represent creative market-based strategies with the potential to strengthen a community's economic vitality (see, for example, Brown and Miller, 2008; Hancharick and Kiernan, 2008; Bagdonis et al, 2009). Proponents of these practices often speak of the resiliency that they add to a region's food provisioning capabilities (see, for example, McCullum et al, 2005; King, 2008). I do not disagree with these positions. But they paint an incomplete picture of what short-chain food systems provide. We can look to the marketplace and quantify the economic value of such things as CSA, heritage seed banks and backyard chicken coops. We can physically count how much food they provide us and the dollars they generate. Yet, is that where their value ends? Is there anything about them that is non-quantifiably valuable?

I have heard people try to paint the Local Food and Slow Food Movements as elitist and unrealistic. As mentioned in Chapter 6, for example, from an ecological standpoint sometimes local food is terribly unsustainable. And as far as price is concerned, food from a local artisan is rarely sold for less than its cheap industrial counterpart. Yet, these critiques are more hand-wavy than anything and have the tendency of side-tracking the debate (much like a magician uses excessive hand movements to distract the audience from the hand they ought to be watching). Affordable, resilient, sustainable and just food systems have nothing to do with either/or logic. I have no interest in debates about whether the food system ought to be entirely either 'local' or 'global', 'slow' or 'fast', or 'organic' or 'non-organic'. One-size-fits-all thinking is one of the things this book rails against.

Nothing is ever as black and white as our handy dichotomies would have us believe. Probyn (2001) makes this point beautifully when discussing McDonald's. Eating at McDonald's for many is an expression of care: of caring for one's family and providing for one's children. More than that, Probyn notes, McDonald's provides a warm and welcoming family space, where everyone can eat and play together without the distractions of ringing doorbells and televisions. All this 'caring' occurs, however, at the expense of care for the environment, animals and social justice. At McDonalds:

> 'Two worlds collide' in terms of a vision of care, when, on the one hand, McDonald's stitches us all together through our stomachs, and on the other, a politics that directly equates the desire for burgers with the destruction of the rain forest, and the exploitation of workers and children. (Probyn, 2001, p36).

In the same non-essentializing spirit, Julie Guthman (2003, p56) reminds us that categories such as slow, fast, organic or conventional seldom tell us much about what a particular food item truly *is* in terms of its real costs:

> Fast food is often pitched to healthy eaters (e.g. Subway's advertising campaign suggesting you can lose weight and cut fat by eating fast food) and slow food is often made tasty by slavish uses of salt and butter ...

> Little is it considered that organic production depends on the same systems of marginalized labor as does fast food. Or that organic salad mix led the way in convenience packaging, and is often grown out of place and out of season. Or that fast food serves women who work outside the home who are then blamed for depending on it to manage family and work. Or that slow food presumes a tremendous amount of unpaid feminized labor.

When discussing the value of more local, slow and whole foods, it is important not to see localness, slowness and/or wholeness as ends in themselves. The end is affordable food, a state predicated on food production and consumption that is just, equitable, sustainable, nourishing and ethical. The means to this end is not paved with either/or logic. The path, rather, is as colourful as a rainbow.

Focusing on Capacities Instead of Barriers

The point was earlier made that equal rules for unequal players are unequal rules. I fear we have gotten so hung up on reducing trade barriers that we have lost sight of what is really important: real food security. Part of the problem is that the term 'free trade' is used without much thought. Let's face it: many throw the phrase around, having little more than a caricature understanding of the concept. I have found Isaiah Berlin's dual understanding of freedom extremely useful for making sense of this overused, under-understood, concept.

Conventional understandings of 'free trade' are dangerously narrow, as evidenced by the simple fact that free trade is rarely ever fair. Most free trade talk is heavily infused with, to draw from Berlin, 'freedom *from*' talk: freedom *from* domestic protectionist policies, freedom *from* other countries' production-oriented subsidies, and so forth. Isaiah Berlin (1969, p127) describes this as 'negative freedom', which refers to the 'absence of interference'. It seems as though when most people are asked about 'freedom', negative freedom is the freedom most have in mind – a freedom where we are allowed to pursue actions unimpeded (see also Bell and Lowe, 2000, pp287–289).

In addition to its negative component, freedom also has a positive side. 'Positive freedom' – or freedom *to* – refers to the ability 'to lead one prescribed form of life ... that derives from the wish on the part of the individual to be his [or her] own master' (Berlin, 1969, p127). Positive freedom is essential, for without it most people would hardly feel free. In the words of the old English aphorism, 'freedom for the pike is death for the minnow'. Without the active pursuit of positive freedom for all – pikes and minnows alike – the pikes of the world would be clearly advantaged. To put it plainly: positive freedom allows us to talk about *constraints* in the context of freedom without any sense of paradox. In order for me to be and feel free (and trade freely), I need some assurances that the 'pikes' of the world will not freely have their way with me. This is why freedom – and, yes, free trade – *requires* some level of state intervention.

Many cheap food proponents decry calls that involve building real food security capacity in many parts of the world. They say redistributions of wealth, capital and expertise represent the antithesis of freedom, or at least most certainly the antithesis of free trade. I would contend that these individuals are not interested in freedom at all (or perhaps they are only interested in freedom for the economic pikes). Freedom, *real* freedom – which is premised on negative *and* positive freedoms – requires material capabilities that allow all the freedom to pursue trade, innovation and the developmental trajectories of their choosing. Cheap food policies do not allow most this choice.

The importance of positive freedoms for development has been forcefully articulated by Nobel Prize-winning economist Amartya Sen. Sen (2001, pp13–15) famously notes that development will never be complete without attention to a society's and person's *capabilities*, which speaks to a person's freedom to choose between different ways of living. The one-size-fits-all policies and practices of cheap food fundamentally ignore these capabilities by forcing countries down a single developmental and food security trajectory, with very 'lumpy' results. Real food security requires policies that *afford* people (and nations) more alternatives, not less.

Thinking about free trade through the lens of Berlin's dual understanding of freedom profoundly changes the structure of the debate. For one thing, it gives force to arguments that favour opening up the middle of the 'hourglass' figure detailed in Chapter 9 to allow for greater competition between firms. Indeed, market monopolies and oligopolies are the inevitable outcome of food policies overly influenced by one-sided understandings of freedom – namely, freedom *from*. Until we do something about this market concentration, we will continue to have an economic environment where the pikes of the food chain not only are able to 'eat' their minnow competitors, but dictate price as well.

This dual understanding of freedom also justifies massive spending initiatives to create food security capabilities in less developed countries – what some have likened to an international New Deal (Haq, 1976). Significant sums of money are already being spent on agriculture. Perhaps we could just spend some of it differently. For example, the world's top ten crop-science companies spend approximately US$3 billion a year on biotech research (Karapinar and Temmerman, 2008, p192). Compare this to the US$30 million spent annually by the Consultative Group on International Agricultural Research (CGIAR) on biotech research aimed at benefiting farmers in the developing world. A sizeable chuck of this research defect could be made up by simply altering our understanding of 'food aid', from being a Band-Aid type of policy to something interested in creating actual long-term food security. The US Agency for International Development (USAID) food aid budget in 2005 was US$1.6 billion. Of that, only 40 per cent (US$654 million) went to paying for food. The remainder was spent on overland transportation (US$141 million), ocean shipping (US$341 million), transportation and storage in destination countries (US$410 million), and administrative costs (US$81 million) (Dugger, 2005). Redirecting this money to, say, crop R&D specifically for the world's poor farmers would produce funds equal to more than half of what the world's top ten crop-science companies spend on biotech research.

Admittedly, more money will not, in itself, solve anything for the world's smallholders. As noted in a recent report by the UK-based Overseas Development Institute (Compton et al, 2010), assistance directed at the rural poor needs to be vigilantly monitored. Helping the rural poor increase their food production is a sensible policy response. Yet, as the report goes on to argue, assistance to producers, unless carefully targeted, risks benefiting the people least in need of help – namely, those with access to land, labour, water, credit and markets – the very things many of the rural poor lack.

Finally, revisiting foundational concepts that underlie understandings of 'growth' and 'development' will hopefully cause us to re-evaluate whether conventional economic measures adequately encourage behaviours directed towards ends that we as a society value. For instance, in growing our own food (with seeds saved from the previous year in soil fertilized with homemade compost), we add no 'value' to the world, at least according to those who measure gross domestic product (GDP); if anything, that action *takes value away* because it results in fewer trips to the grocery store. There is an old joke among economists that when a man or woman marries their housekeeper, the economy – and, thus, the overall well-being of society – shrinks. There is something fundamentally wrong-headed with this logic.

Besides, where does it say that food security *must* come under the exclusive purview of the market? Over the last century we have witnessed the transfer of responsibility for food security from the state to the market – a process that has been called the 'marketization of food security' (Zerbe, 2009, p172). Even in cases where the state could effectively provide food to its people, and seed, fertilizer and credit to its farmers (as was the case in the Philippines, as discussed in Chapter 3), the need for small government took precedence over the need for real food security.

The Efficiency of Small Farms

We need to be committed to a model of 'efficiency' that will feed us well into the future. The model that we are currently following only endangers food security by producing a system of limited redundancies, highly concentrated markets and trade dependencies. What we have in cheap food is *relative* efficiency. What we need is *absolute* efficiency (see, for example, Jackson, 2009). I am all for cost minimization; that is, after all, what this book is all about. But I am only for cost minimization in the absolute sense. Being able to selectively pick and choose what costs 'count' just seems like dishonest book-keeping to me.

Broadening our understanding of efficiency places the small farm in noticeably different light. Rather than backward and unproductive, small farms can be shown to be considerably more efficient and, thus, productive than larger farms (see Box 10.6). For instance, small diverse (e.g. polyculture) farms that raise grains, fruits, vegetables and livestock easily out-produce, in terms of harvestable products per unit area, large specialized (monoculture) operations. Yield advantages of small diverse farms can range from 20 to 60 per cent (Altieri and Nicholls, 2008, p474). One study found that a 1.73ha plot of land in Mexico needs to be planted with corn in order to produce as much food as a 1ha plot planted with a mixture of corn, squash and beans. Moreover, the 1ha polyculture plot produces twice as much dry matter, which eventually gets ploughed back into the soil (Gliessman, 1998, p10). In Brazil, polycultures containing 12,500 maize plants per hectare and 150,000 soybean plants per hectare exhibited a 28 per cent yield advantage over soybean monocultures (Altieri, 1999, p200).

Box 10.6 A review of the organic agriculture literature

Examining close to 300 studies from all around the world, Badgley and colleagues (2007) conclude that, contrary to what most cheap food proponents proclaim, organic agriculture has the potential to produce a lot of food, perhaps even enough to feed the world. Comparing yields between organic and conventional systems, it was found that properly managed organic systems could produce as much, if not more, than conventional farms. It was also discovered that yield ratios between organic and conventional farms were higher in developing countries, which is to say that organic systems compared even more favourably in less affluent nations. Although, as others have pointed out, this 'should not have been surprising since in developing countries the "conventional" farms are far from efficient' (e.g. fertilizer is applied when available rather than during optimal application windows) (Perfecto et al, 2009, p68). If this is correct, we might want to think twice about forcing farmers in developing economies into production regimes that require the purchase of expensive inputs when less expensive agro-ecological and/or organic techniques could produce the same (or perhaps even better) yields.

The inverse relationship between farm size and productivity has been attributed to a more (absolute) efficient use of land, water, biodiversity and other agricultural resources by small farmers. Polycultures not only minimize losses due to weeds, insects and diseases – by keeping intact ecological internal controls – but also more efficiently utilize water, light and nutrients (Altieri and Nicholls, 2008, p474). Small farms are also overwhelmingly owner operated, which, as discussed in Chapter 8, benefits rural communities and their economies. Small diverse operations are less vulnerable to catastrophic loss due to environmental events – a point of great significance in light of climate change-induced weather risks (Alexander, 2008). Polycultures also display better yield stability during extreme weather events than monocultures (Altieri and Nicholls, 2008, p475). In India, for example, the number of varieties that a farmer grows increases with the variability of conditions (e.g. atmospheric, agronomic, pest ecology). Thus, the low terraces, which are wetter and prone to flooding, are planted with indigenous long-growing rice varieties. The upper terraces, in contrast, dry out more rapidly after the rains, so they are planted with drought-resistant fast-growing varieties. In total, a small Indian rice farmer can plant up to ten different rice varieties. Just how many varieties end up in small rice farms in India is determined by on-farm conditions – just the opposite of a one-size-fits-all approach (Holdrege and Talbott, 2008, p22).

Small farm efficacy can be attained with surprisingly small levels of investment, especially when compared to the massive capital expenditures of industrial farms. In Peru, for example, several non-governmental organizations (NGOs) and government agencies have created initiatives to restore abandoned terraces and build new ones by offering peasant communities low-interest loans or seeds and other inputs. Terraces reduce losses associated with frost and drought, minimize soil erosion, improve water efficiency, and increase crop yields by 45 to 65 per cent compared to crops raised on sloping land (Altieri and Nicholls, 2008, p476). And let's not forget: these crops are being produced at 4160m (13,648 feet).

This last point raises a question that I never hear asked: just what do we mean by 'arable land' anyway? 'Arable' comes from the Latin word *arare*, which means 'to plough'. More generally, arable land refers to land where soil and climate are suitable for agriculture. But what's suitable for small-scale agriculture may not be suitable for industrial agriculture – at least, not without a massive injection of inputs and energy. High-altitude land, such as that found at 4100m, is completely off the radar of cheap food proponents, given its steep slopes and relatively shallow soils (Lal, 1995, p5). Having a prejudice towards large-scale industrial farms marginalizes (from a food system standpoint) those communities that live where industrial monocultures can't.

Then there is the issue of land access. The future viability of small farms hinges on the future availability of small plots of land. Until the world's rural poor have greater access to land, global food security, and most certainly national and household food sovereignty (see Box 10.7), will only be a tenuously held state at best.

Precisely *how* we go about improving land access will vary from situation to situation and country to country. To take just one example: the Landless Workers' Movement in Brazil, also known as MST after its Portuguese title *Movimento dos Trabalhadores Rurais Sem Terra* (see Wright and Woldford, 2003). According to the MST's website (www.mstbrazil.org), it is the largest social movement in Latin American, with an estimated 1.5 million members. The impetus of this movement centres on the perceived unjust distribution of land throughout the region and especially in Brazil, where 3 per cent of the population own and controls two-thirds of the country's arable land.

Under Brazilian law, a landowner risks losing title to their land by letting it lie idle, especially after someone else settles the land and begins using it productively. Under such a scenario, the latter party (the 'squatter') could challenge

Box 10.7 A declaration on food sovereignty

The World Forum on Food Sovereignty was held in Nyéléni village in Mali, Africa, from 23 to 27 February 2007. The meeting brought together 600 delegates from five continents to reaffirm the right to food sovereignty and to begin an international drive to reverse the worldwide decline in local community production of food. The forum was organized through an alliance of social movements: Friends of the Earth International, Via Campesina, the World March of Women, the Network of Farmers' and Producers' Organizations of West Africa (ROPPA), the World Forum of Fish Harvesters and Fish Workers (WFF) and the World Forum of Fisher Peoples (WFFP). Collectively, this group wrote the following statement, known as the Declaration of Nyéléni:[4]

We, more than 500 representatives from more than 80 countries, of organizations of peasants/family farmers, artisanal fisherfolk, indigenous peoples, landless peoples, rural workers, migrants, pastoralists, forest communities, women, youth, consumers, and environmental and urban movements have gathered together in the village of Nyéléni in Sélingué, Mali, to strengthen a global movement for food sovereignty ...

Food sovereignty is the right of peoples to healthy and culturally appropriate food produced through ecologically sound and sustainable methods, and their right to define their own food and agriculture systems. It puts the aspirations and needs of those who produce, distribute and consume food at the heart of food systems and policies rather than the demands of markets and corporations. It defends the interests and inclusion of the next generation. It offers a strategy to resist and dismantle the current corporate trade and food regime, and directions for food, farming, pastoral and fisheries systems determined by local producers and users. Food sovereignty prioritizes local and national economies and markets and empowers peasant and family farmer-driven agriculture, artisanal-fishing, pastoralist-led grazing, and food production, distribution and consumption based on environmental, social and economic sustainability. Food sovereignty promotes transparent trade that guarantees just incomes to all peoples as well as the rights of consumers to control their food and nutrition. It ensures that the rights to use and manage lands, territories, waters, seeds, livestock and biodiversity are in the hands of those of us who produce food. Food sovereignty implies new social relations free of oppression and inequality between men and women, peoples, racial groups, social and economic classes and generations ...

What are we fighting against?

- Imperialism, neo-liberalism, neo-colonialism and patriarchy, and all systems that impoverish life, resources and eco-systems, and the agents that promote the above such as international financial institutions, the World Trade Organization, free trade agreements, transnational corporations, and governments that are antagonistic to their peoples.

- The dumping of food at prices below the cost of production in the global economy.
- The domination of our food and food producing systems by corporations that place profits before people, health and the environment.
- Technologies and practices that undercut our future food producing capacities and that damage the environment and put our health at risk. These include transgenic crops and animals, terminator technology, industrial aquaculture and destructive fishing practices, the so-called 'White Revolution' of industrial dairy practices, the so-called 'old' and 'new' Green Revolutions, and the 'Green Deserts' of industrial bio-fuel monocultures and other plantations ...

Now is the time for food sovereignty!

the landholder's title to the land in court. MST has sought to utilize this law to the advantage of landless peasants. To give their movement further legitimacy, they also point to a passage in Brazil's constitution that states that that land should serve a 'larger *social* function', which they interpret to mean it should not be viewed through solely an economic lens. Members of the MST thus seek to identify and occupy idle land and make it agriculturally productive, arguing that such actions met the constitutional requirement of a 'social function' (imagine if this legal threshold existed in other countries where land speculation is rife and where precious arable land is bought up and held until a 'suitable' return can be obtained for the investment). While doing this, the MST is also pushing for broader land reform throughout the region. It is worth mentioning, too, that agro-ecological principles are becoming more widely used in MST settlements, involving things such as tree planting, organic farming, and the use of crop rotations and manures to build soil fertility (Boyce et al, 2007, p135).

Rather than providing the rural poor with access to already 'improved' arable land, the Brazilian government has been attempting to lure people to the Amazon, where they would have to 'improve' the land themselves. History suggests, however, that promises of land are no substitute for real land reform, where laws and policies are rewritten to better serve the rural poor. For the last few centuries the Brazilian government promised land to those who cleared the Atlantic Coast rainforest (about 93 per cent has been cut). And those who came received and maintained their claims to the new land for a short period of time. Yet, in the end, they 'were nearly always displaced as the more powerful land barons moved in ... [to] become, once again, hopelessly indebted dependents of wealthy landowners' (Perfecto et al, 2009, p112). MST wants desperately to avoid having the landless poor fall for these empty promises again.

Those in the West concerned about the affordability of food and their abilities as agents for creating social change towards this end should take heart knowing that others – particularly in the developing world – are doing so. MST is just one example. Other small farm, peasant-focused organizations include, for example, La Via Campesina (a movement seeking to defend the peasant way of life and livelihood with subnational organizations in Latin America, North America, Asia, the Caribbean, African, Europe and Asia) (Borras, 2008, p132) and the International Planning Committee for Food Sovereignty (a self-managed global network of more than 45 people's movements and NGOs and at least 800 organizations from around the world).[5]

It's Not a Choice If You Don't Have One

Maybe it's because I live in the US, where so many are hypnotized by the fiction of consumer sovereignty. Whatever the reason, I feel compelled to return to the issue of consumer 'choice' one last time. I hear it often: 'What right does anyone have to tell me what I can and cannot eat!?' Yet, when you look at the incentive structure driving consumer and farmer behaviour, there really isn't much 'choice' out there. Making food affordable is a step towards broadening consumer and producer choice, not restricting it. Some examples will help to illustrate my point.

Taxpayers for cheesier pizzas. As detailed in a recent article in *The New York Times*, Domino's Pizza, whose sales were tumbling up until recently, turned to an organization called Dairy Management (Moss, 2010). Together they developed a new line of pizzas with 40 per cent more cheese (each slice contains roughly two-thirds of the daily recommended amount of saturated fat) and promoted them with a US$12 million marketing campaign. The cheesier pizzas were a hit, ultimately saving the company. So who is Dairy Management? They are not a consulting agency or marketing firm, but a product of taxpayer dollars. Dairy Management is a child of the USDA, who out of the other side of its mouth is discouraging the consumption of the very fatty foods that they are helping Dominos push upon the public. With an annual budget of close to US$140 million, Dairy Management gets its money from a government-mandated fee on the dairy industry, plus a couple of million dollars more from the USDA. Dairy Management, the USDA and Pizza Hut also teamed up for the 'Summer of Cheese' promotion in 2002, which resulted in the consumption of 102 million *additional* pounds of cheese.

Meeting nutritional guidelines is not one of your choices. A study just released by Yale University's Rudd Center for Food Policy and Obesity examined 12 popular restaurant chains and found only 12 out of more than 3000 kids' meal combinations met the nutritional guidelines for preschool-aged kids. According to the study, the fast-food industry spent US$4.2 billion on advertising in 2009 and found that 40 per cent of preschool-aged children ask to go to McDonald's on a weekly basis, and 15 per cent ask on a daily basis (Yale Rudd Center, 2010).

Consumer choice or media influence? An article recently published in the *Journal for the American Dietetic Association* found that a diet based on observed food items advertised on television would provide 2560 per cent of the recommended daily servings for sugars, 2080 per cent for fat, 40 per cent for vegetables and 27 per cent for fruits (Mink et al, 2010). Perhaps it's no coincidence, then, that the average American diet greatly exceeds the recommended daily servings for sugars and fat while being deficient in vegetables and fruit (Nestle, 2007, pp8–10).

For most farmers the choice has already been made for them. As taxpayer monies incentivize the production of certain crops, the production of non-incentivized commodities on farms has plummeted. Moreover, as the middle of the 'hourglass' narrows – yet another factor beyond a farmer's control – so too narrows farmers' choice in terms of what they can produce (at least, that is, if they are interested in selling what they raise). Table 10.3 offers a historical depiction of this narrowing of the commodity basket for the average farm in the 'heartland' of the US – namely, Iowa. The table shows the number of commodities produced for sale in at least 1 per cent of all Iowa farms for various years from 1920 to 2007 (commodities in bold indicate that they are produced in at least 50 per cent of all farms). Note the diversity and subsequent commodity 'shrinkage' with the birth and growth of cheap food policy, which has narrowed farmer sovereignty to the point that farmers are now, essentially, told what to produce.

I realize, given the entrenched interest involved, that subsidies for corn, cotton, rice, soybeans and the like will not be going away anytime soon. They should, for reasons discussed in earlier chapters. Yet, if politicians are too afraid to touch the sacred cow of traditional agricultural subsidies, they should at least be amenable to drastically increasing support for things such as small farms, the production of fruits and vegetables, and more sustainable forms of food production. I recently came across an article in *The Denver Post* proudly proclaiming that Feed Denver – an urban farming organization – is in the running for a US$4000 grant (Robles, 2010). The fact that a US$4000 grant is a big deal for an urban agriculture organization – when colleagues of mine in my university's College of Agriculture routinely obtain grants worth hundreds of thousands of dollars – says a lot about food choice in the US (and I know the story is the same in other countries) (see Box 10.8).

I also think it is disingenuous to say that we cannot use policy to alter the consumption end of the equation. This is because we have been using policy to shape consumer behaviour as it pertains to food for generations. By socializing many of its costs, we are making cheap food attractive to consumers. In contrast, pricing these foods in ways that more accurately reflect their real cost would probably reduce demand for them. A new study in the *American Journal of Public Health* finds that sales of sugary carbonated drinks fell by 26 per cent after a 35 per cent rise in their price (Block et al, 2010). The researchers also conducted a four-week educational campaign that focused on the health benefits of reduced soda consumption. By itself, the educational component had no measurable effect on consumption. However, when price hike and education

Table 10.3 *Number of commodities produced for sale in at least
1 per cent of all Iowa farms during various years, from 1920 to 2007*

1920	(%)	1935	(%)	1945	(%)	1954	(%)	1964	(%)
Horses	(95)	Cattle	(94)	Cattle	(92)	Corn	(91)	Corn	(87)
Cattle	(95)	Horse	(93)	Chicken	(91)	Cattle	(89)	Cattle	(81)
Chicken	(95)	Chicken	(93)	Corn	(91)	Oats	(83)	Hogs	(69)
Corn	(94)	Corn	(90)	Horses	(84)	Chicken	(82)	Hay	(62)
Hogs	(89)	Hogs	(83)	Hogs	(81)	Hogs	(79)	Soybeans	(57)
Apples	(84)	Hay	(82)	Hay	(80)	Hay	(72)	Oats	(57)
Hay	(82)	Potatoes	(64)	Oats	(74)	Horses	(42)	Chicken	(48)
Oats	(81)	Apples	(56)	Apples	(41)	Soybeans	(37)	Horses	(26)
Potatoes	(62)	Oats	(52)	Soybeans	(40)	Potatoes	(18)	Sheep	(17)
Cherries	(57)	Cherries	(34)	Grapes	(23)	Sheep	(16)	Potatoes	(06)
Wheat	(36)	Grapes	(28)	Potatoes	(23)	Ducks	(05)	Wheat	(03)
Plums	(29)	Plums	(28)	Cherries	(20)	Apples	(05)	Sorghum	(02)
Grapes	(28)	Sheep	(21)	Peaches	(16)	Cherries	(04)	Red clover	(02)
Ducks	(18)	Peaches	(16)	Sheep	(16)	Peaches	(04)	Apples	(02)
Geese	(18)	Pears	(16)	Plums	(15)	Goats	(04)	Ducks	(02)
Strawberry	(17)	Mules	(13)	Pears	(13)	Grapes	(03)	Goats	(02)
Pears	(17)	Ducks	(12)	Red clover	(10)	Pears	(03)	Geese	(01)
Mules	(14)	Wheat	(12)	Mules	(06)	Plums	(03)		
Sheep	(14)	Geese	(11)	Strawberry	(06)	Wheat	(03)		
Timothy	(10)	Sorghum	(09)	Ducks	(06)	Red clover	(03)		
Peaches	(09)	Barley	(09)	Wheat	(04)	Geese	(03)		
Bees	(09)	Red clover	(09)	Timothy	(04)	Popcorn	(02)		
Barley	(09)	Strawberry	(08)	Geese	(03)	Timothy	(02)		
Raspberry	(07)	Soybeans	(08)	Rye	(02)	Sweet potato	(02)		
Turkeys	(07)	Raspberry	(06)	Popcorn	(02)	Sweet corn	(01)		
Watermelon	(06)	Bees	(05)	Sweet corn	(02)	Turkeys	(01)		
Sorghum	(06)	Timothy	(05)	Raspberry	(02)				
Gooseberry	(03)	Turkeys	(04)	Bees	(02)				
Sweet corn	(02)	Rye	(02)	Sorghum	(01)				
Apricots	(02)	Popcorn	(02)						
Tomatoes	(02)	Sweet corn	(02)						
Cabbage	(01)	Sweet clover	(01)						
Popcorn	(01)	Goats	(01)						
Currents	(01)								
n = 34		n = 33		n = 29		n = 26		n = 17	

Source: prepared by author based on US Census of Agriculture, 1920–2007

were combined, an additional 18 per cent reduction (beyond that obtained by price rise alone) in soft drink consumption was observed. We must always be careful, however, to make sure the effects of such price hikes do not disproportionally fall on the shoulders of low-income and minority groups, especially those living in a food desert where 'cheap' food might be their only option. Perhaps some of the money generated from these 'sin' taxes could be invested in local/urban food systems that are most insecure to help bring affordable food to those least able to afford it.

In closing, I would like to make a final point about the notion of food 'affordability'. I strongly believe cheap food policy is obsessed to a fault with retail price. This obsession leads to, among other things, the various cost externalizations discussed throughout this book. To be clear: I think we need to be extremely mindful of retail price. But affordability is a relative term – a US$100 grocery bill affects the households of Bill Gates and my students very differently. Cheap food policy has helped to spur a type of 'race to the bottom'

1978	(%)	1987	(%)	1997	(%)	2002	(%)	2007	(%)
Corn	(90)	Corn	(79)	Corn	(68)	Corn	(58)	Corn	(54)
Soybeans	(68)	Soybeans	(65)	Soybeans	(62)	Soybeans	(54)	Soybeans	(45)
Cattle	(60)	Cattle	(47)	Hay	(42)	Hay	(37)	Hay	(28)
Hay	(56)	Hay	(46)	Cattle	(42)	Cattle	(35)	Cattle	(32)
Hogs	(50)	Hogs	(35)	Hogs	(19)	Horses	(13)	Horses	(11)
Oats	(34)	Oats	(25)	Oats	(12)	Hogs	(11)	Hogs	(09)
Horses	(13)	Horses	(10)	Horses	(11)	Oats	(08)	Oats	(03)
Chicken	(09)	Sheep	(08)	Sheep	(04)	Sheep	(04)	Sheep	(04)
Sheep	(08)	Chicken	(05)	Chicken	(02)	Chicken	(02)	Goats	(02)
Wheat	(01)	Ducks	(01)	Goats	(01)				
Goats	(01)	Goats	(01)						
Ducks	(01)	Wheat	(01)						
	n = 12		n = 12		n = 10		n = 9		n = 9

environment that is negatively affecting millions of people around the world. So we have cheap food ... but at what, and *whose*, expense?

Affordable food might in the end cost, retail price-wise, a little more, since by definition its retail price will represent a fairer reflection of its real cost. But if your income is rising faster than the price of food, what does it matter? Boosting household incomes around the world therefore must be a top priority as we move towards a more affordable food system. Adopting policies more interested in 'affordability' rather than 'cheapness' will help to push global incomes upward as international agricultural trade begins to operate on the premise of fairness rather than just freeness. Thus, in the end, I do not see affordable food costing consumers any more, as a fraction of total annual disposable income, than what they are paying today. Remember also that waste is anathema to any affordable food system. Drastically reducing food waste, improving irrigation efficiencies, boosting overall yields per unit of land utilizing polycultures, and better managing our animals (and our appetite for their

Box 10.8 Get big or get organized

In a recent issue of the journal *Gastronomica* you'll find an article titled 'A tale of two dairies' (Estabrook, 2010). The article begins with a report about the suicide of a small dairy farmer in New York State, a sad act emblematic of the dairy crisis in the US. The crisis is a product of basic dairy economics. As described by Bill Rowell, a medium-sized dairy farmer in northern Vermont:

> When milk is paying good, the signal to the farmer is, you better put on more cows to make more milk to make up what you've lost. So what happens? You quicken the pace toward oversupply and a deteriorating price (Estabrook, 2010, p50).

To combat this vicious cycle, some dairies in the US looked north to Canada and its production-limiting models. Again, Bill Rowell: 'So our idea was, if there is a surplus of milk, why wouldn't you have some sort of mechanism during a receding market to slow your production and balance it with demand rather than overproducing to the point where you spoil your price and you spoil your industry?' (Estabrook, 2010, p50).

Dairies in Canada, you see, have been following a production-limiting system since the 1970s (milk supply management systems are also operated in the European Union, Switzerland, Norway and Japan, where it is operated by co-operatives) (OECD, 2005, p50). The goal of these schemes: to maintain a sufficient milk inventory to meet demand but avoid a market glut that drops prices below the cost of production. Under such an arrangement, dairies collectively organize and agree to production limits set by a national board of farmers, processors, consumers and dairy economists.

Corporate dairy and milk processors have likened such systems to socialism. Call them what you want: they work. In Canada, consumers pay the same price for their milk as consumers in the US. Yet, Canadian dairy farmers receive often twice to three and a half times what US farms get. Understandably, big processors, who currently profit widely from milk prices below the cost of production (as discussed previously in Chapter 9), hate such schemes. Raising the farm gate price for milk helps small- and mid-sized farms, while cutting into the bloated profits of large dairy processors. But does it affect consumers? The evidence suggests that it does not. As described in 'A tale of two dairies', the price for 1 gallon of milk in a small town due north of Syracuse, New York, in Canada, is nearly identical to 1 gallon of milk in Vermont.

protein) will save consumers and taxpayers (as well as future generations) a significant sum of money.

The next time you hear someone using the cheap food argument to shut down debate about alternative food trajectories, push the discussion. Don't shrink from it; get them to talk about the *real* cost of cheap food. For when all the costs are tallied, cheap food is neither fiscally or environmentally responsible, nor just or sustainable. We just can't afford it.

Notes

1 It has been estimated that a US$1 increase per bushel of corn increases average dairy cow ration costs between US$0.27 and $0.34 per day (this quickly adds up when talking about sizeable LMC farms) (Garcia et al, 2007).
2 This case study draws upon Bohlen et al (2009).
3 See www.nyc.gov/html/dep/html/about_dep/bureaus.shtml#WaterSupply.
4 Translated from www.nyeleni.org.
5 See www.foodsovereignty.org/Aboutus/WhatisIPC.aspx.

References

Achard, F., H. Eva, H.-J. Stibig, P. Mayaux, J. Gallego, T. Richards and J.-P. Malingreau (2002) 'Determination of deforest rates of the world's tropical forests', *Science*, vol 297, no 5583, pp999–1002

Alexander, W. (2008) *Resiliency in Hostile Environments: A Comunidad Agricola in Chile's Norte Chico*, Rosemount Publishing, Cranbury, NY

Allman, W. (1995) *Stone Age Present: How Evolution Has Shaped Modern Life*, Simon and Schuster, New York, NY

Altieri, M. (1999) 'Applying agroecology to enhance the productivity of peasant farming systems in Latin America', *Environment, Development and Sustainability*, vol 1, pp197–217

Altieri, M. and C. Nicholls (2008) 'Scaling up agroecological approaches for food sovereignty in Latin America', *Development*, vol 51, no 4, pp472–480

Badgley, C., J. Moghtader, E. Quintero, E. Zakem, M. Chappell, K. Aviles-Vazquez, A. Samulon and I. Perfecto (2007) 'Organic agriculture and the global food supply', *Renewable Agriculture and Food Systems*, vol 22, pp86–108

Bagdonis, J., C. Hinrichs and K. Schafft (2009) 'The emergence and framing of farm-to-school initiatives: Civic engagement, health and local agriculture', *Agriculture and Human Values*, vol 26, pp207–219

Balmford, A., R. Green and J. Scharlemann (2005) 'Sparing land for nature: Exploring the potential impact of changes in agricultural yield on the area needed for crop production', *Global Change Biology*, vol 11, pp1594–1605

Beddington, J. (2010) 'Food security: Contributions from science to a new and greener revolution', *Philosophical Transactions of the Royal Society B*, vol 365, pp61–71

Bell, M. and P. Lowe (2000) 'Regulated freedoms: The market and the state, agriculture and the environment', *Journal of Rural Studies*, vol 16, pp285–294

Berlin, I. (1969) (originally published 1958) *Four Essays on Liberty*, Oxford University Press, London

Beman, M., K. Arrigo and P. Matson (2005) 'Agricultural runoff fuels large phytoplankton blooms in vulnerable areas of the ocean', *Nature*, vol 434, pp211–214

Block, J., A. Chandra, K. McManus and W. Willett (2010) 'Point-of-purchase price and education intervention to reduce consumption of sugary soft drinks', *American Journal of Public Health*, vol 100, no 8, pp1427–1433

Bohlen, P., S. Lynch, L. Shabman, M. Clark, S. Shukla and H. Swain (2009) 'Paying for agricultural services from agricultural lands: An example from the northern Everglades', *Frontiers in Ecology and the Environment*, vol 7, no 1, pp46–55

Borras, S. (2008) *Competing Views and Strategies on Agrarian Reform: International Perspective*, Ateneo De Manila University Press, Manila, the Philippines

Boyce, J., P. Rosset and E. Stanton (2007) 'Land reform and sustainable development', in J. Boyce, S. Narain and E. Stanton (eds) *Reclaiming Nature: Environmental Justice and Ecological Restoration*, Anthem Press, New York, NY, pp127–150

Brown, C. and S. Miller (2008) 'The impacts of local markets: A review of research on farmers' markets and community supported agriculture', *American Journal of Agricultural Economics*, vol 90, no 5, pp1296–1302

Bruinsma, J. (2009) *The Resource Outlook to 2050: By How Much do Land, Water and Crop Yields Need to Increase by 2050?*, FAO Expert Meeting on How to Feed the World in 2050, Rome, ftp://ftpfaoorg/docrep/fao/012/ak971e/ak971e00pdf, last accessed 7 November 2010

Brummer, E. C. and K. Moore (2000) 'Persistence of perennial cool-season grass and legume cultivars under continuous grazing by beef cattle', *Agronomy Journal*, vol 92, pp466–471

Carolan, M. (2009) 'The costs and benefits of biofuels: A review of recent peer-reviewed research and a sociological look ahead', *Environmental Practice*, vol 11, pp17–24

Carolan, M. (2011) *Embodied Food Politics*, Ashgate, Surrey, UK

Cartwright, J. (2000) *Evolution and Human Behavior: Darwinian Perspectives on Human Nature*, MIT Press, Cambridge, MA

Chedd, G. (1975) 'Cellulose from sunlight', *New Scientist*, vol 65, no 939, pp572–575

Compton, J., S. Wiggins and S. Keats (2010) *Impact of the Global Food Crisis on the Poor: What Is the Evidence?*, Overseas Development Institute, London, www.odi.org.uk/resources/download/5187.pdf, last accessed 31 December 2010

Connolly, W. (2003) *Neuropolitics: Thinking, Culture, Speed*, University of Minnesota Press, Minneapolis, MN

Cox, J. (2009) 'What is vertical farming?', *On Earth*, 6 November, www.onearth.org/community-blog/what-is-vertical-farming, last accessed 4 December 2010

Dartt, B., J. Lloyd, B. Radke, J. Black and J. Kaneene (1999) 'A comparison of profitability and economic efficiencies between management-intensive grazing and conventionally managed dairies in Michigan', *Journal of Dairy Science*, vol 82, pp2412–2420

Deckers, J. (2010) 'Should the consumption of farmed animal products be restricted, and if so, by how much?', *Food Policy*, vol 35, no 6, pp497–503

Dempster, Q. (2007) *Carbon Farming*, Australian Broadcasting Corporation, Sydney, 2 September, www.abc.net.au/stateline/nsw/content/2006/s1844608.htm

Despommier, D. (2010) *The Vertical Farm: Feeding the World in the 21st Century*, Thomas Dunne Books, New York, NY

Despommier, D. and E. Ellingsen (2008) 'The vertical farm: The sky-scraper as vehicle for a sustainable urban agriculture', Paper presented to the CTBUH 8th World

Congress, Dubai, 3–5 March, www.ctbuh.org/Portals/0/Repository/
T7_DespommierEllingsen.b8a44415-acfe-44b7-9d2d-c31c028f88ea.pdf, last
accessed 4 December 2010

Dugger, C. (2005) 'Africa food for Africa's starving is road blocked in Congress', *The
New York Times*, 12 October, www.nytimes.com/2005/10/12/international/
africa/12memo.html?ex=1286769600&en=0de1afa6dd7990e7&ei=5090&partner
=rssuserland&emc=rss, last accessed 12 November 2010

Dybas, C. (2009) 'Report from the (2009) AIBS Annual Meeting: Ensuring a food
supply in a world that's hot, packed, and starving', *BioScience*, vol 59, no 8,
pp640–646

Ehrenberg, R. (2008) 'Let's get vertical: City buildings offer opportunities for farms to
grow up instead of out', *Science News*, vol 174, no 8, pp16–20

Eshel, G. and P. Martin (2006) 'Diet, energy, and global warming', *Earth Interactions*,
vol 10, pp1–17

Estabrook, B. (2010) 'A tale of two dairies', *Gastronomica*, vol 10, no 4, pp48–52

FAO (2006) *Livestock's Long Shadow: Environmental Issues and Options*, Food and
Agriculture Organization of the United Nations, Rome, Italy,
ftp://ftp.fao.org/docrep/fao/010/a0701e/a0701e.pdf, last accessed March 14, 2011

FAO (Food and Agriculture Organization of the United Nations) (2009) *Food
Production Will Have to Increase by 70 Percent – FAO Convenes High-Level
Expert Forum*, FAO, United Nations, Rome,
www.fao.org/news/story/0/item/35571/icode/en, last accessed 6 November 2010

Fargione, J., J. Hill, D. Tilman, S. Polasky and P. Hawthorne (2008) 'Land clearing
and the biofuel carbon debt', *Science*, vol 319, pp1235–1238

Ferraro, P. and A. Kiss (2002) 'Direct payments to conserve biodiversity', *Science*, vol
298, no 5599, pp1718–1719

Finetto, C., C. Lobascio and A. Rapisarda (2010) 'Concept of LUNAR farm: Food
and revitalization module', *Acta Astronautica*, vol 66, no 9–10, pp1329–1340

Forster, J. (2010) 'Turning to the sea: Can seaweed culture conserve other marine feed
ingredients and help us raise more food more efficiently?', *The Advocate*, vol 13,
no 5, pp96–100

Friends of the Earth (2010) *Healthy Planet Eating: How Lower Meat Diets Can Save
Lives and the Planet*, Friends of the Earth, London, www.foe.co.uk/resource/
reports/healthy_planet_eating.pdf, last accessed 4 November 2010

Garcia, A., K. Kalscheur, A. Hippen and R. Schafer (2007) 'High priced corn and
dairy cow rations', *The DairySitecom*, www.thedairysite.com/articles/832/high-
priced-corn-and-dairy-cow-rations, last accessed 3 November 2010

Gliessman, S. (1998) *Agroecology: Ecological Process in Sustainable Agriculture*, Ann
Arbor Press, Ann Arbor, MI

Gray, L. (2010) 'Cows absolved of causing global warming with nitrous oxide', *The
Telegraph*, 8 April, www.telegraph.co.uk/earth/environment/climatechange/
7564682/Cows-absolved-of-causing-global-warming-with-nitrous-oxide.html

Guthman J. (2003) 'Fast food/organic food: Reflexive tastes and the making of
"yuppie chow"', *Social and Cultural Geography*, vol 4, no 1, pp45–58

Hancharick, A. and N. Kiernan (2008) 'Improving agricultural profitability through
an income: Opportunities for Rural Areas Program', *Journal of Extension*, vol 46,
no 5, www.joe.org/joe/2008october/a3.php

Haq, M. ul (1976) *The Poverty Curtain: Choices for the Third World*, Columbia
University Press, New York, NY

Harrison, P. and C. Dunmore (2010) 'EU report signals U-turn on biofuels target',
Reuters, 25 March, www.reuters.com/article/idUSTRE62O3O420100325, last

accessed 29 December 2010

Harvey, D. (2004) 'Policy dependency and reform: Economic gains versus political pains', *Agricultural Economics*, vol 31, pp265–275

Hayes-Conroy, A. and J. Hayes-Conroy (2008) 'Taking back taste: Feminism, food, and visceral politics', *Gender, Place and Culture*, vol 15, no 5, pp461–473

Hey, D. (2002) 'Nitrogen farming: Harvesting a different crop', *Restoration Ecology*, vol 10, no 1, pp1–10

Holdrege, C. and S. Talbott (2008) *Beyond Biotechnology: The Barren Promise of Genetic Engineering*, University of Kentucky Press, Lexington, KY

Honore, C. (2004) *In Praise of Slow*, Orion, London

Howlett, C. (2010) 'Carbon farming is on-track', *G Magazine*, 29 October, www.gmagazine.com.au/news/2313/carbon-farming-track, last accessed 9 November 2010

Humphries, J. (2010) 'EU biofuels effects food production', *New Generation Food*, 15 February, www.nextgenerationfood.com/news/eu-biofuels-effects-food-production, last accessed 29 December 2010

InterAcademy Council (2003) *Realising the Promise and Potential of African Agriculture: Science and Technology Strategies for Improving Agricultural Productivity and Food Security in Africa*, Amsterdam, The Netherlands, www.cgiar.org/pdf/agm04/agm04_iacpanel_execsumm.pdf, last accessed 7 November 2010

International Livestock Research Institute (2010) *Greener Pastures and Better Breeds Could Reduce Carbon 'Hoofprint'*, International Livestock Research Institute, September, Nairobi, Kenya, www.ilri.org/aggregator/sources/79, last accessed 4 November 2010

Jackson, T. (2009) *Prosperity without Growth: Economics for a Finite Planet*, Earthscan, London

Karapinar, B. and M. Temmerman (2008) 'Benefiting from biotechnology: Pro-poor IPRS and public–private partnerships', *Biotechnology Law Report*, vol 27, no 3, pp189–202

Khan, S. and M. Hanjra (2009) 'Footprints of water and energy inputs in food production – global perspectives', *Food Policy*, vol 34, pp130–140

King, C. (2008) 'Community resilience and contemporary agri-ecological systems: Reconnecting people and food, and people with people', *Systems Research and Behavioral Science*, vol 25, no 1, pp111–124

Koning, N., M. Van Ittersum, G. Becx, M. Van Boekel, W. Brandenburg, J. Van den Broek, J. Goudriaan, G. Van Hofwegen, R. Jongeneel, J. Schiere and M. Smies (2008) 'Long-term global availability of food: Continued abundance or new scarcity?', *NJAS Wageningen Journal of Life Sciences*, vol 53, no 3, pp229–292

Knopman, D. (2009) 'Mediterranean diet and late-life cognitive impairment: A taste of benefit', *JAMA*, vol 302, no 6, pp686–687

Knowler, D. and B. Bradshaw (2007) 'Farmers' adaption of conservation agriculture: A review and synthesis of recent research', *Food Policy*, vol 32, no 1, pp25–48

Koning, N., M. Van Ittersum, G. Becx, M. Van Boekel, W. Brandenburg, J. Van den Broek, J. Goudriaan, G. Van Hofwegen, R. Jongeneel, J. Schiere, and M. Smies (2008) 'Long-term global availability of food: Continued abundance or new scarcity?' *NJAS Wageningen Journal of Life Sciences, vol* 53, no 3, pp229–292

Kriegl, T. and G. Frank (2005) *A Ten Year Economic Look at Wisconsin Dairy Systems*, University of Wisconsin-Madison, Madison, WI, http://cdp.wisc.edu/pdf/Ten%20Yr%20COP3.pdf, last accessed 3 November 2010

Kumar, D. and O. P. Singh (2005) 'Virtual water in global food and water policy making: Is there need for rethinking?, *Water Resources Management,* vol 19, pp759–789

Lal, R. (1995) *Tillage Systems in the Tropics: Management Options and Sustainability Implications,* FAO, Rome, Italy

Liebig, M., S. Gross, R. Kronberg, L. Phillips and J. Hanson (2010) 'Grazing management contributions to net global warming potential: A long-term evaluation in the Northern Great Plains', *Journal of Environmental Quality,* vol 39, pp799–809

MacKenzie, A. (2002) *Transductions,* Continuum, London

Mariola, M. J., K. Stiles and S. Lloyd (2005) *The Social Implications of Management Intensive Rotational Grazing: An Annotated Bibliography,* Center for Integrated Agricultural Systems, University of Wisconsin-Madison, Madison, WI

McCullum, C., E. Desjardins, V. Kraak, P. Ladipo and H. Costello (2005) 'Evidence-based strategies to build community food security', *Journal of the American Dietetic Association,* vol 105, no 2, pp278–283

McMichael, A., J. Powles, C. Butler and R. Uauy (2007) 'Food, livestock production, energy, climate change, and health', *The Lancet,* vol 370, pp1253–1263

Miele, M. and J. Murdoch (2002) 'The practical aesthetics of traditional cuisines: Slow food in Tuscany', *Sociologia Ruralis,* vol 42, no 4, pp312–328

Mink, M., A. Evans, C. Moore, K. Calderon and S. Deger (2010) 'Nutritional imbalance endorsed by televised food advertisement', *Journal for the American Dietetic Association,* vol 110, no 6, pp904–910

Moss, M. (2010) 'Watch your weight and pass the cheese', *The New York Times,* Sunday, 7 November

Mougeot, L. (2005) 'Introduction: Urban agriculture and Millennium Development Goals', in L. Mougeot (ed) *Agropolis: The Social, Political and Environmental Dimensions of Urban Agriculture,* Earthscan, London, pp1–29

Myers, N. and J. Kent (2001) *Perverse Subsidies: How Tax Dollars Can Undercut the Environment and Economy,* Island Press, Washington, DC

National Research Council (2010) *Towards Sustainable Agricultural Systems in the 21st Century,* The National Academy Press, Washington, DC

Nestle, M. (2007) *Food Politics: How the Food Industry Influences Nutrition and Health,* University of California Press, Berkeley, CA

New York City (undated) *New York City's Waste Water Treatment System,* Bureau of Wastewater Treatment, New York City, www.nyc.gov/html/dep/pdf/wwsystem.pdf, last accessed 10 November 2010

OECD (Organisation for Economic Co-operation and Development) (2005) *Dairy Policy Reform and Trade Liberalization,* OECD Publishing, Paris, France

Oltenacu, P. and D. Broom (2010) 'The impact of genetic selection for increased milk yield on the welfare of dairy cows', *Animal Welfare,* vol 19, supplement, pp39–49

Pagiola, S., E. Ramírez, J. Gobbi, C. de Haan, M. Ibrahim, E. Murgueitio and J. P. Ruíz (2007) 'Paying for the environmental services of silvopastoral practices in Nicaragua', *Ecological Economics,* vol 64, no 2, pp374–85

Patnaik, U. (2009) 'Origins of the food crisis in India and developing countries', *Monthly Review* July/August, http://monthlyreview.org/090727patnaik.php, last accessed 14 March 2011

People's Daily (2009) 'China allocates $1048 bln for rural development', *People's Daily Online* (English), 26 December, http://english.peopledaily.com.cn/90001/90778/90860/6852750.html, last accessed 9 November 2010

Perfecto, I., J. Vandermeer and A. Wright (2009) *Nature's Matrix: Linking Agriculture, Conservation, and Food Sovereignty,* Earthscan, London

Peterson, E. W. (2009) *A Billion Dollars a Day: The Economics and Politics of Agricultural Subsidies*, Wiley-Blackwell, Malden, MA

Petrini, C. and W. McCuaig (2004) *Slow Food: The Case for Taste*, Columbia University Press, New York, NY

Petrini, C. and B. Watson (2001) *Slow Food: Collected Thoughts on Taste, Tradition, and the Honest Pleasures of Food*, Chelsea Green, White River, VT

Pope, K. (2010) *Corn-Fed Beef is a Nutritious Friend to Environment*, Nebraska Corn Board, Press Release, 13 August, www.nebraskacorn.org/editorials-letters/corn-fed-beef-is-a-nutritious-friend-to-environment, last accessed 4 November 2010

Popp, A., H. Lotze-Campen and B. Bodirsky (2010) 'Food consumption, diet shifts, and associated non-CO_2 greenhouse gases from agricultural production', *Global Environmental Change*, vol 20, pp451–462

Probyn, E. (2001) *Carnal Appetites: Food, Sex, Identities*, Taylor and Francis, New York, NY

Rabobank (2010) *Sustainability and Security of the Global Food Supply Chain*, Rabobank Group, The Netherlands, www.rabobank.com/content/images/Rabobank_IMW_WB_report-FINAL-A4-total_tcm43-127734.pdf, last accessed 23 May 2011

Ravallion, M., S. Chen and P. Sangraula (2007) *New Evidence on the Urbanization of Global Poverty*, Policy Research Working Paper 4199, World Bank, Washington, DC, http://siteresources.worldbank.org/INTWDR2008/Resources/2795087-1191427986785/RavallionMEtAl_UrbanizationOfGlobalPoverty.pdf, last accessed 8 November 2010

Redwood, M. (2010) 'Commentary: Food price and volatility and the urban poor', in C. Pearson (ed) *Urban Agriculture: Diverse Activities and Benefits for City Society*, Earthscan, London, pp5–6

Rice, T. (2010) *Meals Per Gallon: The Impact of Industrial Biofuels on People and Global Hunger*, Actionaid UK, London, www.actionaid.org/micrositeAssets/eu/assets/aa_biofuelsreportweb100210.pdf, last accessed 29 December 2010

Roberto, K. (2003) *How to Hydroponics,* 4th edition, Futuregarden, Lindenhurst, NY

Robles, Y. (2010) 'Urban-farming group in running for $4,000 grant', *The Denver Post*, 30 September, www.denverpost.com/news/ci_16211365?source=email, last accessed 11 November 2010

Rogers, S. (2009) 'Egypt regrets killing trash-eating pigs', *The Mother Nature Network*, 22 September, www.mnn.com/lifestyle/health-well-being/stories/egypt-regrets-killing-trash-eating-pigs#, last accessed 5 November 2010

Ruhl, J. B. (2008) 'Agriculture and ecosystem services: Strategies for state and local governments', *NYU Environmental Law Journal*, vol 17, pp424–459

Schade, C. and D. Pimentel (2010) 'Population crash: Prospects for famine in the twenty-first century', *Environment, Development and Sustainability*, vol 12, no 2, pp245–262

Sen, A. (2001) *Development as Freedom*, Oxford University Press, New York, NY

Stack, L. (2009) 'For Egypt's Christians, pig cull has lasting effects', *The Christian Science Monitor*, 3 September, www.csmonitor.com/World/Middle-East/2009/0903/p17s01-wome.html

Stengers, I. (2002) 'A cosmo-politics: Risk, hope, and change', in M. Zournazi (ed) *Hope: New Philosophies for Change*, Routledge, New York, NY, pp244–272

Stehfest, E., L. Bouwman, D. van Vuuren, M. den Elzen, B. Eickhout and P. Kabat (2009) 'Climate benefits of changing diet', *Climate Change*, vol 95, pp83–102

Stokstad, E. (2010) 'Could less meat mean more food?', *Science*, vol 327, no 5967, pp810–811

Taylor, J. and S. Neary (2008) *How Does Managed Grazing Affect Wisconsin's Environment*, Center for Integrated Agricultural Systems, University of Wisconsin-Madison, Madison, WI, www.ciaswisc.edu/wp-content/uploads/2008/10/grzgenvweb.pdf, last accessed 2 November 2010

Thompson, C. (2007) *Africa: Green Revolution or Rainbow Evolution? Foreign Policy in Focus*, 17 July, Washington, DC, www.fpif.org/articles/africa_green_revolution_or_rainbow_evolution, last accessed 7 November 2010

Thornton, P. and M. Herrero (2010) 'Climate mitigation and food production in tropical landscapes special feature: potential for reduced methane and carbon dioxide emissions from livestock and pasture management in the tropics', *Proceedings of the National Academy of Sciences*, vol 107, no 46, pp19667–19672

Tilman, D., J. Fargione, B. Wolff, C. D'Antonio, A. Dobson, R. Howarth, D. Schindler, W. H. Schlesinger, D. Simberloff and D. Swackhamer (2001) 'Forecasting agriculturally driven global environmental change', *Science*, vol 292, pp281–284

Tudge, C. (2010) 'How to raise livestock – and how not to', in J. D'Silva and J. Webster (eds) *The Meat Crisis: Developing More Sustainable Production and Consumption*, Earthscan, London, pp9–21

UNICEF, WTO, and WFP (2007) *Preventing and Controlling Micronutrient Deficiencies in Populations Affected by an Emergency*, Joint Statement, Geneva, Switzerland, www.helid.desastres.net/en/d/Js13449e/1.html, last accessed 9 March 2010

Vance, C. (2001) 'Symbiotic nitrogen fixation and phosphorus acquisition: Plant nutrition in a world of declining renewable resources', *Plant Physiology*, vol 127, no 2, pp390–397

Vidal, J. (2010) 'EU biofuels significantly harming food production in developing countries', *The Guardian*, 15 February, www.guardian.co.uk/environment/(2010)/feb/15/biofuels-food-production-developing-countries, last accessed 29 December 2010

Wansink, B. (2007) *Mindless Eating: Why We Eat More Than We Think*, Bantam Books, New York

Watkiss, P., T. Downing, C. Handley and R. Butterfield (2005) *The Impacts and Costs of Climate Change*, European Commission DG Environment, Brussels, http://ec.europa.eu/clima/studies/package/docs/final_report2.pdf, last accessed 23 May 2011

Webster, J. (2010) *The Meat and Dairy Crisis*, Earthcast, 11 October, sponsored by Earthscan, www.earthscan.co.uk/Earthcasts/tabid/101760/Default.aspx, last accessed 3 November 2010

Wirsenius, S. and F. Hedenus (2010) 'Policy strategies for a sustainable food system: Options for protecting the climate, in J. Webster and J. D'Silva (eds) *The Meat Crisis: Developing More Sustainable Production and Consumption*, Earthscan, London, pp237–253

Wolf, B., X. Zheng, N. Brüggemann, W. Chen, M. Dannenmann, X. Han, M. A. Sutton, H. Wu, Z. Yao and K. Butterbach-Bahl (2010) 'Grazing-induced reduction of natural nitrous oxide release from continental steppe', *Nature*, vol 464, pp881–884

Wolin, S. (1997) 'What time is it?', *Theory and Event*, vol 1, p1

World Bank (2008) *Agriculture for Development: The World Development Report 2008*, World Bank, Washington, DC, http://siteresources.worldbank.org/INTWDR2008/Resources/WDR_00_book.pdf, last accessed 6 November 2010

Wright, A. and W. Woldford (2003) *To Inherit the Earth: The Landless Movement and the Struggle for a New Brazil*, Food First, Oakland, CA

Wunder S., S. Engel and S. Pagiola (2008) 'Taking stock: A comparative analysis of payments for environmental services programs in developed and developing countries', *Ecological Economics*, vol 65, no 4, pp834–852

Yale Rudd Center (2010) *Evaluating Fast Food Nutrition and Marketing to Youth*, Yale Rudd Center, Yale University, New Haven, CT, www.fastfoodmarketing.org/media/FastFoodFACTS_Report.pdf, last accessed 12 November 2010

Yersin, H. and W. Finkenzeller (2007) 'Triplet emitters for organic light-emitting diodes: Basic properties', in H. Yersin (ed) *Highly Efficient OLEDs with Phosphorescent Materials*, Wiley, Regensburg, Germany, pp1–20

Zerbe, N. (2009) 'Setting the global dinner table: Exploring the limits of the marketization of food security' in J. Clapp and M. Cohen (eds) *The Global Food Crisis: Governance Challenges and Opportunities*, Wilfrid University Press, Waterloo, Ontario, pp161–177

Zezza, A. and L. Tasciotti (2010) 'Urban agriculture, poverty and food security: Empirical evidence from a sample of developing countries', *Food Policy*, vol 35, pp265–273

Index